Plant Allelopathy: Mechanisms and Applications in Regenerative Agriculture

Plant Allelopathy: Mechanisms and Applications in Regenerative Agriculture

Guest Editors

Margot Schulz
Vincenzo Tabaglio

MDPI

Basel • Beijing • Wuhan • Barcelona • Belgrade • Novi Sad • Cluj • Manchester

Guest Editors
Margot Schulz
Institute of Molecular Physiology and Biotechnology of Plants
University of Bonn
Bonn
Germany

Vincenzo Tabaglio
Department of Sustainable Crop Production
Section Agronomy and Plant Biotechnologies
Università Cattolica del Sacro Cuore
Piacenza
Italy

Editorial Office
MDPI AG
Grosspeteranlage 5
4052 Basel, Switzerland

This is a reprint of the Special Issue, published open access by the journal *Plants* (ISSN 2223-7747), freely accessible at: https://www.mdpi.com/journal/plants/special_issues/Plant_Allelopathy_Mechanisms.

For citation purposes, cite each article independently as indicated on the article page online and as indicated below:

Lastname, A.A.; Lastname, B.B. Article Title. *Journal Name* **Year**, *Volume Number*, Page Range.

ISBN 978-3-7258-3495-2 (Hbk)
ISBN 978-3-7258-3496-9 (PDF)
https://doi.org/10.3390/books978-3-7258-3496-9

Cover image courtesy of Margot Schulz and Vincenzo Tabaglio

Contents

About the Editors

Margot Schulz

Margot Schulz works since many years in the field of allelopathy, focusing on biochemical responses of target plants, allelochemical detoxification of benzoxazinoids, and glucosinolates by plants and microorganisms, and on molecular effects of terpenoids. She published more than 150 articles, most of them dealing with allelopathy, book chapters and conference abstracts. After studying chemistry and biology and finishing her PhD at the University of Cologne, Germany, she had a postdoc position at the University of Wisconsin, Madison, USA, before starting her habilitation at the University of Bonn. Germany. She got her qualification for a teaching career in higher education in 1993. Presently, she is involved in projects addressing the role of microorganisms in allelopathy and plant-microbe interactions.

Vincenzo Tabaglio

Vincenzo Tabaglio, working at the DI.PRO.VE.S., University of Cattolica del Sacro Cuore, Piacenza, Italy, is an established scientist in the field of allelopathic weed control, agroecology, crop residue decomposition, soil fertility and health, conservation tillage systems, carbon farming, nitrogen fertilization and mitigation of greenhouse gasses. He is engaged in projects and research for promoting of the rural development in India, in sustainable crop production for food security in Africa and low-income countries. He has published more than 90 articles, book chapters, and conference abstracts/papers, eleven of them dealing with allelopathy.

Preface

The use of allelopathic interaction for weed control is a promising component of integrated crop management in sustainable agriculture. Our conviction, that the use of allelopathy can mitigate environmental risks and contribute to maintaining soil fertility and biodiversity, motivated us to design this Special Issue. The functional agents in plant-plant, plant-microbe interactions are allelochemicals/secondary metabolites, which are entirely produced by the organisms and released into the environment. The isolation of the compounds from allelopathic plants and other organisms and the structural identification of the compounds is of outstanding importance to develop efficient application methods and to elucidate the specific mode of action on the physiological, biochemical, and molecular level, including changed protein abundancy and altered gene expression in target plants. This Special Issue contains 15 contributions, 10 research articles and 5 review articles. Altogether, 75 Authors from Brazil, China, Colombia, Egypt, France, Germany, Japan, Iran, Italy, Jordan, Poland, Spain, Tunisia, and USA contribute to the success of the Special Issue, demonstrating a word-wide interest in the field of allelopathy. We think that the theme will find a broad audience at relevant Universities and Research Stations, and we hope that the Special Issue motivates young scholars engaged in science and agriculture to continue research on allelopathy for improving its applications and promotion of Green Technologies development. Finally, we want to thank all Authors for all their work and the Editorial Office, particularly Bobo Wang, for their help and support to complete this Special Issue.

Margot Schulz and Vincenzo Tabaglio

Guest Editors

plants

Editorial

Allelopathy: Mechanisms and Applications in Regenerative Agriculture

Margot Schulz [1,*] and Vincenzo Tabaglio [2]

1 IMBIO Institute of Molecular Physiology and Biotechnology of Plants, University of Bonn, Karlrobert-Kreiten Str. 13, 53115 Bonn, Germany
2 Department of Sustainable Crop Production DI.PRO.VE.S., Section Agronomy and Plant Biotechnologies, Università Cattolica del Sacro Cuore, Via Emilia Parmense 84, 29122 Piacenza, Italy; vincenzo.tabaglio@unicatt.it
* Correspondence: ulp509@uni-bonn.de

Allelopathy is an important mechanism in plant communication and interference, involving the release of plant/microorganism self-produced, special featured organic molecules into the environment. These molecules (allelochemicals) inhibit or stimulate the growth of neighbored plants and microorganisms (targets), depending on their type and dosage. They are therefore regarded as natural plant growth modulators, with the effects also influenced by the developmental and other characteristics of the target plants. When used as a form of negative interaction between species, allelopathy in agriculture serves as a proven, sustainable tool for integrated weed control in organic farming, thereby mitigating environmental risks, a key objective of the Green Deal and Farm to Fork Strategy of the EU Agricultural Policy. Allelopathy can reduce the use of synthetic herbicides or of chemically upgraded natural compounds which cannot be no longer considered natural products.

The adoption of allelopathy (plant-plant and plant-microbe interactions) in regenerative agricultural practices still requires thorough multidisciplinary research (encompassing chemistry, genetics, metabolism, and agronomy) to address two complementary areas of focus. First, screening is needed to identify new allelopathic plants and their allelochemicals. Second, it is crucial to investigate the agronomic feasibility of effectively using these plants and their metabolites in conventional agroecosystems through field studies. For the latter, it is important to assess the potential for incorporating these allelopathic plants into crop rotations, often as cover crops, without adversely affecting subsequent crops.

This special issue on allelopathy aims to gather current experimental findings from around the world, covering both basic and applied aspects. It serves as a convenient and accessible resource for researchers working in this field and may contribute to promote future research on allelopathic interactions.

It compiles fourteen articles (nine original research and five original reviews) covering different aspects of allelopathy. The original research articles present recent investigations on allelochemical identification, on tritrophic interaction, on molecular effects in plant growth modulation and on undesired pathogen propagation via allelochemicals.

As can be observed, there is a lack of articles in the more strictly agronomic significance, specifically regarding the application of these results in open-field conditions. This highlights the need for greater research efforts in this important area, which is certainly more costly from an experimental perspective but is absolutely necessary for transitioning from basic knowledge to applied knowledge.

Following, some key points of the research articles are listed.

Vieites-Álvarez et al. [1] characterize the chemical profile of shoots, roots, and root exudates of four buckwheat accessions, emphasizing on polyphenol accumulation and exudation. The buckwheat Gema (accession number 01Z5000112) appeared to be most promising for sustainably control of the two herbicide-resistant target weeds *Lolium rigidum*

Citation: Schulz, M.; Tabaglio, V. Allelopathy: Mechanisms and Applications in Regenerative Agriculture. *Plants* **2024**, *13*, 3301. https://doi.org/10.3390/plants13233301

Received: 17 November 2024
Accepted: 19 November 2024
Published: 25 November 2024

Gaud. and *Portulaca oleracea* L. Espinosa-Colín et al. [2] present phytotoxic propiophenone, 4-ethylacetophenone, and 2,4-dimethylacetophenone obtained from *Cistus ladanifer*. Using *Lactuca sativa* and *Allium cepa* as target plants, dosage dependency of inhibitory and hormetic effects of the compounds were investigated.

The study of Amri et al. [3] addresses the chemical composition of essential oils from *Eucalyptus falcata*, *E. sideroxylon* and *E. citriodora*. They determined remarkable antioxidant, antimicrobial and phytotoxic properties of the essential oils. Germination and growth of *Sinapis arvensis*, *Phalaris canariensis*, and *Triticum durum* seedlings were inhibited. Marques et al. [4] contribute with a work on phytotoxic, water-soluble quillaic acid-based triterpene saponins isolated from the leaves of *Quillaja lancifolia*. Kato-Noguchi et al. [5] investigated the hitherto unknown allelopathic properties of *Osmanthus fragrans*, *Osmanthus heterophyllus* and of a hybrid of these two species, *Osmanthus* × *fortunei* Carrière. Main allelopathic substances of *O.* × *fortunei* and *O. fragrans* were identified as (+)-pinoresinol and 10-acetoxyligustroside, which are released into the environment during decomposition.

The contribution of Preusche et al. [6] demonstrated the concentration-dependent influence of the *Mentha x piperita* volatilome, *Mentha* essential oil fraction and limonene on the expression profile of distinct TCP transcription factors that modulate *Arabidopsis thaliana* and cabbage leaf growth under different culture conditions. Staszek et al. [7] investigated the phytotoxic effect of the digestive fluid of *Nepenthes x ventrata* on tomato seed germination and seedlings root growth. Oxidative stress elicited by the digestive fluid is supposed to inhibit tomatoes. A study dealing with the trophic interaction between *Peudosphinx tetrio* L. larvae and the toxic Apocynaceae *Allamanda cathartica* L. is presented by Matignon et al. [8]. The chemical profiles of the leaves of healthy and herbivorous *A. cathartica* and of the excretions of the caterpillars was analyzed. Larval excretion of the bioactive compounds from the host plant is one possibility to escape the harmful effects of these compounds.

Xin et al. [9] present investigations on the interactions of the allelochemicals and pathogens of potato including *A. solani*, *B. cinerea*, *F. solani*, *F. oxysporum*, *C. coccodes*, and *V. dahlia*. 7-Methoxycoumarin had inhibitory effects but, for instance, caffeic acid and chlorogenic acid promoted the pathogens. The results led to the conclusion that autotoxic allelopathy and promotion of pathogens caused by the accumulation of distinct allelochemicals can result in replant problems of potato.

The five original reviews summarize important features of *Cyperus esculentus*, *Solidago species*, *Abutilon theophrasti*, and microalgae allelopathy, whereas the contribution of Lewerenz et al. [10] attends to translocation of allelochemicals between plants, emphasizing on pyrrolizidine alkaloids. Zhang et al. [11] present the current state of *Cyperus esculentus* L. including the allelopathic potential and antibacterial, antioxidant and insecticidal activities of this plant. Kato-Noguchi and Kato [12] give an overview of *Solidago canadensis* L. and *Solidago altissima* L. allelopathy, two species which are aggressive invaders in many parts of the world. Tabaglio et al. [13] thematize properties and microorganisms possibly involved in the resilience of velvetleaf (*Abutilon theophrasti* Medik.) against allelochemicals, such as benzoxazinoids. Casanova et al. [14] reviewed the great variety of allelochemicals found in microalgae. These compounds may be valuable to biocontrol weeds, insects, phytopathogenic fungi and bacteria.

This special issue demonstrates that the sustainable use of allelopathy in agriculture demands in-depth knowledge about the bioactive molecules responsible for influencing the complex interactions between organisms. It is imperative that microorganisms must be considered in allelopathic interactions. This knowledge is the first step in designing smart agroecosystems that ensure both the sustainable intensification of agricultural production, as advocated by the FAO to meet global food demands and combat poverty, and the protection of the environment and species diversity, as outlined in the European Green Deal.

Author Contributions: The authors contribute to this article in equal shares. All authors have read and agreed to the published version of the manuscript.

Funding: This research received no external funding.

Conflicts of Interest: The authors declare no conflict of interest.

References

1. Vieites-Álvarez, Y.; Otero, P.; López-González, D.; Prieto, M.A.; Simal-Gandara, J.; Reigosa, M.J.; Hussain, M.I.; Sánchez-Moreiras, A.M. Specialized Metabolites Accumulation Pattern in Buckwheat Is Strongly Influenced by Accession Choice and Co-Existing Weeds. *Plants* **2023**, *12*, 2401. [CrossRef] [PubMed]
2. Espinosa-Colín, M.; Hernandez-Caballero, I.; Infante, C.; Gago, I.; García-Muñoz, J.; Sosa, T. Evaluation of Propiophenone, 4-Methylacetophenone and 2′,4′-Dimethylacetophenone as Phytotoxic Compounds of Labdanum Oil from *Cistus ladanifer* L. *Plants* **2023**, *12*, 1187. [CrossRef] [PubMed]
3. Amri, I.; Khammassi, M.; Ben Ayed, R.; Khedhri, S.; Mansour, M.B.; Kochti, O.; Pieracci, Y.; Flamini, G.; Mabrouk, Y.; Gargouri, S.; et al. Essential Oils and Biological Activities of *Eucalyptus falcata, E. sideroxylon* and *E. citriodora* growing in Tunisia. *Plants* **2023**, *12*, 816. [CrossRef]
4. Marques, M.E.M.; de Carvalho, A.C.; Yendo, A.C.A.; Magedans, Y.V.S.; Zachert, E.; Fett-Neto, A.G. Phytotoxicity of Quillaja lancifolia Leaf Saponins and Their Bioherbicide Potential. *Plants* **2023**, *12*, 663. [CrossRef]
5. Kato-Noguchi, H.; Hamada, Y.; Kojima, M.; Kumagai, S.; Iwasaki, A.; Suenaga, K. Allelopathic Substances of *Osmanthus* spp. for Developing Sustainable Agriculture. *Plants* **2023**, *12*, 376. [CrossRef] [PubMed]
6. Preusche, M.; Vahl, M.; Riediger, J.; Ulbrich, A.; Schulz, M. Modulating Expression Levels of TCP Transcription Factors by *Mentha x piperita* Volatiles—An Allelopathic Tool to Influence Leaf Growth? *Plants* **2022**, *11*, 3078. [CrossRef] [PubMed]
7. Staszek, P.; Piekarniak, M.; Wal, A.; Krasuska, U.; Gniazdowska, A. Is the Phytotoxic Effect of Digestive Fluid of *Nepenthes* x *ventrata* on Tomato Related to Reactive Oxygen Species? *Plants* **2023**, *12*, 755. [CrossRef] [PubMed]
8. Matignon, L.; Lo, M.M.; Monpierre, M.; Correia, M.V.; Valencia, D.P.; Palmeira-Mello, M.V.; Sylvestre, M.-N.; Pruneau, L.; Sylvestre, M.; Domenech, A.; et al. Phytochemical and Biological Study of Trophic Interaction between *Pseudosphinx Tetrio* L. Larvae and *Allamanda Cathartica* L. *Plants* **2023**, *12*, 520. [CrossRef]
9. Xin, A.; Jin, H.; Yang, X.; Guan, J.; Hui, H.; Liu, H.; Cui, Z.; Dun, Z.; Qin, B. Allelochemicals from the Rhizosphere Soil of Potato (*Solanum tuberosum* L.) and Their Interactions with the Soilborne Pathogens. *Plants* **2022**, *11*, 1934. [CrossRef]
10. Lewerenz, L.; Abouzeid, S.; Yahyazadeh, M.; Hijazin, T.; Selmar, D. Novel Cognitions in Allelopathy: Implications from the "Horizontal Natural Product Transfer". *Plants* **2022**, *11*, 3264. [CrossRef] [PubMed]
11. Zhang, S.; Li, P.; Wei, Z.; Cheng, Y.; Liu, J.; Yang, Y.; Wang, Y.; Mu, Z. Cyperus (*Cyperus esculentus* L.): A Review of Its Compositions, Medical Efficacy, Antibacterial Activity and Allelopathic Potentials. *Plants* **2022**, *11*, 1127. [CrossRef] [PubMed]
12. Kato-Noguchi, H.; Kato, M. Allelopathy and Allelochemicals of *Solidago canadensis* L. and *S. altissima* L. for Their Naturalization. *Plants* **2022**, *11*, 3235. [CrossRef] [PubMed]
13. Tabaglio, V.; Fiorini, A.; Sterling, T.M.; Schulz, M. Abutilon theophrasti's Resilience against Allelochemical-Based Weed Management in Sustainable Agriculture—Due to Collection of Highly Advantageous Microorganisms? *Plants* **2023**, *12*, 700. [CrossRef] [PubMed]
14. Casanova, L.M.; Macrae, A.; de Souza, J.E.; Neves Junior, A.; Vermelho, A.B. The Potential of Allelochemicals from Microalgae for Biopesticides. *Plants* **2023**, *12*, 1896. [CrossRef] [PubMed]

plants

MDPI

Article

Specialized Metabolites Accumulation Pattern in Buckwheat Is Strongly Influenced by Accession Choice and Co-Existing Weeds

Yedra Vieites-Álvarez [1], Paz Otero [2], David López-González [1], Miguel Angel Prieto [2], Jesus Simal-Gandara [2,3], Manuel J. Reigosa [1,3], M. Iftikhar Hussain [1] and Adela M. Sánchez-Moreiras [1,3,*]

1 Departamento de Bioloxía Vexetal e Ciencia do Solo, Facultade de Bioloxía, Campus Lagoas-Marcosende s/n, Universidade de Vigo, 36310 Vigo, Spain; yedra.vieites.alvarez@uvigo.es (Y.V.-Á.); davidlopez@uvigo.es (D.L.-G.); mreigosa@uvigo.es (M.J.R.); iftikhar@uvigo.es (M.I.H.)

2 Nutrition and Bromatology Group, Department of Analytical and Food Chemistry, Faculty of Food Science and Technology, Universidade de Vigo, Ourense Campus, 32004 Ourense, Spain; potero@uvigo.es (P.O.); mprieto@uvigo.es (M.A.P.); jsimal@uvigo.es (J.S.-G.)

3 Instituto de Agroecoloxía e Alimentación (IAA), Universidade de Vigo, Campus Auga, 32004 Ourense, Spain

* Correspondence: adela@uvigo.es; Tel.: +34-6-5059-9150

Abstract: Screening suitable allelopathic crops and crop genotypes that are competitive with weeds can be a sustainable weed control strategy to reduce the massive use of herbicides. In this study, three accessions of common buckwheat *Fagopyrum esculentum* Moench. (Gema, Kora, and Eva) and one of Tartary buckwheat *Fagopyrum tataricum* Gaertn. (PI481671) were screened against the germination and growth of the herbicide-resistant weeds *Lolium rigidum* Gaud. and *Portulaca oleracea* L. The chemical profile of the four buckwheat accessions was characterised in their shoots, roots, and root exudates in order to know more about their ability to sustainably manage weeds and the relation of this ability with the polyphenol accumulation and exudation from buckwheat plants. Our results show that different buckwheat genotypes may have different capacities to produce and exude several types of specialized metabolites, which lead to a wide range of allelopathic and defence functions in the agroecosystem to sustainably manage the growing weeds in their vicinity. The ability of the different buckwheat accessions to suppress weeds was accession-dependent without differences between species, as the common (Eva, Gema, and Kora) and Tartary (PI481671) accessions did not show any species-dependent pattern in their ability to control the germination and growth of the target weeds. Finally, Gema appeared to be the most promising accession to be evaluated in organic farming due to its capacity to sustainably control target weeds while stimulating the root growth of buckwheat plants.

Keywords: buckwheat; polyphenols; sustainable weed control; allelopathy; competition

Citation: Vieites-Álvarez, Y.; Otero, P.; López-González, D.; Prieto, M.A.; Simal-Gandara, J.; Reigosa, M.J.; Hussain, M.I.; Sánchez-Moreiras, A.M. Specialized Metabolites Accumulation Pattern in Buckwheat Is Strongly Influenced by Accession Choice and Co-Existing Weeds. *Plants* **2023**, *12*, 2401. https://doi.org/10.3390/plants12132401

Academic Editor: Ippolito Camele

Received: 27 April 2023
Revised: 17 May 2023
Accepted: 15 June 2023
Published: 21 June 2023

1. Introduction

Pests and weeds are probably the primary biotic limitation that farmers face when attempting to increase the yield of their crops. Therefore, research, development, and the use of chemical products exclusively produced for the control of weeds/pests in the field has increasingly grown in the last decades. However, these synthetic chemical products, which have been extensively applied in the field, are dangerous to terrestrial environment and to human health [1], and increase weed resistance [2]. Several synthetic herbicides, including glyphosate, exhibit significant potential for soil adsorption and cannot move around freely in the environment. The effects of herbicides on human health depend on the concentration, length, and frequency of exposure, and often lead to cytotoxic and DNA damage and carcinogenicity [3]. Reducing their use is an increasing necessity in order for more sustainable food production.

There are several recent reports that highlight the development of resistance in different *Lolium* sp., including *Lolium rigidum* Gaudin, the species used in this study, which is likely the cause of the significant loss in cereal crop yield in Mediterranean countries and Australia [4,5]. In the US, populations of Italian ryegrass (*L. perenne* ssp. *multiflorum* (Lam.) Husn.) have also been found to be resistant to glufosinate [6]. As a result of the quick evolution of resistance in *L. rigidum*, comprehensive weed management measures, including crop allelopathic varieties, are required to slow down this rapid evolution and sustainable control. In several European countries, including Spain and Portugal, resistance to acetyl-CoA carboxylase (ACCase), acetolactate synthase (ASL), photosystem II (PSII), 5-enolpyruvylshikimate-3-phosphate synthase (EPSPS), glutamine synthase, very long-chain fatty acid (VLCFA) synthesis, and protoporphyrinogen oxidase (PPO)-inhibiting herbicides has been reported in *L. rigidum*, which is mainly associated with winter cereals, vineyards, and orchards [7,8]. The other species used in this study, *Portulaca oleracea* L., commonly known as purslane, is a weed belonging to the Portulacaceae family that has been ranked ninth out of the world's worst weeds, and has been recorded in 45 crops in 81 countries [9]. According to the International Herbicide-Resistant Weed Database, *P. oleracea* has been already found to develop resistance to Group 5 (Legacy C1 C2) herbicides, which are referred to as PSII inhibitors—Serine 264 Binders. These specific biotypes are resistant to the atrazine and linuron herbicides, and may also be cross-resistant to other members of the Group 5 (Legacy C1 C2) family. Other researchers have also reported *P. oleracea* as a resistant weed to linuron [10]. This is a common phenomenon in many weeds due to the massive use of synthetic herbicides.

Several integrated weed management strategies can be employed to limit herbicide use, such as harrowing or hoeing, flames or hot water, allelopathic cover crops, smother crops, and green mulching, which are crucial for controlling the weed seed population in the soil for green economy, plant biodiversity, and environmental sustainability [11]. The circular economy, promoted by the European Commission, defends the reuse of various types of organic biomass and organic waste, thus transforming waste management into economic opportunities [12,13]. In this context, screening suitable allelopathic crops and crop genotypes that are competitive with weeds can be a secure and environmentally responsible weed control strategy [14]. By releasing specialized metabolites into the surrounding environment that may act as phytotoxins to suppress weeds, they fall in the framework of a process known as allelopathy and provide themselves a competitive edge [15,16]. By encouraging diversification of the agricultural system and reducing the reliance on herbicides, this approach can also reduce the risk of the development of herbicide resistance in monocot and dicot weeds populations. Weed management must protect environmental quality and human health, and allelochemicals released from live crops and crop residues can be used in this way to manage weeds and improve crop performance [17].

The *Fagopyrum* genus is a dicotyledonous pseudo-cereal crop of the family Polygonaceae. It is comprised of both perennial and annual species with diploid (2n = 2x = 16) and tetraploid (2n = 4x = 32) chromosome numbers with a haploid genome size of −1.2 Gb. Out of the 34 reported species to date, this genus is recognized primarily by two cultivated species, common buckwheat (*F. esculentum* Moench.) and Tartary buckwheat (*F. tataricum* (L.) Gaertn.), along with the wild species *F. cymosum* (Trev.) Meisn. [18]. Buckwheat, which is known for its high nutritional value and bioactive components, is cultivated mainly for the production of food and pharmacological products for humans, and is increasingly considered as a promising emergent crop, as it can be an important source of proteins with well-balanced amino acid composition, dietary fibre, and phenolic substances [19]. Buckwheat is rich in phenolic acids and flavonoids, and exhibits antioxidative properties [20–22]. Besides well-known antioxidative effects, it has recently been suggested that buckwheat, as a source of the flavonoid quercetin, may prevent health problems in patients with diabetes [19].

Regarding its allelochemical potential, Kumar et al. [23] demonstrated that buckwheat residues in the soil were able to reduce powell amaranth, shepherd's purse, and corn chamomile growth. There are several research and review manuscripts published advocating that buckwheat tissues possess abundant specialized metabolites, especially phenolic acids and flavonoids, with a significant allelopathic potential [24–26] that could be responsible for the weed-suppressive activity of buckwheat [23,27,28]. In addition, it has been reported that the suppression of some weeds is caused by the light competition of rapidly growing buckwheat plants [29]. However, previous studies have shown that common buckwheat, thanks to allelochemicals, can also significantly reduce the biomass of certain weed species such as *Thlaspi arvense* L., *Cirsium arvense* (L.) Scop., and *Plantago lanceolata* L. [30].

Phenolic compounds, abundant metabolites in buckwheat [31,32], are well known for their allelochemical properties. Phenolic acids such as cinnamic, dihydroxybenzoic, ferulic, *p*-coumaric, phthalic, syringic, *m*-toluic, or protocatechuic have been related to the allelochemical-induced growth inhibition of several weed species [33]. These compounds can be found in different parts of the buckwheat plant, such as the leaves, stems, and roots [34]. When buckwheat residues, such as decomposed plant material or root exudates, are present in the soil, the phenolic acids released can exert inhibitory effects on weed germination, growth, and development [23,29]. The specific mechanism through which phenolic acids inhibit weeds is not fully understood, but it is believed to involve interference with weed seed germination, root elongation, and nutrient uptake. The research has demonstrated the allelopathic effects of buckwheat on various weed species. For example, studies have shown that buckwheat residues can inhibit the germination and growth of weeds such as common lambsquarters (*Chenopodium album*), pigweed (*Amaranthus* spp.), and barnyard grass (*Echinochloa crus-galli*) [23,34]. It is important to note that the allelopathic effects can vary depending on factors such as the buckwheat cultivar and the weed species in question.

Both competition and allelopathy, as mechanisms of plant interference, have been well documented under controlled conditions [35]. The combined effects of allelopathy and crop competition determine the total weed-suppressive potential of a given variety, and research groups worldwide have been working to improve both traits simultaneously in order to achieve maximum gains in weed suppression [36,37], particularly in cereal crops. The root length of ryegrass, mustard, and lettuce was significantly reduced when their seeds co-germinated with buckwheat [30]. Although more recent studies are mostly focused on the allelopathic action of common buckwheat on weeds, Tartary buckwheat has also been studied for years regarding its potential for weed management [38].

However, little is known about the direct relation among allelochemicals exudation or accumulation and weed control by common or Tartary buckwheat. Therefore, the objective of this research was to examine the impact of three accessions of common buckwheat *Fagopyrum esculentum* Moench. (Gema, Kora, and Eva), and one accession of Tartary buckwheat, *Fagopyrum tataricum* (L.) Gaertn. (PI481671), on monocot (*Lolium rigidum* Gaud.) and dicot (*Portulaca oleracea* L.) weeds under laboratory-based germination and seedling growth bioassays. The potential of allelopathic compounds was carried out through identifying and quantifying a broad polyphenols profile (phenolic acids and flavonoids) in the roots, shoots, and root exudates of the different buckwheat accessions to compare the abundance of these allelochemicals with their weed suppressive capacity. The specific objectives of this study include the elucidation of the allelopathic potential of selected buckwheat accessions and their possible use to promote plant-based herbicides and their use in sustainable weed management strategies.

2. Results

2.1. Germination and Seedling Growth

Our results show that the germination rate of both weed species, *L. rigidum* and *P. oleracea*, was significantly decreased following being co-cultured with the four tested buckwheat accessions (Gema, Kora, Eva, and PI481671), which showed the strong capacity to control weed germination following a 7-day co-culture and it was species-specific

(Table 1). Only *P. oleracea* could maintain its germination unaltered in the presence of the Eva accession. Although different buckwheat accessions behaved differently, they induced significant inhibition of germination when compared with the control, with Gema (19%) and PI481671 (35%) being the most inhibitory accessions for *L. rigidum* followed by Eva (50%) and Kora (54%). On the contrary, Kora (41%) was the most inhibitory accession for *P. oleracea* (41%), followed by PI481671 (55%) and Gema (59%).

Table 1. Allelopathic impact of different buckwheat genotypes (Gema, Kora, Eva, and PI481671) on the germination (Germ: number of seeds as % of the control), fresh plant weight (PW; g given as % of the control) (i.e., weed species grown alone), and shoot and root lengths (cm given as % of the control) in monocot (*Lolium rigidum*) and dicot (*Portulaca oleracea*) weeds. Grey cells indicate significant inhibition, while bold numbers indicate significant stimulations compared with the control. The asterisk indicates statistical significant differences compared to the control at $p < 0.05$.

	Lolium rigidum				*Portulaca oleracea*			
Accession	Germ	PW	SL	RL	Germ	PW	SL	RL
PI481761	35 *	68 *	106	78	55 *	**162** *	**186** *	134
Eva	50 *	52 *	**149** *	122	73	107	**146** *	90
Kora	54 *	73 *	127	119	41 *	91	100	126
Gema	19 *	79	135	29 *	59 *	107	**158** *	**166** *

Regarding the impact of the presence of buckwheat accessions on weed growth and development, we obtained a species dependent behaviour again, with *L. rigidum* as the most sensitive weed species to buckwheat co-cultivation (Table 1). The most inhibited parameter was the fresh plant weight for *L. rigidum* in the presence of Kora (73%), PI481761 (68%), and Eva (52%), while the most stimulated parameter was the shoot length of *P. oleracea* in the presence of PI481761 (186%), Gema (158%) and Eva (146%). *L. rigidum* was especially sensitive to Eva, as shoot length significantly increased (149%) while reducing fresh plant weight (52%), which resulted in longer but weaker plants. Something similar was found for *P. oleracea* in front of Gema accession, which significantly increased both, shoot (158%) and root (166%) length while maintaining unaltered the total weight of the plant (107%).

Regarding the germination and growth parameters of the different buckwheat accessions in front of the two tested weeds (Table 2), our results show that the shoot and root lengths of Eva were significantly inhibited in front of *P. oleracea,* as well as the root length in front of *L. rigidum*. In fact, Eva was the only accession where the shoot and root length were statistically inhibited in the presence of the weeds, while Gema and PI481671 also showed significant stimulation of root length when co-cultured with *P. oleracea*. PI481671 showed a stimulation of leaf weight in front of *P. oleracea* when compared with the buckwheat plants growing alone. Finally, the root weight was only inhibited in Kora in the presence of the monocot weed *L. rigidum*. From the parameters measured in the different accessions, we could see also that common buckwheat accessions (Eva, Gema, and Kora) grew much more than Tartary buckwheat (PI481671), which showed shoot length values that were two times lower than common buckwheat and root lengths values that were even three times lower than Eva and Kora and two times lower than Gema. These differences were even stronger in the case of leaf weight, where Tartary buckwheat values were ten times under common buckwheat accessions.

Table 2. Mean values and standard deviation of growth parameters for different buckwheat accessions (Gema, Kora, Eva, and PI481671) when growing alone or in co-culture with the monocot weed *Lolium rigidum* (LR) or the dicot weed *Portulaca oleracea* (PO). The parameters measured are plant and root weight (g), and shoot and root lengths (cm). Grey cells indicate significant inhibition, while bold numbers indicate significant stimulations compared with the control.

		Leaf Weight	Root Weight	Shoot Length	Root Length
Eva	Alone	3.12 ± 0.70	1.01 ± 0.24	13.3 ± 2.09	27.2 ± 0.50
	LR	2.97 ± 0.54	0.71 ± 0.10	12.8 ± 3.11	20.6 ± 4.95
	PO	2.43 ± 0.21	0.87 ± 0.02	9.84 ± 0.83	17.3 ± 3.19
Gema	Alone	1.68 ± 0.22	0.37 ± 0.15	9.81 ± 1.83	13.8 ± 1.82
	LR	1.65 ± 0.17	0.53 ± 0.10	9.14 ± 3.97	14.9 ± 2.73
	PO	1.76 ± 0.34	**0.59 ± 0.05**	11.9 ± 2.9	**20.3 ± 2.85**
Kora	Alone	2.69 ± 0.25	1.20 ± 0.28	12.6 ± 1.07	21.4 ± 0.62
	LR	1.99 ± 1.04	0.50 ± 0.20	10.7 ± 4.33	18.7 ± 6.58
	PO	1.91 ± 0.44	0.76 ± 0.19	12.5 ± 2.46	19.6 ± 1.27
PI481761	Alone	0.25 ± 0.07	0.56 ± 0.34	4.18 ± 1.12	6.63 ± 2.36
	LR	0.33 ± 0.10	0.43 ± 0.29	4.04 ± 0.04	8.07 ± 2.26
	PO	**0.95 ± 0.07**	0.46 ± 0.17	4.43 ± 0.60	**12.7 ± 4.56**

2.2. Polyphenols Profile in Different Buckwheat Accessions

The identification and quantification of different polyphenols (phenolic acids and flavonoids), such as 4-chlorobenzoic acid (4-CBA), vanillic acid (VA), cinnamic acid (CA), dihydroxybenzoic acid (DA), ferulic acid (FA), p-coumaric acid (P-CA), phthalic acid (PA), syringic acid (SA), m-toluic acid (M-TA), luteolin (LU), syringaldehyde (SY), protocatechuic acid (PTA), quercetin (QE), vanillin (VN), salicylic acid (SAA), 4-hydroxyacetophenone (4-HA), catechin (CAT), epicatechin (ECAT), orientin (OR), vitexin (VIT), hypericin (HYP), and rutin (RU) showed different profiles for each of the four tested buckwheat accessions. In fact, while DA, FA, P-CA, SA, LU, QE, VN, 4-HA, ECAT, OR, VIT, and RU were present in the four buckwheat accessions (Gema, PI481671, Kora, and Eva), VA, SY, and SAA were only present in Gema, PI481671, and Eva; 4-CBA and PTA in Gema and Eva; CAT in Gema, Kora, and Eva; PA in Eva and PI481671; and HYP only in Gema, while M-TA was missed in this buckwheat accession (Table 3).

From all of the polyphenols measured, DA, P-CA, LU, 4-HA, OR, and VIT were the most commonly accumulated compounds, especially in the roots, for most of the four buckwheat accessions in the presence of both weeds, *L. rigidum* and *P. oleracea*, while DA, LU, 4-HA, CAT, ECAT, and OR were the most altered compounds in the roots, shoots, and root exudates of the different buckwheat accessions (Table 3). Although in the presence of *L. rigidum* buckwheat tended to accumulate the compounds indistinctly in roots or shoots, depending of the buckwheat accession, there was a clear trend in the four buckwheat accessions to accumulate the compounds in the roots when co-grown with *P. oleracea*, as DA, P-CA, LU, OR, and VIT were significantly accumulated in the roots in three of the four buckwheat accessions tested when *P. oleracea* was present in the medium. On the contrary, *L. rigidum* was the weed that more root exudates induced from the buckwheat accessions tested, especially for Eva (Table 3). Finally, the Tartary buckwheat accession (PI481671) showed a generally reduced root exudation of polyphenols in comparison with the three common buckwheat accessions (Gema, Kora, and Eva), which reacted more to the presence of weeds exuding more or less phenolic acids and flavonoids, but also exuding more polyphenols when growing alone.

Table 3. Significant increases and decreases ($p \leq 0.05$) for different polyphenols (PP) in the shoots, roots, or root exudates of the buckwheat accessions Gema (G), Kora (K), Eva (E), and PI481671 (PI) after co-culture with *L. rigidum* (LR) or *P. oleracea* (PO) compared to the control (each accession growing alone). The PP detected were 4-chlorobenzoic acid (4-CBA), vanillic acid (VA), dihydroxybenzoic acid (DA), ferulic acid (FA), *p*-coumaric acid (*P*-CA), syringic acid (SA), luteolin (LU), syringaldehyde (SY), protocatechuic acid (PTA), quercetin (QE), rutin (RU), vanillin (VN), salicylic acid (SAA), 4-hydroxyacetophenone (4-HA), catechin (CAT), epicatechin (ECAT), orientin (OR), vitexin (VIT), phthalic acid (PA), *m*-toluic acid (*M*-TA), and hypericin (HYP).

INCREASED		**4-CBA**	**VA**	**DA**	**FA**	***P*-CA**	**SA**	**LU**	**SY**	**PTA**	**QE**	**RU**
Co-cultured with LR	Shoots		E	G; E	E	E		E	E	E	E; PI	
	Roots					E		E	E	E		G
	Root exudates		E	E; PI	E	E	E; PI	E	E	E	E; PI	E
Co-cultured with PO	Shoots	E	E	PI	PI					G		
	Roots		E; PI	G; K; PI	G; PI	G; E; PI	PI	G; K; E	E; PI	E	G	G; PI
	Root exudates			K	K	K				E	G	
INCREASED		VN	SAA	4-HA	CAT	ECAT	OR	VIT	PA	*M*-TA	HYP	
Co-cultured with LR	Shoots	E		K; E						K; PI		
	Roots			G; E	E	E	E; PI	E				
	Root exudates	E	E	E	G; E	G; E; PI	G; K; E	E	E			
Co-cultured with PO	Shoots		G; PI	G; E; PI		PI						
	Roots	PI	PI	G; K; PI	E	E; PI	G; K; E	G; K; E		K		
	Root exudates			K; E	E		K	PI		K		
DECREASED		4-CBA	VA	DA	FA	*P*-CA	SA	LU	SY	PTA	QE	RU
Co-cultured with LR	Shoots			K		K	K				K	K
	Roots			K	K	K		K			K	K
	Root exudates	E		K	K	K	K	K			K	K
Co-cultured with PO	Shoots		PI	G; K; E	K; E	K; E	K	G; K; E	G	E	K; E	E; K
	Roots						K					K
	Root exudates	G;		G; E	G	ECAT	PI		G; E		K	G
DECREASED		VN	SAA	4-HA	CAT	ECAT	OR	VIT	PA	*M*-TA	HYP	
Co-cultured with LR	Shoots	G		PI	G; K;	G; K; E						
	Roots			PI			G; K	G; K	PI	PI		
	Root exudates	PI		K	K	K		K	PI	E		
Co-cultured with PO	Shoots	G	E	K	K; E	G; K; E	G; K; E	G; K; E	E; PI	K		
	Roots	G			G; K	G; K	PI		PI	PI		
	Root exudates	K; PI			K; G	G			PI			

During the comparison of the polyphenol content of shoots, roots, and root exudates of the four buckwheat varieties (Eva, Gema, Kora, and PI481671), the generated score plot of the PCA (Figure 1A) revealed a complex accession distribution pattern that, in the largest distance, was not mainly accession-dependent, but organ-dependent as the root exudates showed strong dissimilarity with shoots and roots, especially for Gema, Eva, and PI491671. The supervised PLS-DA analysis (Figure 1B) confirmed the separation, previously observed with the PCA. The separation was achieved by virtue of the first two principal components, which explained a total variance of 60.2%. Component 2 explained the highest variance (40.5%), while component 1 explained 19.7% of the total variance. The hierarchical cluster showed similarity in the root and shoot polyphenol contents of each accession, being the root and shoot contents of the four accessions grouped together in the same cluster of the dendrogram, but each species constituting and independent subclusters of the tree (Figure 1C). Surprisingly, this division was not species dependent, as there was more distance among the common accessions Eva and Gema with Kora than among the common accessions Gema and Eva with the Tartary accession PI481671.

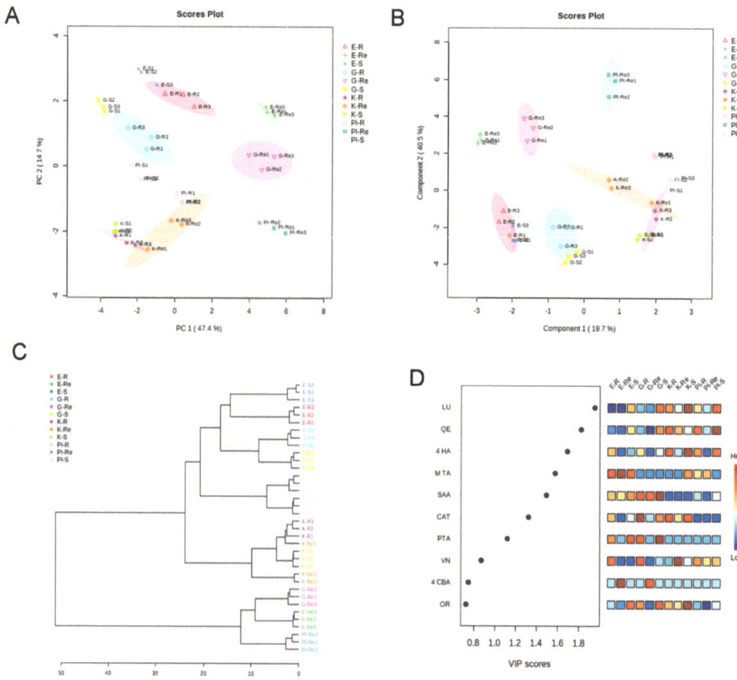

Figure 1. Analyses of the polyphenol content in the shoots (S), roots (R), and root exudates (Re) in the three common buckwheat accessions Eva (E), Gema (G), and Kora (K) and the Tartary buckwheat accession PI481671 (PI). (**A**) PCA scores plot between the selected PCs; (**B**) PLS-DA scores plot between the selected PCs, the explained variances of PCA and PLS-DA are shown in brackets; (**C**) clustering result shown as a dendrogram (distance measured using Euclidean, and clustering algorithm using ward). (**D**) Important features identified by PLS-DA. E−S (chemical profile of Eva shoots; dark blue), E-R (chemical profile of Eva roots; red), E-Re (chemical profile of Eva root exudates; dark green), G-S (chemical profile of Gema shoots; yellow), G-R (chemical profile of Gema roots; light blue), G-Re (chemical profile of Gema root exudates; dark pink), K-S (chemical profile of Kora shoots; light green), K-R (chemical profile of Kora roots; violet), K-Re (chemical profile of Kora root exudates; orange), PI-S (chemical profile of PI481671 shoots; malva), PI-R (chemical profile of PI481671 roots; light pink), and PI-Re (chemical profile of PI481671 root exudates; dark green). The coloured boxes on the right indicate the relative concentrations of the corresponding polyphenol in each group under study. N = 3.

In contrast, the root exudates of Gema, Eva, and PI481671 were grouped in a totally independent cluster, showing its dissimilarity from the rest of the samples and indicating a characteristic behaviour of exudation in buckwheat. Moreover, although the content of root exudates of common and Tartary accessions was grouped in the same cluster of the dendrogram, Gema and Eva were grouped in a different subcluster than PI481671, suggesting a species-dependent dissimilarity among accessions at this level of distance. The compounds that had stronger weight in this classification were LU, QE, 4-HA, M-TA, SAA, CAT, and PTA (over 1.0 VIP score). LU, QE, and 4-HA were characterized to be increased in Gema, Kora, and PI481671 shoots, roots, and root exudates in comparison with Eva, while M-TA was increased in Eva shoots, roots, and root exudates in comparison with Gema, Kora, and PI481671 (Figure 1D).

2.2.1. Chemical Profile of EVA Polyphenols

When Eva was co-cultured with *L. rigidum*, the concentration of most of the polyphenols was significantly increased in the shoots, roots, and especially root exudates, following a co-culture of 7 days (Table 4).

Table 4. Quantification of polyphenols (PP) from shoots, roots, and root exudates from buckwheat accession Eva, alone (control) or in association with *L. rigidum* (LR) or *P. oleracea* (PO). Quantities are expressed in μg kg^{-1} dry weight (shoots and roots) and μg L^{-1} (root exudates). Means followed by different lowercase letters within a column show significant differences among treatments ($p \leq 0.05$). Bold numbers indicate a significant increase when compared with the control, while grey cells indicate a significant decrease.

Sample	**Treatment**	**4-CBA**	**VA**	**DA**	**FA**	***P*-CA**	**PA**	**SA**
Shoot	Alone	<0.01 b	<0.01 b	2472 b	83.41 b	553.2 b	350.9 a	3.187 a
	+LR	<0.01 b	**238.8 a**	**3956 a**	**289.1 a**	**998.4 a**	539.4 a	28.13 a
	+PO	**5.173 a**	**108.5 a**	131.4 c	28.69 c	98.88 c	31.55 b	6.337 a
Root	Alone	<0.01 a	471.9 b	928.9 a	152.8 a	49.71 b	147.4 a	291.0 a
	+LR	<0.01 a	892.9 ab	1122 a	242.9 a	**122.6 a**	186.2 a	578.7 a
	+PO	<0.01 a	**1067 a**	1214 a	326.9 a	**145.0 a**	218.1 a	222.5 a
Root exudates	Alone	4.533 a	<0.01 b	14.47 b	<0.01 b	123.4 b	28.02 b	<0.01 b
	+LR	<0.01 b	**345.5 a**	**6924 a**	**511.3 a**	**1171 a**	**448.0 a**	**47.23 a**
	+PO	3.980 a	<0.01 b	<0.01 c	<0.01 b	<0.01 c	10.83 b	<0.01 b
Sample	Treatment	*M*-TA	LU	SY	PTA	QE	VN	RU
Shoot	Alone	418.8 a	57.66 b	3.513 b	366.8 b	1174 b	<0.01 b	9011 a
	+LR	553.5 a	**205.5 a**	**14.07 a**	**1384 a**	**3211 a**	**1.948 a**	9164 a
	+PO	775.9 a	1.031 c	2.987 b	55.64 c	299.4 c	<0.01 b	283.2 b
Root	Alone	855.3 a	<0.01 b	2.658 b	219.1 b	122.0 a	106.1 a	409.9 a
	+LR	775.1 a	**3.070 a**	**31.83 a**	**2105 a**	160.6 a	197.2 a	899.9 a
	+PO	627.2 a	**1.810 a**	**33.92 a**	**1589 a**	173.7 a	116.2 a	791.9 a
Root exudates	Alone	1089 a	<0.01 b	2.143 b	<0.01 c	12.54 b	<0.01 b	29.08 b
	+LR	323.0 b	**199.9 a**	**10.54 a**	**2837 a**	**3817 a**	**22.61 a**	**13583 a**
	+PO	900.2 a	<0.01 b	<0.01 c	**3.710 b**	30.44 b	<0.01 b	23.66 b
Sample	Treatment	SAA	4-HA	CAT	ECAT	OR	VIT	
Shoot	Alone	25.94 a	0.603 b	260.5 a	3143 a	4220 a	3078 a	
	+LR	42.41 a	**2.590 a**	365.4 a	1573 b	3547 a	3504 a	
	+PO	16.16 b	**2.958 a**	41.35 b	67.32 c	314.7 b	549.9 b	
Root	Alone	23.76 a	26.59 b	532.3 b	33.95 b	34.92 b	30.78 b	
	+LR	35.64 a	**137.3 a**	**2616 a**	**414.5 a**	**241.9 a**	**449.0 a**	
	+PO	25.20 a	59.02 b	**1782 a**	**291.6 a**	**299.3 a**	**430.0 a**	
Root exudates	Alone	18.20 b	<0.01 b	<0.01 c	<0.01 b	20.36 b	54.87 b	
	+LR	**39.66 a**	**5.327 a**	**346.1 a**	**1034 a**	**4400 a**	**4481 a**	
	+PO	17.34 b	**6.415 a**	**2.110 b**	<0.01 b	14.40 b	24.27 b	

4-chlorobenzoic acid (4-CBA); vanillic acid (VA); dihydroxybenzoic acid (DA); ferulic acid (FA); *p*-coumaric acid (*P*-CA); syringic acid (SA); *m*-toluic acid (*M*-TA); luteolin (LU); syringaldehyde (SY); protocatechuic acid (PTA); quercetin (QE); vanillin (VN); rutin (RU); salicylic acid (SAA); 4-hydroxyacetophenone (4-HA); catechin (CAT); epicatechin (ECAT); orientin (OR); vitexin (VIT).

Different specialized metabolites such as VA, DA, FA, *P*-CA, LU, SY, PTA, QE, VN, and 4-HA were significantly increased in the **shoot tissues** after the co-culture of Eva with *L. rigidum* (Table 4). Even VA and VN were found in high concentrations after co-growth with annual ryegrass, although could not be detected in Eva shoots when growing alone.

In contrast, there was a clear trend to reduce most of the detected polyphenols (DA, FA, *P*-CA, PA, LU, PTA, QE, SAA, CAT, ECAT, OR, VIT, and RU) in Eva shoot tissues when co-cultured with *P. oleracea*, especially LU, ECAT, and RU, which were reduced for more than 57 times, 47 times, and 31 times, respectively, followed by OR and PA, with reductions of more than 12 times. There was also a significant reduction in PTA (6.6 times), and QE (4 times). In contrast, a significant increase was only obtained in 4-CBA, VA, and 4-HA levels in the shoot tissues of Eva when co-cultured with *P. oleracea* (Table 4).

In **root tissues**, the significant increase in the content of phenolic acids and flavonoids, such as *P*-CA, LU, SY, PTA, 4-HA, CAT, ECAT, OR, and VIT when Eva was co-cultured with *L. rigidum* was coherent with the increase observed in the shoots and root exudates of this buckwheat accession in the presence of annual ryegrass. Especially interesting were the results of SY, ECAT, and VIT that increased their contents by more than 12 times, and of LU, which could not be detected in the control, although it appeared in Eva when co-growing with *L. rigidum*. In Eva roots co-cultured with *P. oleracea*, there was also an increase in VA, *P*-CA, LU, SY, PTA, CAT, ECAT, OR, and VIT following the pattern previously described for Gema and Kora. Surprisingly, LU was found to be increased after co-culture with both weeds, but was absent in the root tissues of the control plants of Eva. As a result, Eva root tissues accumulated in general significantly strong levels of polyphenols (Table 4).

In contrast, Eva **root exudates** were strongly increased only in the presence of *L. rigidum*, with 18 significantly increased exuded compounds (VA, DA, FA, *P*-CA, PA, SA, LU, SY, PTA, QE, VN, RU, SAA, 4-HA, CAT, ECAT, OR, and VIT) out of 20 compounds detected (Table 4). In fact, nine of the polyphenols significantly increased after *L. rigidum* co-culture did not practically appear in the control samples (when Eva was growing alone), which could not be detected in the analyses. Especially interesting is the exudation of PTA (2836-times), ECAT (1034-times), FA (511-times), DA (477-times), RU (467-times), CAT (346-times), VA (345-times), QE (304-times), OR (216-times), LU (200-times), VIT (81.6-times), SA (47-times), and VN (22.6-times). In contrast, in general, root exudates significantly decreased following the co-cultivation of Eva with *P. oleracea* and even the polyphenols DA, *P*-CA, and SY practically disappeared from the medium, while the content was only significantly increased for the polyphenols PTA, 4-HA, and CAT, which could not be detected in the root exudates when the Eva buckwheat plants were growing alone (Table 4).

2.2.2. Chemical Profile of GEMA Polyphenols

As shown in Table 5, the distribution of polyphenols in the shoots, roots, and root exudates of Gema was different depending on whether this accession was grown with *L. rigidum* or with *P. oleracea*.

Compared with the control, where Gema plants were growing alone, only the concentration of DA was significantly increased in **shoot tissues** when Gema was co-cultured with *L. rigidum* (Table 5), while the level of PTA, SAA, and 4-HA significantly increased in the shoot tissues when Gema was co-cultured with *P. oleracea*. Although the increases in PTA and SAA were not especially relevant, 4-HA increased by 24 times its concentration in the shoot tissues in the presence of *P. oleracea*. In addition, there were more polyphenols inhibited in the shoots when Gema was co-cultured with *P. oleracea* (DA, LU, SY, VN, ECAT, OR, and VIT) than with *L. rigidum* (VN, CAT, and ECAT). Curiously, the concentration of the polyphenol VN almost disappeared in the shoot tissues when Gema was co-cultured with either weed species. Similar to the shoot analyses, Gema increased significantly more polyphenols in **root tissues** when co-cultured with *P. oleracea* (DA, *P*-CA, LU, 4-HA, RU, OR, and VIT) than with *L. rigidum*, where only 4-HA and RU showed a significantly increased content. In fact, a significantly strong accumulation in the amounts of DA (56-times), LU (98-times), OR (5-times), VIT (4-times), and RU (3-times) was found in the **roots** of Gema when co-cultured with *P. oleracea* (Table 5).

Table 5. Quantification of polyphenols (PP) from shoots, roots, and root exudates from buckwheat accession Gema, alone (control) or in association with *L. rigidum* (LR) or *P. oleracea* (PO). Quantities are expressed in μg kg^{-1} dry weight (shoots and roots) and μg L^{-1} (root exudates). Means followed by different lowercase letters within a column show significant differences among treatments ($p \leq 0.05$). Bold numbers indicate a significant increase when compared with the control, while grey cells indicate a significant decrease.

Sample	Treatm.	4-CBA	VA	DA	FA	*P*-CA	SA	PA
Shoot	Alone	<0.01 a	432.7 a	22123 b	1004 a	1317 a	92.40 a	<0.01 a
	+*LR*	<0.01 a	499.3 a	**33720 a**	933.7 a	944.7 a	131.7 a	<0.01 a
	+*PO*	<0.01 a	395.1 a	417.3 c	605.0 a	2057 a	132.7 a	<0.01 a
Root	Alone	<0.01 a	232.5 ab	286.2 b	125.3 b	35.47 b	223.3 a	<0.01 a
	+*LR*	<0.01 a	343.1 a	352.8 b	203.7 ab	53.13 b	197.0 a	<0.01 a
	+*PO*	<0.01 a	140.4 b	**16210 a**	**297.7 a**	**103.0 a**	276.3 a	<0.01 a
Root exudates	Alone	2.713 ab	<0.01 a	3.397 a	7.573 a	<0.01 a	<0.01 a	2.710 ab
	+*LR*	1.133 a	<0.01 a	2.770 a	3.797 ab	<0.01 a	<0.01 a	1.130 a
	+*PO*	0.640 b	<0.01 a	<0.01 b	<0.01 b	<0.01 a	<0.01 a	0.640 b
Sample	Treatm.	LU	SY	PTA	QE	VN	SAA	RU
Shoot	Alone	146.8 a	19.64 a	593.7 b	2208 ab	22.73 a	183.7 b	399.8 a
	+*LR*	94.40 a	17.77 a	636.3 b	1447 b	<0.01 b	113.8 b	544.9 a
	+*PO*	2.870 b	<0.01 b	**2827 a**	2720 a	<0.01 b	**265.0 a**	411.1 a
Root	Alone	2.517 b	21.00 a	244.0 a	293.0 ab	159.9 a	37.53 a	475.0 b
	+*LR*	2.193 b	34.80 a	236.3 a	247.7 b	47.73ab	43.73 a	**1275 a**
	+*PO*	**245.7 a**	38.23 a	490.0 a	454.3 a	33.77 b	48.40 a	**1617 a**
Root exudates	Alone	0.560 ab	2.331 a	<0.01 a	12.73 b	<0.01 a	26.47 a	16.41 a
	+*LR*	0.420 b	1.227 a	<0.01 a	13.77 b	<0.01 a	29.47 a	17.64 a
	+*PO*	1.070 a	<0.01 b	<0.01 a	**28.77 a**	<0.01 a	25.33 a	<0.01 b
Sample	Treatm.	4-HA	CAT	ECAT	OR	VIT	HYP	
Shoot	Alone	9.240 b	1070 a	951.8 a	1642 a	7825 a	0.097 a	
	+*LR*	14.15 b	268.5 b	61.66 c	3019 a	6542 a	0.120 a	
	+*PO*	**224.0 a**	1842 a	302.0 b	259.1 b	779.9 b	0.093 a	
Root	Alone	14.14 c	2442 a	386.6 a	697.1 b	1619 b	0.033 a	
	+*LR*	**44.00 a**	2134 a	429.7 a	97.16 c	279.1 c	0.050 a	
	+*PO*	**25.96 b**	209.2 b	45.44 b	**3610 a**	**6859 a**	0.073 a	
Root exudates	Alone	<0.01 a	3.910 b	3.197 a	5.523 b	19.05 a	<0.01 a	
	+*LR*	0.167 a	**21.52 a**	4.213 a	**16.51 a**	43.40 a	<0.01 a	
	+*PO*	<0.01 a	<0.01 c	<0.01 b	6.490 b	21.14 a	<0.01 a	

4-chlorobenzoic acid (4-CBA); vanillic acid (VA); dihydroxybenzoic acid (DA); ferulic acid (FA); *p*-coumaric acid (*P*-CA); syringic acid (SA); luteolin (LU); syringaldehyde (SY); protocatechuic acid (PTA); quercetin (QE); vanillin (VN); salicylic acid (SAA); rutin (RU); 4-hydroxyacetophenone (4-HA); catechin (CAT); epicatechin (ECAT); orientin (OR); vitexin (VIT); hypericin (HYP).

Regarding the presence of polyphenols in **root exudates**, the levels of many of the specialised metabolites decreased significantly when Gema plants were grown in co-culture with *P. oleracea* (4-CBA, DA, FA, SY, RU, CAT, and ECAT) and some of them (DA, FA, SY, RU, CAT, and ECAT) practically disappeared in Gema root exudates following co-growth with common purslane. In fact, only one polyphenol (QE) was slightly more exuded in front of common purslane than when Gema plants were grown alone. In contrast, none of the polyphenols was decreased in the presence of annual ryegrass while CAT and OR significantly increased their exudation to the medium with increases of more than five times for CAT and more than three times for OR (Table 5).

2.2.3. Chemical Profile of KORA Polyphenols

When Kora was co-cultured with *L. rigidum* (Table 6), there was a general decrease in polyphenols in the roots (DA, FA, *P*-CA, LU, QE, RU, OR, and VIT) and root exudates (DA, FA, *P*-CA, SA, LU, QE, 4-HA, RU, CAT, ECAT, and VIT) that was not found for common purslane, which showed the most inhibited polyphenols in the **shoots**, with DA, FA, *P*-CA, SA, *M*-TA, LU, 4-HA, CAT, and ECAT practically disappearing from the shoot tissues, and QE, OR, RU, and VIT also significantly reduced in the presence of this weed compared with the control. As a result, no polyphenols were found to be accumulated in the shoots of Kora after co-growth with *P. oleracea*.

Table 6. Quantification of polyphenols (PP) from shoots, roots, and root exudates from buckwheat accession Kora, alone (control) or in association with *L. rigidum* (LR) or *P. oleracea* (PO). Quantities are expressed in μg kg^{-1} dry weight (shoots and roots) and μg L^{-1} (root exudates). Means followed by different lowercase letters within a column show significant differences among treatments ($p \leq 0.05$). Bold numbers indicate a significant increase when compared with the control, while grey cells indicate a significant decrease.

Sample	Treatment	DA	FA	*P*-CA	SA	*M*-TA	LU	QE
Shoot	Alone	1967 a	173.2 a	830.9 a	180.6 a	39.94 b	206.9 a	1022 a
	+*LR*	490.0 b	168.5 a	290.0 b	<0.01 b	**365.5 a**	458.6 a	406.8 b
	+*PO*	<0.01 c	<0.01 b	<0.01 c	<0.01 b	<0.01 c	<0.01 b	14.97 c
Root	Alone	476.3 b	245.3 a	659.2 a	342.1 a	<0.01 b	111.1 b	4785 a
	+*LR*	<0.01 c	21.86 b	35.86 b	147.1 a	<0.01 b	22.04 c	476,5 b
	+*PO*	**7609 a**	230.6 a	674.9 a	<0.01 b	**919.2 a**	**1178 a**	3124 ab
Root exudates	Alone	239.8 b	42.34 b	90.87 b	41.42 a	<0.01 b	18.23 a	2296 a
	+*LR*	<0.01 c	<0.01 c	<0.01 c	<0.01 b	<0.01 b	<0.01 b	13.77 c
	+*PO*	**2373 a**	**238.0 a**	**289.9 a**	24.96 a	**321.1 a**	13.32 a	109.5 b
Sample	Treatment	VN	4-HA	CAT	ECAT	OR	VIT	RU
Shoot	Alone	42.96 a	13,541 b	2230 a	516.8 a	4788 a	5632 a	176.5 a
	+*LR*	34.38 a	**33,077 a**	18.25 b	<0.01 b	2589 ab	4121 a	<0.01 b
	+*PO*	57.69 a	<0.01 c	<0.01 c	<0.01 b	2436 b	31.52 b	1.791 b
Root	Alone	23.95 ab	2214 b	2440 a	463.8 a	472.1 b	793.9 b	1041 a
	+*LR*	65.56 a	2199 b	2264 a	513.6 a	36.16 c	39.10 c	64.32 b
	+*PO*	25.70 b	**38,173 a**	64.30 b	8.760 b	**2650 a**	**5490 a**	<0.01 b
Root exudates	Alone	134.5 a	1412 b	462.6 a	68.51 a	283.0 b	224.2 a	109.1 a
	+*LR*	106.7 ab	<0.01 c	<0.01 c	<0.01 b	**3362 a**	19.57 b	<0.01 b
	+*PO*	66.41 b	**205,000 a**	229.9 b	32.26 a	**1531 a**	269.7 a	132.3 a

Dihydroxybenzoic acid (DA); ferulic acid (FA); *p*-coumaric acid (*P*-CA); syringic acid (SA); *m*-toluic acid (*M*-TA); luteolin (LU); quercetin (QE); vanillin (VN); 4-hydroxyacetophenone (4-HA); catechin (CAT); epicatechin (ECAT); orientin (OR); vitexin (VIT); rutin (RU).

In contrast, once more, different polyphenols were found to be significantly accumulated in the presence of *P. oleracea* in **root tissues**, as *M*-TA (919-times), 4-HA (17-times), DA (16-times), LU (10-times), VIT (7-times), and OR (5.6-times), while most of the polyphenols, were strongly reduced after co-cultivation with *L. rigidum*, as already explained above.

This reduction in the polyphenols in root tissues of annual ryegrass was consistent with the general reduction of polyphenols found in Kora **root exudates** when co-grown with *L. rigidum*, with many of them, such as DA, FA, *P*-CA, SA, LU, 4-HA, CAT, and ECAT, practically disappearing from the root exudates following co-growth with *L. rigidum*. In contrast, Kora strongly exuded a higher amount of several phenolic acids and flavonoids in the presence of *P. oleracea*, such as *M*-TA (321-times), 4-HA (145-times), DA (10-times), FA (5.6-times), OR (5.4-times), and *P*-CA (3.2-times). Finally, OR was the only polyphenol more exuded when Kora was co-cultivated with *L. rigidum* compared with the control (Table 6).

2.2.4. Chemical Profile of PI481671 Polyphenols

In general, the root and shoot tissues of PI481671 were more sensitive to the presence of *P. oleracea* than *L. rigidum* in the medium (Table 7). In fact, while only three polyphenols were altered in **PI481671 shoot tissues** in the presence of annual ryegrass (increases in *M*-TA and QE, and a decrease in PA), the content of seven polyphenols was altered in the presence of common purslane (increases in DA, FA, SAA, 4-HA, and ECAT, and decreases in VA and PA). This pattern was even more obvious in the **root tissues**, where the content of four polyphenols was altered in the presence of annual ryegrass, while the content of up to 14 polyphenols was altered in the presence of common purslane. In fact, after co-growth with *P. oleracea*, the **root tissues** of PI481671 accumulated strong amounts of polyphenols, showing a significant increase in the concentration of different specialized metabolites, such as VA, DA, FA, *P*-CA, SA, VN, SAA, 4-HA, RU, and ECAT. Especially interesting are the strong increases of ECAT (200-times), RU (26-times), and SAA (23-times). In contrast, the results only showed a significant increase in one polyphenol, OR (278-times), in the root tissues after the co-cultivation of PI481671 with *L. rigidum* (Table 7).

Table 7. Quantification of polyphenols (PP) from the shoots, roots, and root exudates from buckwheat accession PI481671, alone (control) or in association with *L. rigidum* (LR) or *P. oleracea* (PO). Quantities are expressed in μg kg^{-1} dry weight (shoots and roots) and μg L^{-1} (root exudates). Means followed by different lowercase letters within a column show significant differences among treatments ($p \leq 0.05$). Bold numbers indicate a significant increase when compared with the control while grey cells indicate a significant decrease.

Sample	Treatment	VA	DA	FA	*P*-CA	PA	SA
Shoot	Alone	911.4 a	1703 b	190.4 b	355.5 ab	74.54 a	112.2 a
	+*LR*	930.7 a	1597 b	190.1 b	293.6 b	61.38 a	136.6 a
	+*PO*	540.5 b	**5940 a**	**480.0 a**	472.6 a	8.597 b	79.17 a
Root	Alone	325.9 b	97.73 b	27.52 b	17.08 b	20.91 a	78.81 b
	+*LR*	368.4 b	71.74 b	71.02 b	22.65 b	3.853 b	84.35 b
	+*PO*	**2146 a**	**276.8 a**	**283.2 a**	**148.9 a**	5.613 b	**360.9 a**
Root exudates	Alone	<0.01 a	<0.01 b	<0.01 a	<0.01 a	26.71 a	<0.01 b
	+*LR*	<0.01 a	**18.48 a**	<0.01 a	<0.01 a	3.300 b	**51.82 a**
	+*PO*	<0.01 a	<0.01 b	<0.01 a	<0.01 a	2.660 b	<0.01 b
Sample	**Treatment**	***M*-TA**	**LU**	**SY**	**QE**	**VN**	**SAA**
Shoot	Alone	<0.01 b	144.9 a	19.62 a	16,993 b	49.77 a	9.467 b
	+*LR*	**3.497 a**	115.5 a	11.41 a	**19,610 a**	37.03 a	14.95 b
	+*PO*	<0.01 b	85.45 a	12.37 a	19,429 ab	25.56 a	**69.19 a**
Root	Alone	13.28 a	21.93 a	7.620 b	4372 a	72.55 b	5.537 b
	+*LR*	<0.01 b	23.68 a	7.293 b	3850 a	37.30 b	8.903 b
	+*PO*	<0.01 b	32.50 a	**30.95 a**	5581 a	**417.6 a**	**131.5 a**
Root exudates	Alone	33.09 a	4.397 a	<0.01 a	535.7 b	57.48 a	<0.01 a
	+*LR*	18.50 a	3.793 a	<0.01 a	**1668 a**	<0.01 b	<0.01 a
	+*PO*	13.27 a	2.467 a	<0.01 a	546.5 b	<0.01 b	<0.01 a
Sample	Treatment	4-HA	ECAT	OR	VIT	RU	
Shoot	Alone	1756 b	330.0 b	98.37 ab	1786 ab	905.3 a	
	+*LR*	1.177 c	175.0 b	101.9 a	2236 a	105.2 a	
	+*PO*	**4260 a**	**6603 a**	38.71 b	1311 b	39.05 a	
Root	Alone	1754 b	23.50 b	22.12 b	193.5 a	57.65 b	
	+*LR*	0.907 c	19.83 b	**6157 a**	229.5 a	31.10 b	
	+*PO*	**4461 a**	**4807 a**	7.701 c	167.1 a	**327.5 a**	
Root exudates	Alone	<0.01 a	1.543 b	0.274 ab	22.57 b	15.84 a	
	+*LR*	<0.01 a	**401.3 a**	0.586 a	25.21 b	18.83 a	
	+*PO*	<0.01 a	1.247 b	<0.01 b	**4404** a	6.040 a	

Vanillic acid (VA); dihydroxybenzoic acid (DA); ferulic acid (FA); *p*-coumaric acid (*P*-CA); phthalic acid (PA); syringic acid (SA); *m*-toluic acid (*M*-TA); luteolin (LU); syringaldehyde (SY); quercetin (QE); vanillin (VN); salicylic acid (SAA); 4-hydroxyacetophenone (4-HA); epicatechin (ECAT); orientin (OR); vitexin (VIT); rutin (RU).

Although there was a general lower **root exudation** in the Tartary buckwheat accession PI481671 when compared with the common buckwheat accessions (Kora, Gema, and Eva), there was a significant increase in the root exudates of DA, SA, QE, and ECAT following the co-cultivation of PI481671 with *L. rigidum*, indicating that the levels of polyphenols in the root exudates had a propensity to rise. In contrast, only VIT increased in the root exudates after co-cultivation with *P. oleracea*.

2.2.5. Multivariate Analyses of Polyphenols' Profile of Shoots, Roots, and Root Exudates of the Four Buckwheat Accessions

When independently comparing the polyphenol profile of the shoots (Figure 2A), roots (Figure 2B) and root exudates (Figure 2C) of Gema, Eva, Kora, and PI481671, the results of the PLS-DA analysis further exacerbated the separation among accessions for both shoots and roots (Figure 2A left, Figure 2B left, respectively), but not for the root exudates. The root exudates of each accession alone or in co-culture with *L. rigidum* or *P. oleracea* were separated and, at the same time, overlapped with the samples of the other accessions (Figure 2C left). PLS-DA explained a total variance of 54.2% for shoots (component 1: 23% and component 2: 31.2%), 46.1% for roots (component 1: 29.4% and component 2: 16.7%), and 60.5% for root exudates (component 1 23.1%: and component 2: 37.4%). In fact, the polyphenol profile of the roots of the accessions Gema (G), Eva (E), Kora (K), and PI481671 (PI) were perfectly separated from each other but perfectly grouped into each accession for all of the plants growing alone (A) or co-cultured with *L. rigidum* (L) or *P. oleracea* (P) (Figure 2B left). The same was found for the shoot polyphenol profile of the accessions Gema, PI481671, and Eva (Figure 2A left). In contrast, no clear groups were found for root exudates, where the polyphenol profile of the plants of the different accessions growing alone or in co-culture with *L. rigidum* or *P. oleracea* were completely overlapped for the four accessions tested in this study (Figure 2C left).

The variable importance in projection (VIP) scores (built on the polyphenols with a VIP score higher than 1.0) revealed that PTA and *M*-TA were the only polyphenols with VIP scores higher than 1.0 that were found in all of the analyses (shoots, roots, and root exudates), while CAT and RU were also common for the shoots and roots. In particular, the compounds with a higher VIP score in the shoots were, in order of importance, CAT, PTA, RU, VN, *M*-TA, 4-HA, and OR, while in the roots they were LU, CAT, PTA, QE, *M*-TA, RU, and PA, and in the root exudates they were OR, 4-CBA, SAA, SY, PTA, PA, and *M*-TA (Figure 2A right, Figure 2B right, Figure 2C right, and Table 8).

Table 8. Variable importance of projection (VIP) features for the groups from the PLS-DA analysis for roots, shoots, and root exudates of the three common buckwheat accessions Eva (E), Gema (G), and Kora (K), and the Tartary accession PI481671 (PI) growing alone or in co-culture with *L. rigidum* or *P. oleracea*. The compounds included in the table are those compounds with a VIP score higher than 1.0 for each type of sample (shoots, roots, and root exudates). Bold numbers indicate compounds with important VIP score in the shoots, roots, and root exudates, while italic numbers indicate compounds with significant VIP score in the shoots and roots.

	Shoots	Roots	Root Exudates
CAT	*1.954*	*1.929*	
PTA	***1.868***	***1.632***	**1.279**
RU	*1.605*	*1.111*	
VN	1.582		
M-TA	***1.572***	***1.343***	**1.183**
4-HA	1.321		
OR	1.293		1.722
LU		1.955	
QE		1.626	
PA		1.047	1.24
4-CBA			1.607

Table 8. *Cont.*

	Shoots	Roots	Root Exudates
SAA			1.521
SY			1.481

Catechin (CAT); protocatechuic acid (PTA); rutin (RU); vanillin (VN); *m*-toluic acid (*M*-TA); 4-hydroxyacetophenone (4-HA); orientin (OR); luteolin (LU); quercetin (QE); phthalic acid (PA); 4-chlorobenzoic acid (4-CBA); salicylic acid (SAA); syringaldehyde (SY).

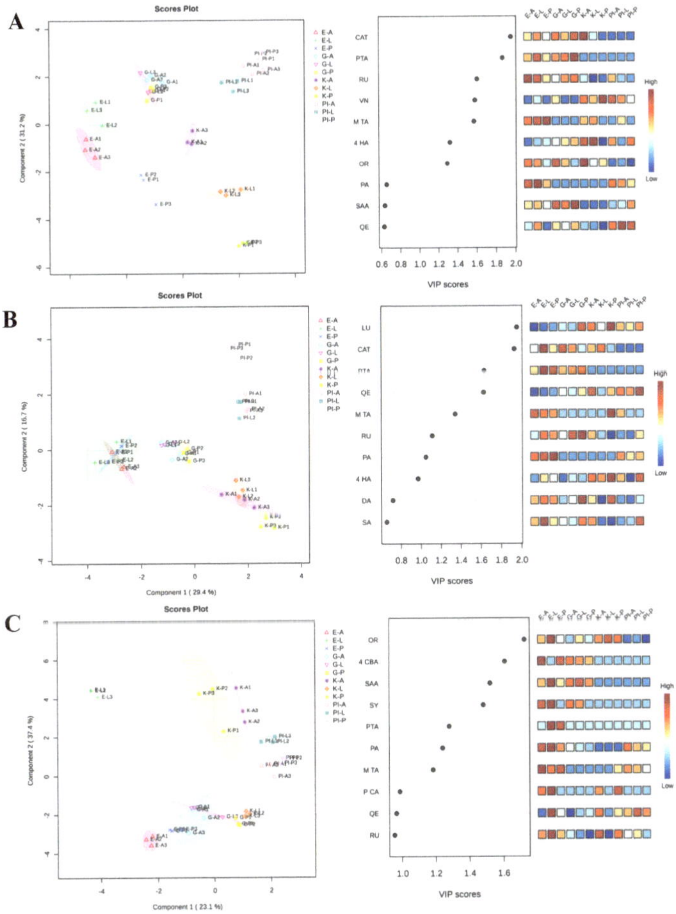

Figure 2. Partial least square discriminant analysis—(PLS-DA) multivariate analysis (left, scores plot between the selected PCs, the explained variances of PLS-DA are shown in brackets) and variable importance of projection (VIP) features (right, the coloured boxes on the right indicate the relative concentrations of the corresponding metabolite in each group under study) for the groups from the PLS-DA analysis, of the polyphenols' profile in the (**A**) shoots, (**B**) roots, and (**C**) root exudates of the three common buckwheat accessions Eva (E), Gema (G), and Kora (K) and the Tartary accession PI481671 (PI) growing alone (A) or in co-culture with *L. rigidum* (L) or *P. oleracea* (P). E-A (Eva growing alone; red colour), G-A (Gema growing alone; light blue), K-A (Kora growing alone; violet), PI-A (PI481671 growing alone; light pink), E-L (Eva growing with *Lolium*; dark green), G-L (Gema growing with *Lolium*; dark pink), K-L (Kora growing with *Lolium*; orange), PI-L (PI481671 growing with *Lolium*; blue green), E-P (Eva growing with *Portulaca*; dark blue), G-P (Gema growing with *Portulaca*; yellow), K-P (Kora growing with *Portulaca*; light green), and PI-P (PI481671 growing with *Portulaca*; malva). N = 3.

3. Discussion

Crop species with allelopathic activity are known as good options for reducing weed damage in sustainable agroecosystems [39]. Although the allelochemical potential of buckwheat crops to manage weeds in the field has not been deeply studied up until now, different papers suggest the presence of bioactive compounds on their extracts and residues that can control the development of different weeds [23,26]. Moreover, different genotypes may produce specialized metabolites differently, opening a wide variety of allelopathic potentials and, consequently, suppressive effects on weeds [40,41]. For this reason, studies such as those carried out in this work, evaluating different varieties of buckwheat that, by themselves, as a crop (alive plants), can control the presence of weeds in their environment, are highly indispensable.

The current study provides further evidence that phenolic compound synthesis, distribution, and exudation vary among buckwheat accessions, and that these compounds play a role in the interference among plants of crops and weeds. Phenolic compounds are specialized metabolites that can behave as phytotoxic when exuded into the medium, affecting the growth of neighbouring plants, as reported for several phenolic acids [42]. Meanwhile, the identification of allelochemical substances and their particular mode of action and interference in different physiological processes is required to make use of the allelopathic capabilities of crop plants in weed control [43]. Additionally, previous studies revealed that production of phenolic compounds differ greatly between buckwheat accessions in the different tissues [40,44].

Phenolic compounds, and especially flavonoids, have been reported for decades as strong antioxidant compounds, behaving as protectors into the plant metabolism against any external biotic or abiotic damage to which the plant can be exposed [45,46]. In fact, flavonoids are the most reported specialized metabolites in the plant defence system [47], so they may play an important role in plant—plant competition other than allelopathy, by accumulating in the different organs of the plant (leaves, roots, stems, seeds, etc.), and make plants more resistant and resilient against external attacks [48].

Moreover, as reported by Uddin et al. [41], different buckwheat cultivars can show different contents of phenolic compounds, and even the same cultivar can show organ-related differences in the phenolic composition. Studying three common buckwheat cultivars (Suwon1, Suwon 2, and Suwon 12), they found that Suwon 1 had the highest levels of catechin and epicatechin, while the greatest amount of 4-hydroxybenzoic acid, chlorogenic acid, and 4-hydroxy-3-methoxybenzoic acid was present in the cultivar Suwon 2. In this context, the Suwon 2 cultivar dominated over the other two cultivars, with the highest phenolic compound content in the stem, flowers, and roots of common buckwheat. Something similar was found in this study, as when comparing the four accessions, the hierarchical cluster showed more dissimilarities between the samples, grouping shoots, and roots separately from the root exudates, than between species, as common and Tartary accessions were grouped together in the same branch of the dendrogram, although each accession was separated from the others in the sub-branches of the tree.

When analysing the response of the different buckwheat accessions to the weeds, our results showed that the most relevant polyphenolic compounds were DA, LU, 4-HA, CAT, ECAT, and OR, as were the compounds whose chemical profile changed more in the roots or shoots and root exudates of the different buckwheat accessions along the study.

4-HA was discovered in buckwheat root exudates and in soil extracts following buckwheat cultivation, in addition to flavonoids and phenolic acids [26]. On the other hand, CAT was found to be highly phytotoxic against *Arabidopsis thaliana* (L.) Heynh. and *Festuca idahoensis* Elmer [49]. According to previous research, the phytotoxicity of CAT on the root cell tissues of *A. thaliana* is caused by the cytoplasm condensing due to the rapid induction of reactive oxygen species, which is followed by an increase in Ca^{2+} and acidification of the cytoplasm, resulting in cell death [50]. In our results, a strong significant increase in the root exudates (346-fold higher than the control) of CAT was observed in buckwheat variety Eva when co-cultured with *L. rigidum*. Golisz et al. [25] also established the effective concentrations for lettuce to be in this range (0.4 mM). These findings suggest that CAT is highly phytotoxic, but less selective against different weeds. In a similar way, Serniak [51] showed, in a comparative allelochemical study, that ECAT exerted strong phytotoxic effects on radish seedling growth. Moreover, ECAT significantly decreased the growth of *Lepidium sativum* L. [52], and radish root growth was also inhibited in vivo as the result of the phytotoxic activity of ECAT [51].

Because of the quick evolution of resistance in *L. rigidum* and *P. oleracea*, comprehensive weed management measures including crop allelopathic varieties are required to slow down this rapid evolution and promote sustainable control [53]. Determination of the mechanism(s) associated with weed suppression is essential to determine if the use of crop varieties for allelopathic and competitive weed suppression in cereal and pseudo-cereal crops is going to provide sustainable solutions for weed management and to overcome resistance problems in weeds.

In this study, the variety Gema showed the strongest crop competitive ability against mono and dicot weeds compared with the other buckwheat accessions of Eva, Kora, and PI481671. This common buckwheat accession greatly inhibited the germination and root length of the monocot *L. rigidum* in more than 70% when compared with the control (i.e., *L. rigidum* growing alone). Strong effects of Gema were also observed on the dicot weed *P. oleracea*, where shoot and root lengths were stimulated, while no increases in fresh plant weight could be detected, resulting in longer, but much weaker, shoots and roots. Gema accumulated more DA, FA, *P*-CA, LU, 4-HA, OR, RU, and VIT in the roots when co-cultured with *P. oleracea*, while QE was the only polyphenol significantly more exuded to the medium after co-growing with this dicot weed. Previous works have reported that buckwheat varieties can accumulate polyphenols in the roots and shoots as a means of defence or protection [27,54]. Our results indicate that strong competition may be taking place between Gema and *P. oleracea*, and that *P. oleracea* might be trying to colonize more space (via longer roots) at the cost of making its roots weaker. The pressure that Gema has on *P. oleracea* can be related to the significant increase in root length of Gema in the presence of this weed, which would be competing with *P. oleracea* by colonizing the medium. In this context, flavonoid accumulation in the roots might be protecting Gema in front of this dicot weed. Root exudation represents a carbon cost to the plant [55]; therefore, the reduced root exudation of polyphenols could save energy that the buckwheat plant could use for defence or protection against *P. oleracea*. Several researchers have reported the antioxidant properties of flavonoids from different buckwheat varieties [45,56]. In this sense, the significant increase in some polyphenols in the roots and shoots of Gema plants could be protecting them from the damage induced by the presence of this dicot weed.

Allelochemical plants, such as buckwheat, have distinct mechanisms for inducing the phytotoxic effects on monocot and dicot weeds, so that biological action on the target weed differs from one weed to another [24,26]. In this sense, the behaviour of Gema with the monocot weed *L. rigidum* was totally different than with *P. oleracea*. In fact, no alterations in leaf and root weight or shoot and root length were observed in Gema plants when co-growing with *L. rigidum*. On the contrary, the germination and root length of *L. rigidum* were strongly inhibited by Gema, with 80% and 70% inhibitions, respectively. Gema increased the exudation of CAT by more than five times and OR by more than three times in the presence of *L. rigidum*, which could be enough to inhibit the germination and growth of *L. rigidum*, as there is no relevant accumulation of polyphenols on the roots or shoots of buckwheat plants and neither growth parameters of Gema plants are affected in front of this monocot weed, which suggests that *L. rigidum* does not represent a threat to Gema plants. The phytotoxic activity of CAT and OR would be enough for Gema to handle *L. rigidum* development [22].

The next accession with a strong capacity to sustainably control weeds was the Eva variety. This common buckwheat accession greatly inhibited the germination and fresh plant weight of the monocot *L. rigidum*, while it strongly stimulated the shoot length of this weed, which resulted in longer but weaker plants. Something similar was observed on the dicot *P. oleracea*, where Eva stimulated the shoot length but did not increase the plant weight. We revealed that this competitive ability of Eva was related to its robust root exudation of different polyphenols (phenolic acids and flavonoids), such as VA, DA, FA, *P*-CA, PA, SA, LU, SY, PTA, QE, VN, 4-HA, RU, CAT, ECAT, OR, and VIT, when co-cultured with both weeds; this was especially true for *L. rigidum*, which induced a significant increase in the exuded content for 16 polyphenols out of the 19 analysed. This increased the production and exudation of different polyphenols, especially PTA (2836-times), ECAT (1034-times) FA (511-times), DA (477-times), RU (467-times), CAT (346-times), VA (345-times), QE (305-times), OR (215-times), and LU (200-times) by Eva compared with the control, which would ensure the success in the inhibition of the germination and development of *L. rigidum* and in the induction of increased weakness in *P. oleracea*. Our results demonstrated that when buckwheat recognizes the presence of the weeds it subsequently changes its root exudation profile to impede their growth. These results are consistent with those found by Gfeller et al. [57] for buckwheat in the presence of redroot pigweed. Moreover, although previous studies [20,54] have suggested the accumulation and exudation of RU as the responsible allelochemical molecule to inhibit the growth of different weeds, our results showed that there are a plethora of compounds participating in this phenomenon, and that there are other polyphenols, such as PTA, ECAT, or FA, that can playing an even more strong allelochemical role than RU. In this sense, in recent research, Krumsri et al. [58] evaluated the phytotoxic potential of *Dalbergia cochinchinensis* Pierre ex Laness. and found that PTA, the most exuded compound by Eva roots, caused growth inhibition on *Echinochloa crus-galli* (L.) P. Beauv. and *L. sativum* at low concentrations. At concentrations greater than 10 mM, ECAT, the second most exuded compound by Eva, significantly decreased the growth of *L. sativum* [52]. Radish root growth was also inhibited in vivo because of the ECAT phytotoxic activity [51]. In another study, Hussain and Reigosa [43] evaluated the effects of FA and DA on the photosynthesis of *Rumex acetosa* L., and found that both compounds behaved as potent inhibitors of photosynthetic traits, leading to weaker plants. The strong increase in the root exudation by living plants indicates that Eva molecules attacked the herbicide-resistant weeds (*L. rigidum* and *P. oleracea*), inhibiting the germination of *L. rigidum* and hindering the development of both monocot and dicot weeds. The exudation of several flavonoids (QE, VN, 4-HA, CAT, ECAT, OR, RU, and VIT) demonstrates that the defence strategy of Eva is alive and working closely with the attacking phenomena to obtain access to the available resources (space and light) for its growth and development. Moreover, polyphenols can also play a role of defence and protection in the plant, as previously demonstrated by several authors [27,54]. In this context, Eva also significantly increased the production of several polyphenols in the shoots (VA, DA, FA, *P*-CA, LU, SY, PTA, QE, VN, and 4-HA) and roots (*P*-CA, LU, SY, PTA, 4-HA, CAT, ECAT, OR, and VIT) after growing with *L. rigidum*, and in root tissues (VA, *P*-CA, LU, SY, PTA, CAT, ECAT, OR, and VIT) after growing with *P. oleracea*. Especially interesting were the polyphenols *P*-CA, LU, SY, PTA, CAT, ECAT, OR, and VIT, which were found to increase in the root tissues in the presence of both weeds in very high concentrations.

Kora was the common buckwheat accession that affected the development of *L. rigidum* and *P. oleracea* less in the present study, although it showed a strong effect on the germination of both monocot and dicot weeds. Although Kora did not exude phenolic compounds and flavonoids after growth with *L. rigidum*, the accumulation in the roots and shoots of different polyphenols, such as *M*-TA, 4-HA, and OR could improve the antioxidant activity in buckwheat plants, providing an advantage in plant−plant competition [59].

In contrast, Kora exuded significant amounts of DA, FA, *P*-CA, *M*-TA, 4-HA, and OR after growth with *P. oleracea*, which were not only related to the strongest decrease in *P. oleracea* germination for the four tested accessions, but also to the chemical control of *P. oleracea* by Kora, where the weed seedlings could normally grow, without affecting the growth and development of Kora plants.

The Tartary buckwheat accession PI481671 followed a similar chemical profile and buckwheat plant development to the common buckwheat accession Gema, although this accession did not affect the weeds in a similar pattern. PI481671 stimulated the weed total biomass and shoot length of *P. oleracea* while inhibiting the total weight of *L. rigidum*. This could be explained by the results previously found by Sijahović et al. [38], who demonstrated that buckwheat–weed interactions are dependent on the type of weed present in the neighbours, resulting in changes in the exudation behaviour of buckwheat plants. The root tissues of PI481671 indeed accumulated considerable levels of polyphenols after co-growing with *P. oleracea*, displaying a marked rise in the content of many specialized metabolites, including VA, DA, FA, *P*-CA, SA, VN, SAA, RU, 4-HA, and ECAT. The significant increases in ECAT (200 times) and SAA (23 times), a well-known defence compound, are particularly fascinating. Competitive genotypes can better access light, nutrients, and water resources in limited space, thus suppressing the growth and reproduction of nearby weed species [37]. Although the Tartary buckwheat (PI481671) showed generally less root exudation than the common buckwheat accessions, PI481671 increased, as Eva, the exudation of DA, SA, and QE, by several folds after co-culture with *L. rigidum*. In a recent study, Šćepanović et al. [60] showed that strong doses of VA, DHA, and *P*-CA, as well as the phenolic acid mixture, inhibited the early growth of *Ambrosia artemisiifolia* L.

As shown by the multivariate analyses, in this study, no differences were found among species (*F. esculentum* and *F. tataricum*), but among the analysed samples. In fact, the roots and shoots were grouped separately from the root exudates. When having a look at the compounds with a higher VIP score (weight) for the comparison of the different samples, the polyphenols PTA and *M*-TA were common to all of the analyses (shoots, roots, and root exudates), while CAT and RU were also common for the shoots and roots. Most of these compounds (i.e., CAT, RU, and PTA) have been shown to have antioxidant properties [61–63] against different stress factors, which could be one main reason for their accumulation in the roots and shoots of common and Tartary buckwheat accessions, giving an advantage to buckwheat plants in front of the surrounding weeds.

Our findings show that different buckwheat accessions have varying capacities to release or accumulate specific metabolites in the presence of surrounding weeds, as well as varying capacities to manage those weeds sustainably. Additionally, each accession exhibited varied the inhibitory capacities and chemical profiles against monocot or dicot weeds, depending on the type of weed in their vicinity [20,25]. These results are consistent with those of Kalinova [30], who found varietal differences for the inhibition of lettuce by three different buckwheat varieties and related these differences to allelochemical action of buckwheat by measuring the production of the known allelochemical compounds ECAT and RU.

The findings of this study indicate that the buckwheat accessions that most significantly impacted the growth of the tested weeds were those with the highest production of allelopathic compounds and their exudation into the rhizosphere. The inhibitory effect on weed germination and growth could be caused by the allelochemicals that were exuded to the medium, because in the current experiments, there was no direct physical contact between the roots or shoots of the buckwheat and weeds. This was particularly true for the variety Eva, which demonstrated a high potential for controlling monocot weed *L. rigidum* trough root exudates. However, the superior competitive ability that the accumulation of polyphenols in shoots and roots provided buckwheat plants in front of weeds could be driving the negative impact of the tested buckwheat accessions on the two target weeds, even though these weeds showed resistance against different herbicides.

The present results highlight the necessity to screen different buckwheat accessions to find the better ones to be used in organic agriculture, due to the variation in the synthesis, distribution, and exudation of polyphenols, which can provide a different allelopathic or competitive ability to different accessions.

4. Materials and Methods

4.1. Germination and Growth Bioassays

Four different buckwheat accessions (Gema, Kora, Eva, and PI481671), previously pre-selected for their potential to be used in organic farming (in the frame of the EU project ECOBREED), were screened for their allelopathic potential against germination and seedling growth of common purslane (*Portulaca oleracea* L.) and annual ryegrass (*Lolium rigidum* Gaud.) using perlite as an inert substrate. Eva is a commercial variety from Kmetijski inštitut Slovenije/Agricultural Institute of Slovenia (KIS), Slovenia. Gema, Kora, and PI481671 were provided by the Czech Gene Bank—Crop Research Institute of Prague (Czech Republic).

Identification Number	Species	Name	Origin
01Z5000095	*Fagopyrum esculentum*	Kora	Poland
01Z5000112	*Fagopyrum esculentum*	Gema	Poland
01Z5100009	*Fagopyrum tataricum*	PI 481671	Bhutan

Seeds of common purslane and annual ryegrass were obtained commercially from "Semillas Cantueso" (Cantueso Natural Seeds, Cordoba, Spain) and Herbiseed (Herbiseed Twyford, Berkshire, UK), respectively. Surface-sterilized seeds of each buckwheat accession were grown alone in individual plastic trays (32 × 20 × 6 cm) filled with a 5 cm deep layer of perlite (500 g/tray), watered with distilled water, and kept in a growth chamber with a day and night temperature of 20 °C and 12/12 h light/dark photoperiod. The light was supplied by cool white fluorescent tubes and irradiance was maintained at 275 μmol m^{-2} s^{-1}. The buckwheat seeds were sown and left to germinate for 5 days before transferring to treatment plastic trays.

Ten buckwheat seedlings per plastic tray and three plastic trays per treatment were used in these experiments for each common and Tartary buckwheat accession. At day 1, similar seedlings for each buckwheat accession were selected and sown in a plastic tray with perlite and were watered each other day. The seedlings were placed in three rows on one half of the tray and placed in a controlled environment growth cabinet with a daily photoperiod of 12L:12D and continuous temperature of 20 °C. After 10 days of growth for the buckwheat accessions, *L. rigidum* or *P. oleracea* seeds were added to the other half of the tray. The arrangement was such that the allelochemicals produced and released by the buckwheat seedlings could diffuse throughout the perlite medium to influence weed germination and growth, but no physical contact was allowed among the roots or shoots of buckwheat seedlings and weeds.

After growing together (buckwheat and weeds) under the same conditions for 7 days, the germination rate, total weight, shoot and root length, and plant height were measured in the two target weeds, *L. rigidum* and *P. oleracea*.

The five treatments established for each buckwheat accession were as follows:

(1) Buckwheat alone (Gema, Kora, Eva, or PI481671);
(2) Buckwheat seedlings (10 days germinated) + *L. rigidum* (10 ungerminated seeds);
(3) Buckwheat seedlings (10 days germinated) + *P. oleracea* (10 ungerminated seeds);
(4) *P. oleracea* alone;
(5) *L. rigidum* alone.

4.2. Plant Harvest and Metabolite Extraction

After harvesting the shoots and roots of the buckwheat seedlings and collecting the root exudates from each treatment tray, the samples were stored at −80 °C until extraction. The samples from each buckwheat accession were separately processed for the identification and quantification of phenolic compounds and flavonoids from the shoots, roots, and root exudates. The plant tissues (shoots and roots) were lyophilized, ground into powder with a mortar and pestle after the addition of liquid nitrogen, and macerated with 9 mL of HCl

(1 mM). Afterwards, the whole solution was transferred into a vial, sonicated for 15 min (Branson SINIFIER 250; microTip limit, output 3), and centrifuged at 20,000 rpm at 10 °C for 15 min (SORVALL RC 5B Plus, Du Pont). The supernatant was collected and extracted three times with diethyl ether (DE). The aqueous layers were discarded, and the corresponding organic layers were combined. The organic phases were evaporated in a multivapor (P-12; Buchi, Switzerland) with 12 simultaneous evaporating positions. The multivapor (P-12) comprised of a vacuum pump (V-700), vacuum controller (V-850), rotavapor (R-210), heating bath (B-491), and recirculating chiller (F-105). The temperature of the recirculating chiller was set to −10 °C. The organic layers (DE) were placed in 15 mL plastic tubes and attached to the multivapor to evaporate the organic solvent under reduced pressure (456 mbar for DE) at 35 °C. Because of the chilling temperature (−10 °C) of recirculating chiller, and the temperature of the heating bath (35 °C), the organic solvent was evaporated and condensed in the attached crystal balloon. The final volume of the residual solution was approximately 1 mL and this solution was further dried with N_2. Methanol was used to dissolve the residual powder and was injected for the LC-MS analysis.

The perlite-based nutrient-free water-growth medium was collected and adjusted to pH 3.0 with 0.06 M HCl. Then, 25 mL of the root exudated water was extracted three times with 25 mL DE. Further treatment of the root exudated water samples was identical to the preparation of shoots or roots extraction as described above.

4.3. Identification of Phenolic Compounds by LC-MS

Shoots, roots, and root exudates were extracted from each buckwheat accession using diethyl ether. Specialized metabolites (phenolic acids and flavonoids) were separated on ultra-high performance liquid chromatography coupled to a quadrupole time-of-flight high-definition mass spectrometry detector (UHPLC-qTOF-MS, Thermo Fisher Scientific Inc., Madrid, Spain) according to Wu et al. [64] with small slight amendments according to Hussain et al. [65].

High Performance Liquid Chromatography−Mass Spectrometry (HPLC-MS, 1260 Series, Agilent, Santa Clara, CA, USA) was performed using a system consisted of compact mass detector equipment (TRIPLE QUAD 3500; AB SCIEX INSTRUMENTS, AB Sciex Pte. Ltd., Framingham, MA, USA). Polyphenols were separated with a C18 column (PHENOMENEX LUNA, 150 mm × 2 mm and 3 µm, Phenomenex, Inc., Torrance, CA, USA) using different chromatographic conditions depending of the compounds. Hypericin was separated at a flow rate of 400 µL min^{-1} with a column temperature of 40 °C, an injection volume of 10 µL, and the column was equilibrated for 6 min between runs. The isocratic elution used was a mixture of two solvents: A, consisting of 5 mM ammonium acetate and 0.1% acetic acid in water, and B, consisting of acetonitrile. The isocratic conditions were 25% A and 75% B for 10 min. The other phenolic acids were separated at a flow rate of 300 µL min^{-1}, the column temperature was 40 °C, the injection volume was 10 µL, and the column was equilibrated for 6 min between runs. The gradient elution used was a mixture of two solvents: A, consisting of 0.1% formic acid in water, and B, consisting of 0.1% formic acid in acetonitrile. Initial conditions (98% A and 2% B) were held for 4 min before ramping to 20% B at 7 min and 90% B at 14 min. Initial conditions were recovered at 15 min and held until 21 min. The instrument parameters were as follows: curtain gas (CUR), 25 psi; collision gas (CAD), 7 psi: ion spray voltage (IS), −4500 V; temperature (TEM), 400 °C; ion source gas 1 (GS1), 55 psi; ion source gas 2 (GS2), 55 psi; interface heater, on.

The quantification of the concentration of the compounds was obtained from calibration curves that related the detector's response to the pure analyte's concentration of those compounds identified in the chemical analyses.

4.4. Data Analyses

The experiments were carried out using a completely randomized design with three replications (each replication was a bulk of 10 buckwheat plants and 10 weed plants). IBM SPSS software (SPSS Inc., Chicago, IL, USA, version 22.0) was used to analyse the data. To detect outliers, an exploratory data analysis was performed. The Kolmogorov–Smirnov test was used to check for deviation from normality, and the Levene test was used to check for homogeneity. Depending on the homoscedasticity of the samples, one-way ANOVA or Kruskal–Wallis tests were performed for germination and seedling growth data to establish the significant effect ($p \leq 0.05$) of the treatments (different accessions). The results are presented in the tables as the percentage of increase or decrease when compared with the control. Different letters represent significant differences in treatment. Polyphenol (phenolic compounds and flavonoids) data were analysed through analysis of variance and the Duncan multiple range test was performed to establish the significant effect ($p \leq 0.05$) within the treatments (alone, co-cultured with *L. rigidum*, or co-cultured with *P. oleracea*).

The identified polyphenols were analysed using the Metaboanalyst 5.0 software. The missing values were replaced with half of the minimum value found, and then data were Log_{10} transformed and Pareto scaled. Data were then analysed through unsupervised principal component analysis (PCA), to visualize group discrimination, and through the supervised partial least square discriminant analysis (PLS-DA). Feature selection with the highest discriminatory power was based on their variable importance in projection (VIP) score > 1.0. To avoid overfitting, the PLS-DA model was validated using Q2 as a performance measure, the 10-fold cross-validation and setting in the permutation test a permutation number of 20 (see tables reported in Supplementary Table S1).

5. Conclusions

We conclude that the selection, evaluation, and development of buckwheat accessions with an increased competitive ability and strong allelopathic potential will be a good option for sustainable weed management. Our results confirm the ability of different buckwheat accessions to suppress monocot and dicot weeds, and this ability clearly appears to be accession dependent. In this regard, Gema appears to be the accession that should be used for growing in organic agriculture due to its capacity to sustainably regulate the germination and growth of the monocot weed *L. rigidum* and the dicot weed *P. oleracea*, while stimulating the root growth of buckwheat plants. Meanwhile, all four buckwheat accessions showed varying degrees of allelochemical production and release to control weeds through affecting multiple processes, such as germination, growth, and weed biomass. We conclude that different buckwheat genotypes may have different capacities to produce and exude several types of specialized metabolites, which lead to a wide range of allelopathic and defence functions in the agroecosystem to sustainably manage the growing weeds in their vicinity.

Supplementary Materials: The following supporting information can be downloaded at: https://www.mdpi.com/article/10.3390/plants12132401/s1. Table S1: Multivariate analyses of the polyphenols quantified in the four buckwheat accessions (Eva, Gema, Kora, and PI481671).

Author Contributions: Conceptualization: A.M.S.-M. and M.J.R.; methodology: A.M.S.-M., M.J.R., J.S.-G., M.I.H., Y.V.-Á., P.O. and M.A.P.; data curation: P.O., M.A.P., A.M.S.-M., Y.V.-Á. and M.I.H.; formal analysis: Y.V.-Á., M.I.H. and D.L.-G.; funding acquisition: M.J.R. and J.S.-G.; investigation: Y.V.-Á., M.I.H. and P.O.; project administration: A.M.S.-M.; supervision: A.M.S.-M., M.A.P., J.S.-G. and M.A.P.; visualization: Y.V.-Á., M.I.H., and A.M.S.-M.; writing—review and editing: A.M.S.-M., Y.V.-Á., M.I.H., D.L.-G. and J.S.-G. All authors have read and agreed to the published version of the manuscript.

Funding: This work was supported by the European Union's Horizon 2020 research and innovation programme under grant agreement no. 771367, ECOBREED. The seed samples of the buckwheat accessions Gema, Kora and PI481671 were provided by the Gene Bank—Crop Research Institute of Prague (Czech Republic), Eva from the Kmetijski inštitut Slovenije/Agricultural Institute of Slovenia (KIS).

Data Availability Statement: The raw data have been deposited at Zenodo webpage (https://zenodo.org/communities/ecobreed/ (accessed on 16 May 2023)), with the following DOI: https://doi.org/10.5281/zenodo.8053668 (accessed on 16 May 2023), as per the rules and regulation.

Acknowledgments: The authors want to thank Alberto Acuña Couñago (Centro de Apoio Científico-Tecnolóxico á Investigación, CACTI, Universidade de Vigo) for their help in the performance of the chemical analyses.

Conflicts of Interest: The authors declare no conflict of interest.

References

1. Gandhi, K.; Khan, S.; Patrikar, M.; Markad, A.; Kumar, N.; Choudhari, A.; Sagar, P.; Indurkar, S. Exposure risk and environmental impacts of glyphosate: Highlights on the toxicity of herbicide co-formulants. *Environ. Chall.* **2021**, *4*, 100–149. [CrossRef]
2. Vázquez-García, J.G.; Alcántara-De La Cruz, R.; Palma-Bautista, C.; Rojano-Delgado, A.M.; Cruz-Hipólito, H.E.; Torra, J.; Barro, F.; De Prado, R. Accumulation of target gene mutations confers multiple resistance to ALS, ACCase, and EPSPS inhibitors in *Lolium* species in Chile. *Front. Plant Sci.* **2020**, *11*, 553948. [CrossRef] [PubMed]
3. Koller, V.J.; Fürhacker, M.; Nersesyan, A.; Mišík, M.; Eisenbauer, M.; Knasmueller, S. Cytotoxic and DNA-damaging properties of glyphosate and Roundup in human-derived buccal epithelial cells. *Arch. Toxicol.* **2012**, *86*, 805–813. [CrossRef] [PubMed]
4. Mahmood, K.; Mathiassen, S.K.; Kudsk, P. Multiple herbicide resistance in *Lolium multiflorum* and identification of conserved regulatory elements of herbicide resistance genes. *Front. Plant Sci.* **2016**, *7*, 1160. [CrossRef] [PubMed]
5. Torra, J.; Montull, J.M.; Taberner, A.; Onkokesung, N.; Boonham, N.; Edwards, R. Target-site and non-target-site resistance mechanisms confer multiple and cross-resistance to ALS and ACCase inhibiting herbicides in *Lolium rigidum* from Spain. *Front. Plant Sci.* **2021**, *12*, 625138. [CrossRef]
6. Tehranchian, P.; Nandula, V.K.; Matzrafi, M.; Jasieniuk, M. Multiple herbicide resistance in California Italian ryegrass (*Lolium perenne* ssp. multiflorum): Characterization of ALS-inhibiting herbicide resistance. *Weed Sci.* **2019**, *67*, 273–280. [CrossRef]
7. Fernández-Moreno, P.T.; Travlos, I.; Brants, I.; De Prado, R. Different levels of glyphosate-resistant *Lolium rigidum* L. among major crops in southern Spain and France. *Sci. Rep.* **2017**, *7*, 13116. [CrossRef]
8. Loureiro, I.; Escorial, C.; Plaza, E.H.; Gonzalez-Andujar, J.L., Chueca, M.C. Current status in herbicide resistance in *Lolium rigidum* in winter cereal fields in Spain: Evolution of resistance 12 years after. *Crop Prot.* **2017**, *102*, 10–18. [CrossRef]
9. Okafor, I.A.; Ayalokunrin, M.B.; Orachu, L.A. A review on *Portulaca oleracea* (purslane) plant—Its nature and biomedical benefits. *Int. J. Biomed. Res.* **2014**, *5*, 75–80. [CrossRef]
10. Masabni, J.G.; Zandstra, B.H. Physiological characteristics of linuron-resistant *Portulaca oleracea* . *Weed Sci.* **1999**, *48*, 420–425. [CrossRef]
11. Liebman, M.; Davis, A.S. Integration of soil, crop and weed management in low-external input farming systems. *Weed Res.* **2000**, *40*, 27–48. [CrossRef]
12. Parisi, M.; Verrillo, M.; Luciano, M.A.; Caiazzo, G.; Quaranta, M.; Scognamiglio, F.; Di Meo, V.; Villani, A.; Cantelli, M.; Gallo, L.; et al. Use of Natural Agents and Agrifood Wastes for the Treatment of Skin Photoaging. *Plants* **2023**, *12*, 840. [CrossRef] [PubMed]
13. Verrillo, M., Koellensperger, G.; Puehringer, M.; Cozzolino, V.; Spaccini, R.; Rampler, E. Evaluation of Sustainable Recycled Products to Increase the Production of Nutraceutical and Antibacterial Molecules in Basil Plants by a Combined Metabolomic Approach. *Plants* **2023**, *12*, 513. [CrossRef]
14. Mwendwa, J.M.; Jeffrey, D.W.; Weston, L.A. The use of allelopathy and competitive crop cultivars for weed suppression in cereal crops. In *Integrated Weed Management for Sustainable Agriculture*; Zimdhal, R., Ed.; Burleigh Dodds Science Publishing: Cambridge, UK, 2018; pp. 361–388.
15. Hussain, M.I.; Araniti, F.; Schulz, M.; Baerson, S.; Vieites-Álvarez, Y.; Rempelos, L.; Bilsborrow, P.; Chinchilla, N.; Macías, F.A.; Weston, L.A.; et al. Benzoxazinoids in wheat allelopathy-from discovery to application for sustainable weed management. *Environ. Exp. Bot.* **2022**, *202*, 104997. [CrossRef]
16. Mahé, I.; Chauvel, B.B.; Colbach, N.; Cordeau, S.; Gfeller, A.; Reiss, A.; Moreau, D. Deciphering field-based evidences for crop allelopathy in weed regulation. A review. *Agron. Sustain. Dev.* **2022**, *42*, 50. [CrossRef]
17. Liebman, M.; Mohler, C.L. Weeds and the soil environment. In *Ecological Management of Agricultural Weeds*; Liebman, M., Mohler, C.L., Staver, C.P., Eds.; Cambridge University Press: Cambridge, UK, 2001; pp. 210–268.
18. Yasui, Y.; Hirakawa, H.; Ueno, M.; Matsui, K.; Katsube-Tanaka, T.; Yang, S.J.; Aii, J.; Sato, S.; Mori, M. Assembly of the draft genome of buckwheat and its applications in identifying agronomically useful genes. *DNA Res.* **2016**, *23*, 215–224. [CrossRef] [PubMed]
19. Luthar, Z.; Zhou, M.; Golob, A.; Germ, M. Breeding buckwheat for increased levels and improved quality of protein. *Plants* **2021**, *10*, 14. [CrossRef]
20. Kalinova, J.; Triska, J.; Vrchotova, N. Biological activity of phenolic compounds present in buckwheat plants. *Allelopath. J.* **2005**, *16*, 123–129.
21. Golisz, A.; Gawronska, H.; Gawronski, S.W. Influence of buckwheat allelochemicals on crops and weeds. *Allelopath. J.* **2007**, *19*, 337–349.
22. Kalinova, J.; Vrchotova, N. Level of catechin, myricetin, quercetin and isoquercitrin in buckwheat (*Fagopyrum esculentum* Moench), changes of their levels during vegetation and their effect on the growth of selected weeds. *J. Agric. Food Chem.* **2009**, *57*, 2719–2725. [CrossRef]

23. Kumar, V.; Brainard, D.C.; Bellinder, R.R. Suppression of Powell amaranth (*Amaranthus powellii*), shepherd's-purse (*Capsella bursa-pastoris*), and corn chamomile (*Anthemis arvensis*) by buckwheat residues: Role of nitrogen and fungal pathogens. *Weed Sci.* **2008**, *56*, 271–280. [CrossRef]
24. Iqbal, Z.; Hiradate, S.; Noda, A.; Isojima, S.I.; Fujii, Y. Allelopathy of buckwheat: Assessment of allelopathic potential of extract of aerial parts of buckwheat and identification of fagomine and other related alkaloids as allelochemicals. *Weed Biol. Manag.* **2002**, *2*, 110–115. [CrossRef]
25. Golisz, A.; Lata, B.; Gawronski, S.W.; Fujii, Y. Specific and total activities of the allelochemicals identified in buckwheat. *Weed Biol. Manag.* **2007**, *7*, 164–171. [CrossRef]
26. Kalinova, J.; Vrchotova, N.; Triska, J. Exudation of allelopathic substances in buckwheat (*Fagopyrum esculentum* Moench). *J. Agric. Food Chem.* **2007**, *55*, 6453–6459. [CrossRef] [PubMed]
27. Kato-Noguchi, H.; Sugimoto, H.; Yamada, M. Buckwheat seedlings may inhibit other plant growth by allelopathic substances. *Environ. Control Biol.* **2007**, *45*, 27–32. [CrossRef]
28. Kalinova, J. Varietal differences in allelopathic potential of common buckwheat (*Fagopyrum esculentum* Moench). *Cereal Res. Commun.* **2008**, *36*, 397–408. [CrossRef]
29. Björkman, T.; Shail, J.W. Using a buckwheat cover crop for maximum weed suppression after early vegetables. *Hort. Technol.* **2013**, *23*, 575–580. [CrossRef]
30. Kalinova, J. Influence of common buckwheat on growth of other plant species. In Proceedings of the 9th International Symposium on Buckwheat, Prague, Czech Republic, 18–22 August 2004; pp. 529–531.
31. Guo, X.D.; Ma, Y.J.; Parry, J.; Gao, J.M.; Yu, L.L.; Wang, M. Phenolics content and antioxidant activity of tartary buckwheat from different locations. *Molecules* **2011**, *16*, 9850–9867. [CrossRef]
32. Vollmannová, A.; Musilová, J.; Lidiková, J.; Árvay, J.; Šnirc, M.; Tóth, T.; Bojňanská, T.; Čičová, I.; Kreft, I.; Germ, M. Concentrations of Phenolic Acids Are Differently Genetically Determined in Leaves, Flowers, and Grain of Common Buckwheat (*Fagopyrum esculentum* Moench). *Plants* **2021**, *10*, 1142. [CrossRef]
33. John, J.; Sarada, S. Role of phenolics in allelopathic interactions. *Allelopath. J.* **2012**, *29*, 215–230.
34. Mohler, C.L.; Dykeman, C.; Nelson, E.B.; Ditommaso, A. Reduction in weed seedling emergence by pathogens following the incorporation of green crop residue. *Weed Res.* **2012**, *52*, 467–477. [CrossRef]
35. Weston, L.A. History and current trends in the use of allelopathy for weed management. *HortTechnology* **2005**, *15*, 529–534. [CrossRef]
36. Bertholdsson, N.O.; Andersson, S.C.; Merker, A. Allelopathic potential of *Triticum* spp., *Secale* spp. and *Triticosecale* spp. and use of chromosome substitutions and translocations to improve weed suppression ability in winter wheat. *Plant Breed.* **2012**, *131*, 75–80. [CrossRef]
37. Worthington, M.; Reberg-Horton, S.C.; Brown-Guedira, G.; Jordan, D.; Weisz, R.; Murphy, J.P. Relative contributions of allelopathy and competitive traits to the weed suppressive ability of winter wheat lines against Italian ryegrass. *Crop Sci.* **2015**, *55*, 57–64. [CrossRef]
38. Sijahović, E.; Đikić, M.; Gadžo, D.; Terzić, I. Allelopathic potential of common (*Fagopyrum esculentum* Moench) and Tartary (*Fagopyrum tataricum* L.) buckwheat for weed control in cereals. *Ege Üniversitesi Ziraat Fakültesi Dergisi* **2012**, *2*, 529–533.
39. Worthington, M.; Reberg-Horton, C. Breeding cereal crops for enhanced weed suppression: Optimizing allelopathy and competitive ability. *J. Chem. Ecol.* **2013**, *39*, 213–231. [CrossRef]
40. Seo, J.M.; Lee, D.B.; Valan, M.; Arasu; Wu, Q.; Suzuki, T.; Yoon, Y.H.; Lee, S.W.; Park, S.U.; Kim, S.J. Quantitative differentiation of phenolic compounds in different varieties of buckwheat cultivars from China, Japan and Korea. *J. Agric. Chem. Environ.* **2013**, *2*, 109–116. [CrossRef]
41. Uddin, M.R.; Li, X.; Park, W.T.; Kim, Y.B.; Kim, S.J.; Kim, Y.S.; Lee, M.Y.; Park, C.H.; Park, S.U. Phenolic compound content in different organs of Korean common buckwheat cultivars. *Asian J. Chem.* **2013**, *25*, 424. [CrossRef]
42. Falquet, B.; Gfeller, A.; Pourcelot, M.; Tschuy, F.; Wirth, J. Weed suppression by common buckwheat: A review. *Environ. Control Biol.* **2015**, *53*, 1–6. [CrossRef]
43. Hussain, M.I.; Reigosa, M.J. Secondary metabolites, ferulic acid and *p*-hydroxybenzoic acid induced toxic effects on photosynthetic process in *Rumex acetosa* L. *Biomolecules* **2021**, *11*, 233. [CrossRef]
44. Danila, A.M.; Kotani, A.; Hakamata, H.; Kusu, F. Determination of rutin, catechin, epicatechin, and epicatechin gallate in buckwheat *Fagopyrum esculentum* Moench by micro-high-performance liquid chromatography with electrochemical detection. *J. Agric. Food Chem.* **2007**, *55*, 1139–1143. [CrossRef]
45. Liu, C.L.; Chen, Y.S.; Yang, J.H.; Chiang, B.H. Antioxidant activity of tartary (*Fagopyrum tataricum* (L.) Gaertn.) and common (*Fagopyrum esculentum* Moench) buckwheat sprouts. *J. Agric. Food Chem.* **2008**, *56*, 173–178. [CrossRef]
46. Sharma, A.; Shahzad, B.; Rehman, A.; Bhardwaj, R.; Landi, M.; Zheng, B. Response of phenylpropanoid pathway and the role of polyphenols in plants under abiotic stress. *Molecules* **2019**, *24*, 2452. [CrossRef]
47. Sugiyama, A.; Yazaki, K. Flavonoids in plant rhizospheres: Secretion, fate and their effects on biological communication. *Plant Biotechnol.* **2014**, *31*, 431–443. [CrossRef]
48. Brunetti, C.; Fini, A.; Sebastiani, F.; Gori, A.; Tattini, M. Modulation of phytohormones signaling: A primary function of flavonoids in plant-environment interactions. *Front. Plant Sci.* **2018**, *9*, 1042. [CrossRef] [PubMed]

49. Kaushik, S.; Bais, H.P.; Biedrzycki, M.L.; Venkatachalam, L. Catechin is a phytototoxin and a pro-oxidant secreted from the roots of *Centaurea stoebe* . *Plant Signal. Behav.* **2010**, *5*, 1088–1098. [CrossRef]
50. Bais, H.P.; Vepachedu, R.; Gilroy, S.; Callaway, R.M.; Vivanco, J.M. Allelopathy and exotic plant invasion: From molecules and genes to species interactions. *Science* **2003**, *301*, 1377–1380. [CrossRef] [PubMed]
51. Serniak, T.L. Comparison of the allelopathic effects and uptake of *Fallopia japonica* phytochemicals by *Raphanus sativus* . *Weed Res.* **2016**, *56*, 97–101. [CrossRef]
52. Okada, S.; Iwasaki, A.; Kataoka, I.; Suenaga, K.; Kato-Noguchi, H. Phytotoxic activity of kiwifruit leaves and isolation of a phytotoxic substance. *Sci. Hortic.* **2019**, *250*, 243–248. [CrossRef]
53. Khamare, Y.; Chen, J.; Marble, S.C. Allelopathy and its application as a weed management tool: A review. *Front. Plant Sci.* **2022**, *13*, 1034649. [CrossRef]
54. Iqbal, Z.; Hiradate, S.; Noda, A.; Isojima, S.I.; Fujii, Y. Allelopathic activity of buckwheat: Isolation and characterization of phenolics. *Weed Sci.* **2003**, *51*, 657–662. [CrossRef]
55. Wen, Z.; White, P.J.; Shen, J.; Lambers, H. Linking root exudation to belowground economic traits for resource acquisition. *New Phytol.* **2022**, *233*, 1620–1635. [CrossRef]
56. Sedej, I.; Sakač, M.; Mandić, A.; Mišan, A.; Tumbas, V.; Čanadanović-Brunet, J. Buckwheat (*Fagopyrum esculentum* Moench) grain and fractions: Antioxidant compounds and activities. *J. Food Sci.* **2012**, *77*, C954–C959. [CrossRef] [PubMed]
57. Gfeller, A.; Glauser, G.; Etter, C.; Signarbieux, C.C.; Wirth, J. *Fagopyrum esculentum* alters its root exudation after *Amaranthus retroflexus* recognition and suppresses weed growth. *Front. Plant Sci.* **2018**, *9*, 50. [CrossRef]
58. Krumsri, R.; Iwasaki, A.; Suenaga, K.; Kato-Noguchi, H. Assessment of allelopathic potential of *Dalbergia cochinchinensis* Pierre and its growth inhibitory substance. *Emir. J. Food Agric.* **2020**, *32*, 513–521. [CrossRef]
59. Liu, X.; Ji, D.; Cui, X.; Zhang, Z.; Li, B.; Xu, Y.; Chen, T.; Tian, S. *p*-Coumaric acid induces antioxidant capacity and defence responses of sweet cherry fruit to fungal pathogens. *Postharvest Biol. Technol.* **2020**, *169*, 111297. [CrossRef]
60. Šćepanović, M.; Košćak, L.; Šoštarčić, V.; Pismarović, L.; Milanović-Litre, A.; Kljak, K. Selected phenolic acids inhibit the initial growth of *Ambrosia artemisiifolia* L. *Biology* **2022**, *11*, 482. [CrossRef] [PubMed]
61. Ismail, H.; Dragišic Maksimovic, J.; Maksimovic, V.; Shabala, L.; Živanovic, B.D.; Tian, Y.; Jacobsen, S.E.; Shabala, S. Rutin, a flavonoid with antioxidant activity, improves plant salinity tolerance by regulating K^+ retention and Na^+ exclusion from leaf mesophyll in quinoa and broad beans. *Funct. Plant Biol.* **2015**, *43*, 75–86. [CrossRef] [PubMed]
62. Popović, B.M.; Štajner, D.; Ždero-Pavlović, R.; Tumbas-Šaponjac, V.; Čanadanović-Brunet, J.; Orlović, S. Water stress induces changes in polyphenol profile and antioxidant capacity in poplar plants (*Populus* spp.). *Plant Physiol. Biochem.* **2016**, *105*, 242–250. [CrossRef] [PubMed]
63. Bernatoniene, J.; Kopustinskiene, D.M. The role of catechins in cellular responses to oxidative stress. *Molecules* **2018**, *23*, 965. [CrossRef]
64. Wu, H.; Haig, T.; Pratley, J.; Lemerle, D.; An, M. Distribution and exudation of allelochemicals in wheat (*Triticum aestivum* L.). *J. Chem. Ecol.* **2000**, *26*, 2141–2154. [CrossRef]
65. Hussain, M.I.; Vieites, Y.; Otero, P.; Prieto, M.A.; Simal-Gandara, J.; Reigosa, M.J.; Sánchez-Moreiras, A.M. Weed pressure determines the chemical profile of wheat (*Triticum aestivum* L.) and its allelochemicals potential. *Pest Manag. Sci.* **2022**, *78*, 1605–1619. [CrossRef] [PubMed]

plants

MDPI

Article

Evaluation of Propiophenone, 4-Methylacetophenone and 2′,4′-Dimethylacetophenone as Phytotoxic Compounds of Labdanum Oil from *Cistus ladanifer* L.

María Espinosa-Colín [1], Irene Hernandez-Caballero [2], Celia Infante [3], Irene Gago [1], Javier García-Muñoz [4] and Teresa Sosa [1,*]

[1] Department of Plant Biology, Ecology and Earth Sciences, Faculty of Science, University of Extremadura, 06006 Badajoz, Spain
[2] Department of Anatomy, Cellular Biology and Zoology, Faculty of Science, University of Extremadura, 06006 Badajoz, Spain
[3] Department of Forest Resources Technology, Center for Scientific and Technological Research of Extremadura (CICYTEX), 06187 Badajoz, Spain
[4] Toxicology Unit, Faculty of Veterinary Medicine, University of Extremadura, 06006 Badajoz, Spain
* Correspondence: tesosa@unex.es

Abstract: This is the first study to evaluate the phytotoxic activity of three phenolic compounds present in the essential oil of the labdanum of *Cistus ladanifer*, an allelopathic species of the Mediterranean ecosystem. Propiophenone, 4′-methylacetophenone, and 2′,4′-dimethylacetophenone slightly inhibit total germination and radicle growth of *Lactuca sativa*, and they strongly delay germination and reduce hypocotyl size. On the other hand, the inhibition effect of these compounds on *Allium cepa* was stronger on total germination than on germination rate, and radicle length compared to hypocotyl size. The position and number of methyl groups will affect the efficacy of the derivative. 2′,4′-dimethylacetophenone was the most phytotoxic compound. The activity of the compounds depended on their concentration and presented hormetic effects. In *L. sativa*, on paper, propiophenone presented greater inhibition of hypocotyl size at greater concentrations, with IC_{50} = 0.1 mM, whereas 4′-methylacetophenone obtained IC_{50} = 0.4 mM for germination rate. When the mixture of the three compounds was applied, in *L. sativa*, on paper, the inhibition effect on total germination and the germination rate was significantly greater compared to the effect of the compounds when they were applied separately; moreover, the mixture inhibited radicle growth, whereas propiophenone and 4′-methylacetophenone did not exert such effect when applied separately. The activity of the pure compounds and that of the mixture also changed based on the substrate used. When the trial was conducted in soil, the separate compounds delayed the germination of the *A. cepa* to a greater extent compared to the trial on paper, although they stimulated seedling growth. In soil, *L. sativa* against 4′-methylacetophenone also showed the opposite effect at low concentrations (0.1 mM), with stimulation of germination rate, whereas propiophenone and 4′-methylacetophenone presented a slightly increased effect.

Keywords: phytotoxicity; phenolic compounds; allelopathy; *Cistus ladanifer*; bioherbicides

Citation: Espinosa-Colín, M.; Hernandez-Caballero, I.; Infante, C.; Gago, I.; García-Muñoz, J.; Sosa, T. Evaluation of Propiophenone, 4-Methylacetophenone and 2′,4′-Dimethylacetophenone as Phytotoxic Compounds of Labdanum Oil from *Cistus ladanifer* L. *Plants* **2023**, *12*, 1187. https://doi.org/10.3390/plants12051187

Academic Editors: Margot Schulz and Vincenzo Tabaglio

Received: 20 January 2023
Revised: 23 February 2023
Accepted: 1 March 2023
Published: 6 March 2023

1. Introduction

In recent years, thanks to different technological innovations and advances, agricultural production has been improved, expanding the irrigated area and thus obtaining greater profit; however, this has also brought major problems related to the deterioration of the natural resources, such as deforestation, soil degradation, biodiversity loss, desertification and contamination due to the use of artificial chemical products [1].

The continuous exponential growth of the human population implies the further exploration of new advances that continue to increase agricultural production. Therefore, it is essential to use pesticides; the question is: where can we find safe products that do not compromise the environment and health? One of the emerging strategies to attain a more sustainable future is to search for solutions in nature itself. Nature has already solved all the challenges we face today; its failures have become fossils, and what surrounds us in the present time is the key to survival. Mimicking these measures is known as biomimicry [2].

In ecosystems where resources are scarce, some plants synthesise phytotoxic compounds, with which they manage to inhibit the growth of other species around them, thereby preventing future competition for resources. This phenomenon is called allelopathy, and it has been widely studied [3]. In addition, other studies have also shown the phytotoxic effect of the essential oils of these allelopathic species on different weeds. This negative effect of essential oils on germination is related to the presence of allelochemical compounds [4–6]. Currently, allelopathic compounds may constitute a natural source of bioherbicides [7], but often these molecules are complex and difficult to reproduce. To avoid this problem, we must focus the search on low molecular weight structures such as phenolic compounds. Although the phytotoxic effect of phenolic compounds is well known [8–10], many of them have not been evaluated to date, due to their diversity and the difficulty in isolating or synthesising them. Acetophenone, due to its chemical, biological, pharmacological, and allelochemical properties, can be a good candidate in different fields such as agronomy. Acetophenone is a common secondary metabolite in plant species [11]. Different studies show that its derivatives are involved in defense mechanisms [11]. Some studies on allelopathic plants have found that acetophenone derivatives such as xanthoxylin and acetosyringone can be good herbicides [12–14]. Another example of an allelopathic species where acetophenone is present is *Cistus ladanifer* Linneo (1753). This shrub is a typical species of the Mediterranean ecosystem that can eventually form monospecific communities known as rock rose thickets. When this species is exploited for the extraction of essential oil and/or labdanum, plant richness increases immediately in the area, although, due to the seed bank preserved in the soil and the regenerative capacity of this species, such plant richness decreases progressively after two years [15]. A large number of studies show that this reduction of biodiversity in rock rose thickets may be due to the allelopathic potential of this species [16–19]. In addition, its essential oil also presents phytotoxic activity [20,21]. The labdanum of its leaves and its essential oil contain a concentrated amount of secondary metabolites with different properties [22]. An important group of this labdanum is constituted of phenolic compounds [21,23–28]. Acetophenone is also found in the essential oil of labdanum. Three derivatives of acetophenone (propiophenone, 4′-methylacetophenone, and 2′,4′-dimethylacetophenone) have been identified in the neutral part of labdanum oil [29,30]. Labdanum oil contains more than 300 constituents, with 186 compounds (154 from the neutral part and 32 from the acidic part) representing about 95% of this oil [30,31].

Propiophenone (Figure 1A), 4′-methylacetophenone (Figure 1B), and 2′,4′-dimethylacet ophenone (Figure 1C) are simple and very similar phenolic compounds. All three of them present an acetophenone backbone, with a methyl group in different positions. Phenolic compounds present variable phytotoxicity depending on their formula, and small differences may result in different phytotoxic activity [9]. The inhibitory action of a phenolic compound may also change based on the concentration applied [19,22]. In some cases, even hormetic effects can be observed [32–35]. Furthermore, in soil, phenolic compounds can undergo processes of transportation, retention, and transformation, which can also influence the final result of bioactivity [36]. Moreover, it is worth highlighting that the action of allelopathic plants is commonly caused by the mixture of several allelochemicals that are found in the soil at very low concentrations [9].

Nowadays, thanks to their commercial availability, propiophenone, 4′-methylacetophe none, and 2′,4′-dimethylacetophenone can be easily evaluated through a standardised phytotoxicity bioassay. To date, no adverse effects have been reported in humans from these three compounds, and 4′-methylacetophenone is used in disinfectants, pest-control products, perfumes, and cosmetics [37].

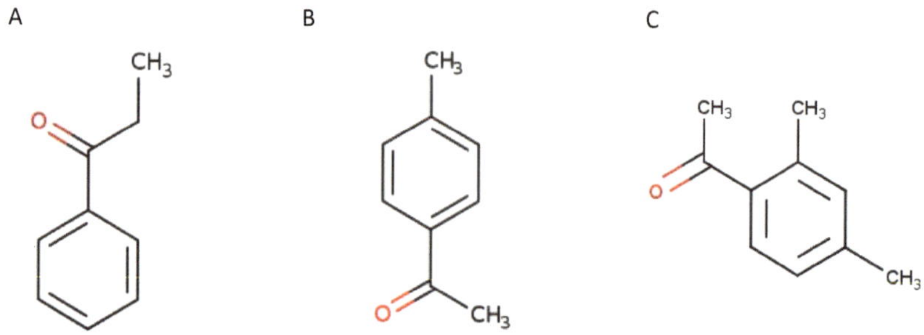

Figure 1. Chemical structure of propiophenone (**A**), 4′-methylacetophenone (**B**), and 2′,4′-dimethylacetophenone (**C**). Source: https://echa.europa.eu/ (accessed on 3 January 2023).

A very important aspect that allows the introduction of allelochemicals in the natural management of weeds is the knowledge of target plants and the exact chemicals responsible for the interaction [38]. Therefore, the aim of this study was to evaluate the phytotoxic activity of propiophenone, 4′-methylacetophenone and 2′,4′-dimethylacetophenone through a standard bioassay that contemplates all the possible variables. To this end, seeds of *Lactuca sativa* L., as a representative of monocotyledons, and *Allium cepa* L., as a representative of dicotyledons, were used, due to their sensitivity. In addition, their rapid, complete, and uniform germination grants reliability and duplicity to the assay [39–41]. These seeds were planted on paper and in soil, and they were watered with different concentrations of each of the compounds separately and a mixture of all three. Their phytotoxic activity was assessed by measuring their effect on germination and seedling growth [42]. The results may contribute to finding possible structures that serve as sources of natural bioherbicides that are safer and less contaminant.

2. Results

2.1. Effect of Propiophenone, 4′-Methylacetophenone and 2′,4′-Dimethylacetophenone on the Germination of Lactuca sativa L.

The results showed that propiophenone, 4′-methylacetophenone, 2′,4′-dimethylacetop henone and the mixture of the three compounds behaved in a similar manner, the three compounds separately, and their mixture inhibited germination rate (%GR) to a greater extent than total germination (%GT), and, lastly, the greater the concentration, the greater the inhibition effect (Figure 2). Both in soil and on paper, the three compounds separately and their mixture at 1 mM slightly inhibited total germination, except when the mixture was applied on paper, which showed full inhibition of total germination. The results for germination rate showed that propiophenone, 2′,4′-dimethylacetophenone and the mixture strongly inhibited %GR on paper and in soil, whereas 4′-methylacetophenone produced a lower effect; moreover, when the soil was watered with 0.1 mM of 4′-methylacetophenone, the opposite effect was obtained, showing significant stimulation of this parameter.

At the three analysed concentrations, propiophenone strongly inhibited germination rate (over 80%) both on paper and in soil, and it also significantly inhibited total germination (over 10%) in soil.

On paper and in soil, 4′-methylacetophenone significantly inhibited total germination at 0.5 mM and 1 mM. On paper, this compound also significantly inhibited germination rate with all analysed concentrations, showing a significant positive correlation (R^2 = 0.99), i.e., the greater the concentration, the greater the inhibition. The effective concentration required to induce half-maximal inhibition (IC_{50}) was 0.4 mM. It is worth highlighting that, when the soil was watered with 1 mM of 4′-methylacetophenone, the germination rate was significantly inhibited, whereas the concentration of 0.1 mM stimulated this parameter.

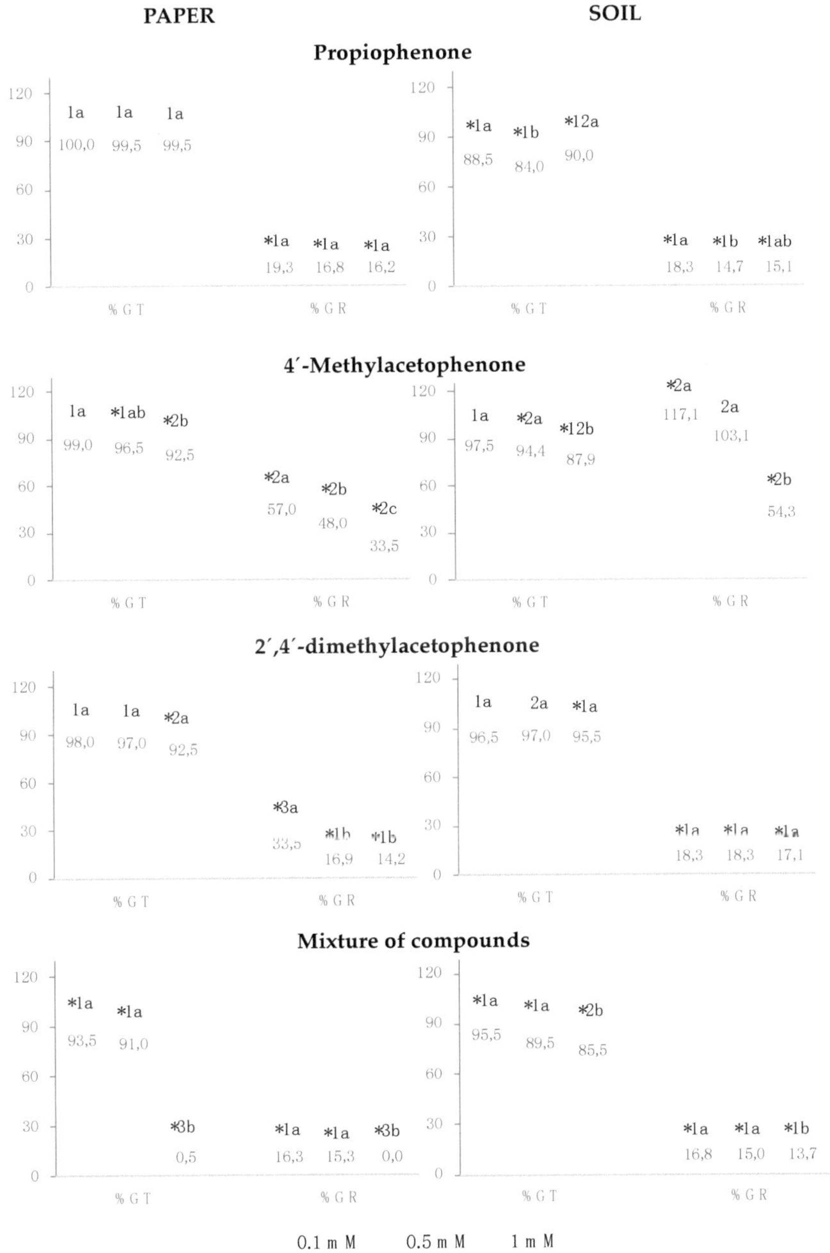

Figure 2. The effect of different concentrations of propiophenone, 4′-methylacetophenone, 2′,4′-dimethylacetophenone, and the mixture of the three compounds on the total germination (%GT) and germination rate (%GR) of *Lactuca sativa*, expressed as the percentage relative to the control. Four replicates of each treatment were performed (n = 4 × 50 = 200 seeds in total for each solution). * Significantly different from the controls; 1, 2, 3: differences numbers indicate significant differences between treatments of the same index and for each concentration. a, b, c: differences in small letters indicate significant differences between concentrations of the same index and for each treatment. $p < 0.05$ (Mann–Whitney U test).

In the assays with 2′,4′-dimethylacetophenone at 1 mM, %GT was significantly inhibited both on paper and in soil. This inhibition was only 7.5% and 4.5% on paper and in soil, respectively; however, as in the case of propiophenone, the germination rate was strongly inhibited at all three concentrations analysed, both on paper and in soil, showing no significant differences between the three concentrations.

The mixture inhibited total germination and germination rate at all analysed concentrations both on paper and in soil. When the mixture of the three compounds was applied on paper at 1 mM, %GT, and %GR were inhibited to a significantly greater extent compared to the inhibition obtained with the compounds separately. However, in soil, the inhibition of %GT produced by the mixture at 1 mM was only significantly greater than that obtained with 2′,4′-dimethylacetophenone separately, whereas %GR was inhibited to a significantly greater extent with the mixture than with 4′-methylacetophenone and 2′,4′-dimethylacetophenone separately.

2.2. *Effect of Propiophenone, 4′-Methylacetophenone and 2′,4′-Dimethylacetophenone on the Seedling Growth of Lactuca sativa L.*

The results for radicle and hypocotyl growth indicate that propiophenone, 4′-methylac etophenone, 2′,4′-dimethylacetophenone, and the mixture played a similar role on paper, the three compounds separately and the mixture inhibited hypocotyl length (%Hypocotyl length) to a greater extent than radicle length (%Radicle length), and greater inhibition was obtained at greater concentrations (Figure 3). Both in soil and on paper, the three compounds separately and their mixture slightly inhibited radicle growth, except for the soil assay with 2′,4′-dimethylacetophenone, which produced an inhibition of over 42%. The three compounds separately and their mixture strongly inhibited hypocotyl growth on paper; on the other hand, in soil, although the effect was greater with 2′,4′-dimethylacetophenone, it decreased with propiophenone and the mixture, obtaining the opposite effect when 4′-methylacetophenone was applied, with significant stimulation of this index at all three concentrations analysed.

When the assay was carried out in soil, propiophenone significantly inhibited radicle and hypocotyl growth at the three analysed concentrations. When the assay was performed on paper, radicle length was not significantly affected, whereas hypocotyl size decreased by over 52%. This inhibition was greater at higher concentrations, being significantly dependent on concentration ($R^2 = 0.98$) and obtaining an IC_{50} of 0.1 mM.

Both on paper and in soil, 4′-methylacetophenone significantly inhibited radicle length at 1 mM, but not at lower concentrations. When the assay was conducted in paper, hypocotyl size was strongly inhibited at all three concentrations. This inhibition showed a positive correlation with concentration ($R^2 = 0.99$), obtaining an IC_{50} lower than 0.1 mM. Nevertheless, when the assay was carried out in soil, hypocotyl growth was stimulated at all three concentrations.

At all concentrations, both on paper and in soil, 2′,4′-dimethylacetophenone significantly inhibited radicle and hypocotyl growth, observing a greater effect for the assays in soil.

The mixture of propiophenone, 4′-methylacetophenone and 2′,4′-dimethylacetophenone inhibited radicle and hypocotyl length at all three concentrations on paper and in soil, except when the paper was watered with 0.1 mM, where radicle length was significantly greater than in the control. On paper, at 1 mM, radicle and hypocotyl length % was cero, since, at that concentration of the mixture, the seeds did not germinate; at 0.5 mM, the mixture inhibited radicle growth to the same extent as 2′,4′-dimethylacetophenone, whereas propiophenone and 4′-methylacetophenone separately did not show such inhibition; at 0.1 mM, the mixture stimulated radicle growth to the same extent as propiophenone alone. On paper, the mixture inhibited %Hypocotyl length to a lesser extent than the isolated compounds. In soil, the inhibitory effect of the mixture on radicle and hypocotyl length was significantly greater than that of the compounds separately, except for 2′,4′-dimethylacetophenone.

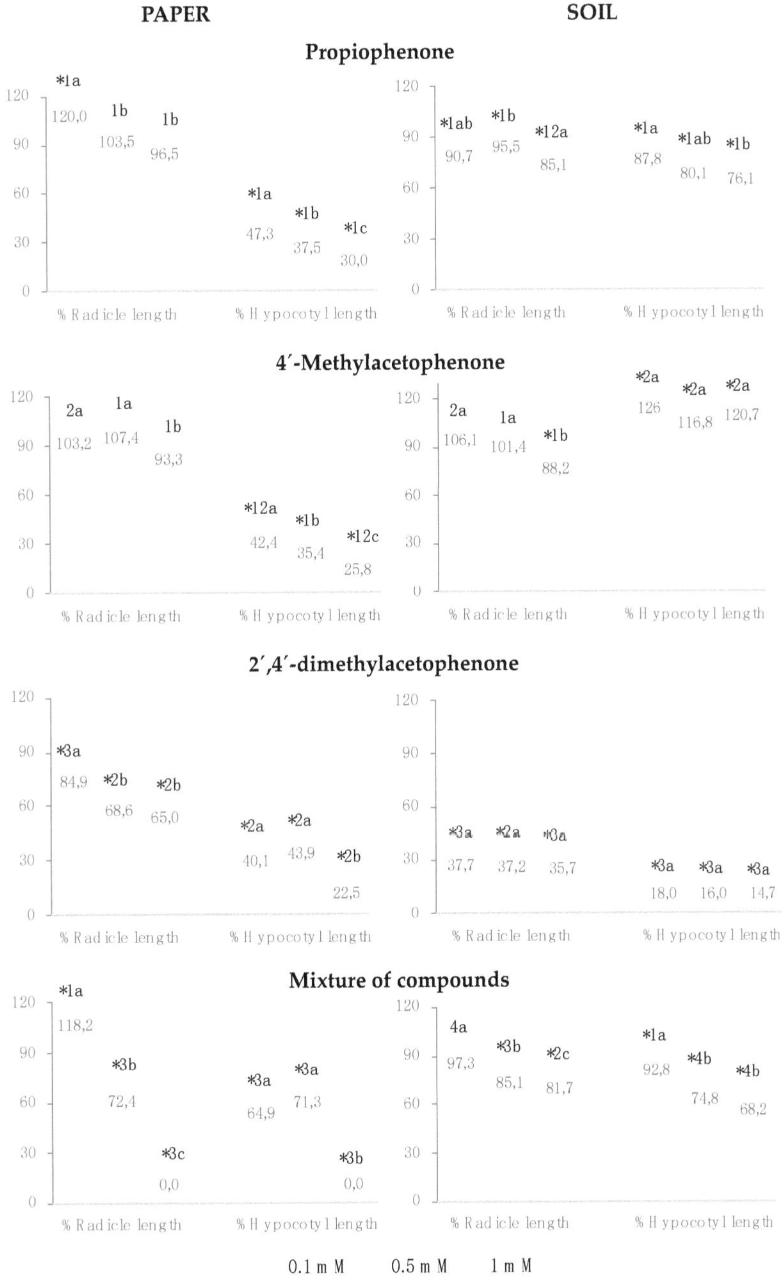

Figure 3. The effect of different concentrations of propiophenone, 4′-methylacetophenone, 2′,4′-dimethylacetophenone, and the mixture of the three compounds on the radicle and hypocotyl length of *Lactuca sativa*, expressed as the percentage relative to the control. Four replicates of each treatment were performed (n = 4 × 50 = 200 seeds in total for each solution). * Significantly different from the controls; 1, 2, 3, and 4: differences numbers indicate significant differences between treatments of the same index and for each concentration. The a, b, c: differences in small letters indicate significant differences between concentrations of the same index and for each treatment. $p < 0.05$ (Mann–Whitney U test).

2.3. Effect of Propiophenone, 4′-Methylacetophenone and 2′,4′-Dimethylacetophenone on the Germination of Allium cepa L.

When the assay was carried out with *Allium cepa*, it was also observed that the inhibition effects of the separate compounds were greater at higher concentrations, although the behaviour was slightly different (Figure 4). Unlike in the case of *Lactuca sativa*, propiophenone, 4′-methylacetophenone, and 2′,4′-dimethylacetophenone inhibited total germination (%GT) to a greater extent than germination rate (%GR). In this case, both in soil and on paper, the three compounds separately at 1 mM strongly inhibited total germination and germination rate, and, in contrast to the results obtained with *Lactuca sativa*, 4′-methylacetophenone did not present hormetic effects, but propiophenone did.

At all three analysed concentrations, propiophenone inhibited %GT more strongly than %GR in soil with respect to the paper assay. Moreover, when the paper was watered with a concentration of 1 mM, the germination rate was significantly inhibited, whereas, at 0.1 mM, it was stimulated by over 20%.

In both paper and soil, 4′-methylacetophenone significantly inhibited total germination and germination rate at all three analysed concentrations. At 1 mM, the inhibition was very strong: over 90% on paper and over 45% in soil. Furthermore, in soil, this inhibition depended on the concentration, presenting a positive correlation (R^2 = 0.99 for %GT, R^2 = 0.98 for %GR). The effective concentration required to induce the half-maximal inhibition of %GT and %GR (IC_{50}) was 0.1 mM.

When the assay was performed with 2′,4′-dimethylacetophenone, a similar behaviour was observed with respect to propiophenone: the inhibition effects of %GT and %GR are greater in soil than on paper, and %GR stimulation was observed when the paper was watered with 0.1 mM. In soil, 2′,4′-dimethylacetophenone strongly inhibited %GT and %GR at the three analysed concentrations. This effect depended on the concentration, presenting a significant positive correlation (R^2 = 0.98 for %GT, R^2 = 0.99 for %GR). The effective concentration required to induce the half-maximal inhibition of %GT (IC_{50}) was 1 mM, and 0.1 mM for %GR.

On its part, the mixture of compounds also presented a different behaviour with respect to that observed in *Lactuca sativa*. On paper, the mixture inhibited total germination and germination rate at the highest concentration; however, in soil, the effect was the opposite. In soil, at the highest and lowest concentrations, %GT and %GR were significantly stimulated by over 26%. When the mixture of the three compounds was applied on paper at 0.5 mM, %GR was inhibited, which was not observed with the separate compounds, and, at 0.1 mM, %GR was inhibited significantly more strongly with the mixture than with propiophenone and 2′,4′-dimethylacetophenone separately.

2.4. Effect of Propiophenone, 4′-Methylacetophenone and 2′,4′-Dimethylacetophenone on the Seedling Growth of Allium cepa L.

The effect on the seedling growth of *Allium cepa*, when watered with propiophenone, 4′-methylacetophenone, 2′,4′-dimethylacetophenone or the mixture, was different from that observed in *Lactuca sativa*. On paper, the three separate compounds and the mixture inhibited radicle length more strongly (%Radicle length) than hypocotyl length (%Cotyledon length), being greater at higher concentrations (Figure 5). However, in soil, the tendency observed was the opposite. The three compounds separately and the mixture stimulated seedling growth at the three analysed concentrations; only at 0.1 mM of propiophenone and 1 mM of 4′-methylacetophenone, hypocotyl length was significantly inhibited.

On paper, radicle and hypocotyl growth were strongly inhibited when watered with concentrations above 0.5 mM, although it was significantly stimulated when watered with a concentration of 0.1 mM. In soil, radicle size was not significantly affected, unlike hypocotyl size, which decreased by over 25% at 0.1 mM and increased by over 20% at higher concentrations.

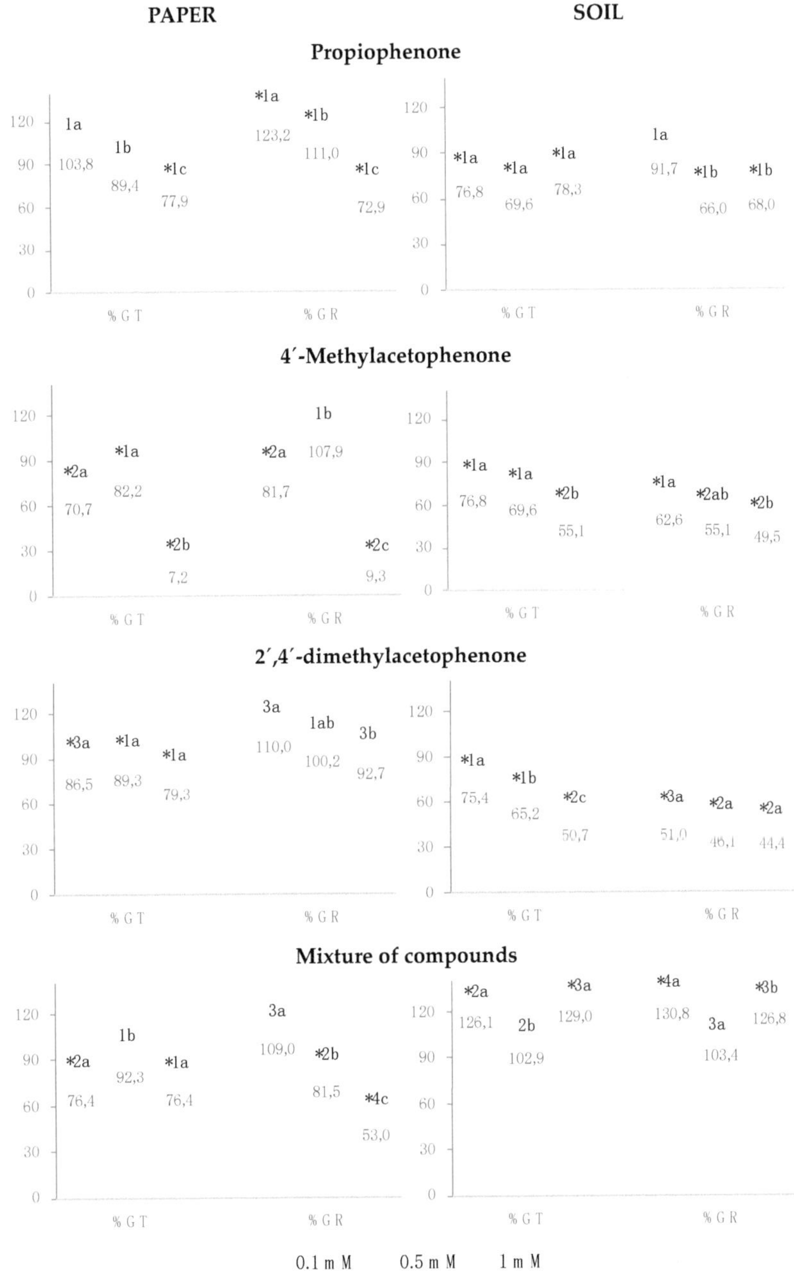

Figure 4. The effect of different concentrations of propiophenone, 4′-methylacetophenone, 2′,4′-dimethylacetophenone, and the mixture of the three compounds on the total germination (%GT) and germination rate (%GR) of *Allium cepa*, expressed as the percentage relative to the control. Four replicates of each treatment were performed (n = 4 × 50 = 200 seeds in total for each solution). * Significantly different from the controls; 1, 2, 3, and 4: differences numbers indicate significant differences between treatments of the same index and for each concentration. The a, b, c: differences in small letters indicate significant differences between concentrations of the same index and for each treatment. $p < 0.05$ (Mann–Whitney U test).

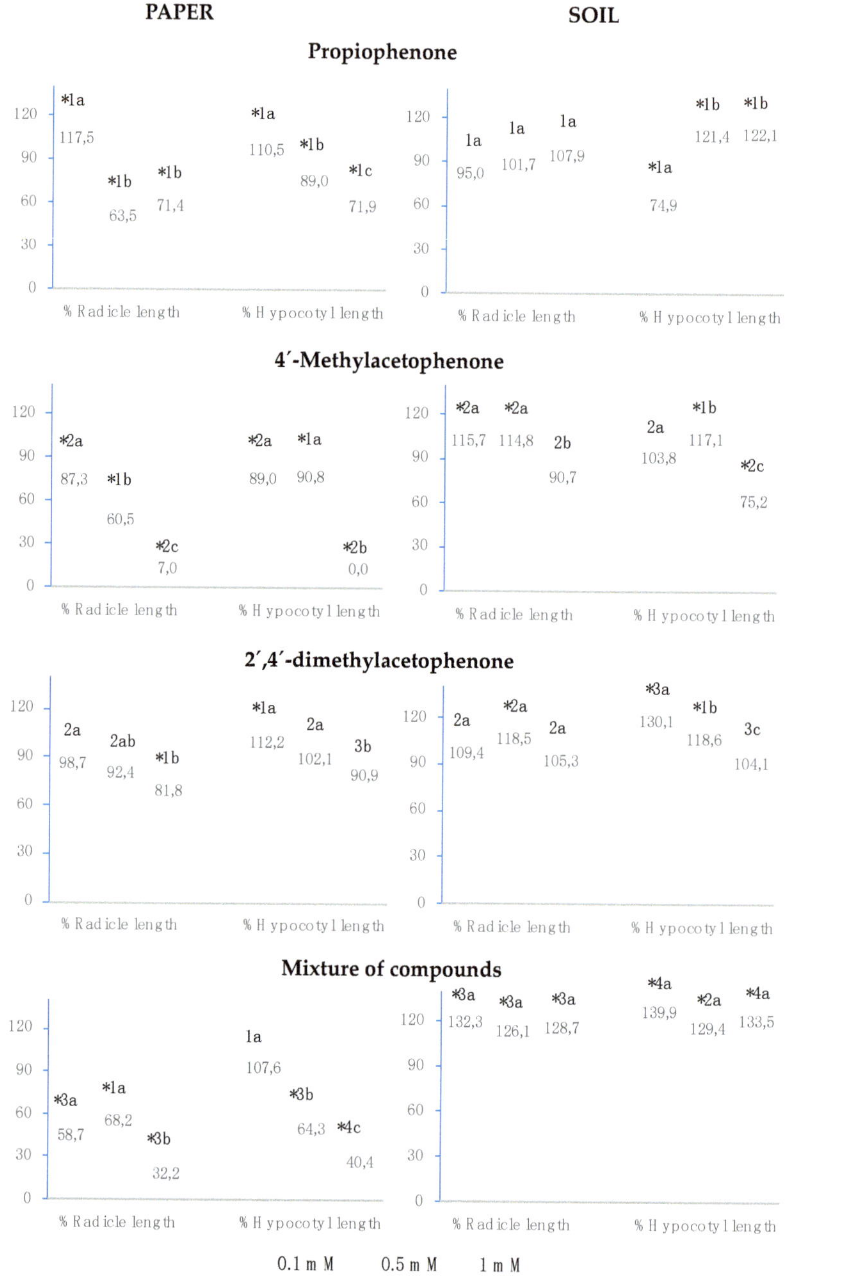

Figure 5. The effect of different concentrations of propiophenone, 4′-methylacetophenone, 2′,4′-dimethylacetophenone, and the mixture of the three compounds on the radicle and hypocotyl length of *Allium cepa*, expressed as the percentage relative to the control. Four replicates of each treatment were performed (n = 4 × 50 = 200 seeds in total for each solution). * Significantly different from the controls; 1, 2, 3, and 4: differences numbers indicate significant differences between treatments of the same index and for each concentration. The a, b, c: differences in small letters indicate significant differences between concentrations of the same index and for each treatment. $p < 0.05$ (Mann–Whitney U test).

The same was observed with 4′-methylacetophenone. When the assays were carried out on paper, radicle, and cotyledon length were significantly inhibited at the three analysed concentrations, whereas the opposite effect was observed in soil.

On paper, 2′,4′-dimethylacetophenone significantly inhibited radicle growth at 1 mM, although at 0.1 mM it significantly stimulated hypocotyl length. In soil, this compound stimulated the seedling growth of *Allium cepa* at all concentrations analysed.

On paper, the mixture of propiophenone, 4′-methylacetophenone and 2′,4′-dimethylac etophenone strongly inhibited radicle and hypocotyl length at all three concentrations; however, at 0.1 mM, hypocotyl length was slightly greater than in the control. The effect of the mixture on radicle and hypocotyl length was significantly greater than that of propiophenone or 2′,4′-dimethylacetophenone separately. On the other hand, in soil, the mixture of compounds significantly stimulated seedling growth at the three concentrations analysed, with this stimulation being greater than that observed when each of the compounds were applied separately.

3. Discussion

The appearance of resistance in weeds and the contamination of natural areas are currently the main problems derived from the massive use of herbicides. To prevent health problems and a decrease in the productivity of crops, it is necessary to find safer and more innovative products with new target sites [43,44]. This study proposes searching for phytotoxic compounds in essential oils of allelopathic species, such as *Cistus ladanifer*. The process of natural selection has given rise to allelochemicals with great structural diversity and adapted to interact with specific target sites. Therefore, through the study of these natural products, it may be possible to develop new tools for the development of new herbicides. However, in many cases, these compounds are complex and difficult to synthesise at an industrial scale. Distillation is the most common method for the extraction of plant essential oils. This method is a process of physical separation, not a chemical reaction. It consists in heating the plant until its most volatile components reach the vapour phase; then, the vapour is cooled down until these components are recovered in liquid form through a condensation process. In this process, complex molecules are hydrolysed into simpler molecules. This may pose a source of simpler derivatives of allelochemicals. If these derivatives preserve their reaction site, their activity may be the same or even greater. Evaluating the phytotoxic activity of simple and low-molecular-weight molecules may help to find a structure that can be used as a backbone for new bioherbicides. An important group of natural phytotoxic compounds is constituted of phenolic compounds [10,12,45]. In the essential oil of *Cistus ladanifer* labdanum, this group of compounds is very abundant and diverse in the exudate of its leaves and photosynthetic stems, although the impossibility of isolating some of them in sufficient amounts has prevented their evaluation. Currently, the commercial availability of three phenolic compounds present in this labdanum oil of *C. ladanifer*, i.e., propiophenone, 4′-methylacetophenone, and 2′,4′-dimethylacetophenone, allows measuring their phytotoxic activity through a germination bioassay in Petri dishes. The results of the present study showed that propiophenone, 4′-methylacetophenone, and 2′,4′-dimethylacetophenone may act as regulators of germination and growth in *Lactuca sativa* L. and *Allium cepa* L.

Although the assay on paper and in soil showed that 4′-methylacetophenone and 2′,4′-dimethylacetophenone slightly inhibited the germination of *L. sativa* at the highest concentration, on paper, all three analysed compounds strongly delayed germination and hypocotyl length; moreover, 2′,4′-dimethylacetophenone also inhibited radicle growth at concentrations of over 0.5 mM. These effects on the germination rate and seedling growth of *L. sativa* could compromise the survival of the seedling and thus be as harmful as the inhibition of germination [46]. This behaviour is similar to that shown by other phenolic compounds analysed in previous studies [10,12,45], which can be attributed to the multiple effects that this group of compounds may have on the processes of photosynthesis, respiration, and transpiration, as well as on the cell cycle [47–50]. Another characteristic of phenols is that their inhibitory action depends on concentration [9,35,51]. In the case of *L. sativa*, the concentration of 4′-methylacetophenone was positively correlated with %GR and %Hypocotyl length, showing that, the greater the concentration, the greater the inhibition of germination rate and hypocotyl growth. The inhibitory effect of propiophenone on hypocotyl size was also positively correlated with concentration. Furthermore, it is known that some phenolic compounds act as regulators of plant growth, presenting hormesis; that is, their effects are either stimulating or inhibitory depending on their concentration [35,52,53]. This behaviour was shown by 4′-methylacetophenone on %GR of *L. sativa*, when the assay was conducted in soil, obtaining inhibitory and stimulating effects at 1 mM and 0.1 mM, respectively.

When the assay was performed with seeds of *Allium cepa*, the results were somewhat different. Unlike in the case of *Lactuca sativa*, propiophenone, 4′-methylacetophenone, and 2′,4′-dimethylacetophenone showed stronger inhibition of total germination (%GT) than of germination rate (%GR), and radicle length (%Radicle length) compared to hypocotyl length (%Cotyledon length). In this case, propiophenone and 2′,4′-dimethylacetophenone presented hormetic effects when the trial was conducted on paper. This species-dependent activity has been observed with other phenolic compounds present in *C. ladanifer* [35,51]. Monocotyledons are usually more resistant to allelochemicals than dicotyledons [54]. Both in soil and on paper, the most phytotoxic compound against *Lactuca sativa* was 2′,4′-dimethylacetophenone. In paper, 1 mM of 4′-methylacetophenone presented a strong inhibition of %GT and %GR of *Allium cepa*; in soil, this compound also strongly inhibited these parameters at the three concentrations analysed, although, in soil, 2′,4′-dimethylacetophenone was the compound that presented the greatest phytotoxic effect on *A. cepa*. The analysed compounds, as in the case of any phenolic compound, have a hydroxyl group bound to a benzene molecule, and they only differ from each other in the binding of methyl groups. Their effects are different due to these small structural differences. In the analysed parameters, 2′,4′-dimethylacetophenone produced the greatest inhibition, which suggests that having two methyl groups may increase the activity of the molecule. Studies with 3,4-dihydroxyacetophenone isolated from *Picea schrenkiana* Fisch. et Mey. have shown that this compound significantly inhibits the germination of lettuce seeds at concentrations of 1 mM and 0.5 mM. In this case, the inhibitory effect of this compound was greater on radicle growth than on hypocotyl growth. Similarly, 3,4-dihydroxyacetophenone also exerted hormesis on the growth of rice radicles and radish hypocotyls [55]. Other studies on acaricide activity with compounds derived from 2′-hydroxy-4′-methylacetophenone identified in *Angelica koreana* Maxim. radicles have shown that, in terms of structure-activity relationships, the activity changed after the introduction of hydroxyl and methyl groups in the acetophenone backbone. In a contact toxicity bioassay, 3′-methylacetophenone was more effective as an acaricide against *Dermatophagoides farinae* than 2′-methylacetophenone and 4′-methylacetophenone [56]. Several studies have shown that 4-hydroxy-(2H)-1,4-benzoxazin-3(4H)-one, or DDIBOA, stands out for its phytotoxic action [57–59], and it may be used as the basis for the development of new herbicides since it is considered to be the most promising structure for the search of new compounds based on the backbones of benzoxazinones [60]. Its structure also presents a benzene ring, and the studies that have been conducted with some of its derivatives, such as 6-Cl-4-hydroxy-(2H)-1,4-benzoxazin-3(4H)-one and 8-Cl-4-hydroxy-(2H)-1,4-benzoxazine-3(4H)-one, show greater activity and selectivity, respectively, exhibiting greater phytotoxicity than the compounds that do not have substituents in their aromatic ring [61,62]. Other studies show that other derivatives such as o-nitroacetophenone inhibit the germination of *Amaranthus tricolor* L. [63]. The herbicide mechanism of benzoxazinoids is still unclear, although it is known that these compounds interfere with electron transport and ATPase activity in mitochondria, as well as with the functions of H^+-ATPase in the plasma membrane [64]. On their part, phenolic compounds increase the production of reactive oxygen species and, consequently, produce oxidative stress [65,66]. Moreover, it has been recently known that carbonyl, aromatic compounds, and azaindole are key interactions of the inhibitors of the enzyme 4-hydroxyphenylpyruvate dioxygenase [63].

In most cases, allelopathy by phenolic compounds is probably never due to a single substance. These compounds are not synthesised separately in plants, and they appear as a mixture. Their individual concentrations in the soil are usually very low, and their phytotoxicity seems to depend on the additive effects of each compound in the mixture [33]. Thus, in order to evaluate the possible interactions among the analysed compounds, they were applied as a mixture, and their effects were analysed. The results showed that, when the mixture of the three compounds was applied to *L. sativa* on paper at 1 mM, the inhibition of %GT and %GR was significantly greater than when the compounds were applied separately. At 0.5 mM, the mixture inhibited radicle growth to the same extent as 2′,4′-dimethylacetophenone, whereas propiophenone and 4′-methylacetophenone separately did not show such inhibition; however, the mixture inhibited % Hypocotyl length to a lesser extent than the isolated compounds. When applied to *Alium cepa* on paper, the mixture showed a greater inhibition effect 2′,4′-dimethylacetophenone, and propiophenone separately, whereas 4′-methylacetophenone showed a greater inhibition effect than the mixture. This behaviour is in line with other studies on volatile compounds that showed a considerable increase in phytotoxicity when applied jointly [67].

Another factor to be considered is the substrate on which the bioassay was carried out. In the soil, phenolic compounds can undergo processes of transportation, retention, and transformation, which, in turn, may influence their availability and action [36].

When the bioassay was conducted on a commercial substrate, in the two species analysed, propiophenone, which did not inhibit germination on paper, did inhibit germination in this substrate at all three analysed concentrations. In *Lactuca sativa*, the inhibition effect of 2′,4′-dimethylacetophenone was greater on all the parameters studied when the assay was carried out in soil; however, in *A. cepa*, although this compound in soil also inhibits %GT and %GR to a greater extent, %radicle length and %Cotyledon length are stimulated. On its part, 4′-methylacetophenone also changed its activity when the assay was performed on paper with *L. sativa*, delaying germination and inhibiting hypocotyl growth at the three analysed concentrations; on the other hand, in soil, this compound stimulated these parameters at the three concentrations, except at 1 mM on %GR. When the assay was conducted with *A. cepa*, 4′-methylacetophenone increased the inhibition of %GT and %GR, although the effect on seedling growth was attenuated and even stimulated at low concentrations. The mixture also seemed to change its action in soil, observing, in the case of *L. sativa*, inhibition of %radicle length at 0.1 mM, whereas, on paper, stimulation was obtained, and, in the case of *A. cepa*, inhibition was observed on paper, while stimulation was observed in soil. These activity changes based on the substrate have also been observed in other phenolic compounds [35,51]. Recent studies show that complex synergistic interactions between volatile and phenolic compounds underlie the efficacy of the allelopathic residues aggregated to the soil for weed control [68]. Propiophenone has moderate mobility in the soil, but also high volatilisation, and, in the vapour phase, it is degraded in the atmosphere by reacting with photochemically-produced hydroxyl radicals with an estimated half-life of approximately 5 days [69]. It is also known that acetophenone can be oxidised to phenol by *Aspergillus* species [70]. This possible, simple, and rapid degradation in the soil of the analysed compounds will influence their availability and concentration, which will affect their activity; however, this also implies that the phytotoxic compound is only found in the environment during the period of time that is adequate for its application, reducing or eliminating the risk of contamination. Therefore, phytotoxicity bioassays must be carried out in the substrates to which they will be applied.

Natural products with high phytotoxic activity and new mechanisms of action could be good options to constitute the basis of new herbicides, although, in most cases, their formulae are too complex to be produced, and the simplification of their structures often reduces their activity [71]. Thus, it is interesting to focus on simple, low-molecular-weight compounds of rapid degradation and low accumulation in the soil, in order to prevent their possible influence on non-target organisms. These characteristics are found in simple phenolic compounds [8–10]. If we search for these types of allelochemicals in essential oils of allelopathic plants, we will find natural products that, due to natural selection, have already been optimised for this activity. It is necessary to evaluate their activity in order to determine the concentration that must be applied in the soil to obtain the desired effect on the target weeds. It is important to highlight that the phytotoxic effects of the allelochemicals observed on paper do not allow extending the results to the field conditions, due to the simultaneous appearance of biotic and abiotic factors that may influence the activity of the mixtures or pure allelopathic compounds [72,73]. Moreover, it is also important to study the effect of formulations of mixtures that allow finding the way of enhancing their activity and attacking several mechanisms of action simultaneously, with the aim of addressing weed resistance.

This is the first study to report that propiophenone, 4′-methylacetophenone, and 2′,4′-dimethylacetophenone have phytotoxic activity and may regulate plant growth depending on the concentration applied and the substrate used. These compounds identified in the essential oil of labdanum from *Cistus ladanifer* may provide relatively simple structures, which allows them to be used as bioagents in weed control. The position and number of methyl groups affected the efficacy of the compound. This finding is consistent with other studies considering the molecular design of new acetophenones that are active against weeds [11,63]. Research on the structure-activity relationship of these types of allelochemicals and their derivatives based on the presence of methyl groups and other functional groups could lead to the development of less polluting and more specific bioherbicides. Furthermore, if their structure is simple, their synthesis and production will be easier.

4. Materials and Methods

4.1. Plant and Substrate Sources

To carry out the bioassay, seeds of *Lactuca sativa* variety romaine verte maraîchère (Vilmorin jardin—CS70110—38291 St Quentin Fallavier Cedex—France) were used as representatives of dicotyledons. This species is ideal for adequately showing the allelochemical effects on the germination processes [40,74]. Moreover, *Lactuca sativa* is recommended by the US EPA (United States Environmental Protection Agency) for phytotoxicity tests, being among the most sensitive species [75]. *Allium cepa* variety Bianca di Maggio (Vilmorin Jardin—CS70110—38291 St Quentin Fallavier Cedex— France) were used as monocotyledons. *Allium cepa* has proved to be the most useful and has repeatedly been suggested as a standard test material in the list of species historically used in plant testing by the International Organization for Standardization [76] and by the Organisation of Economic Cooperation and Development [77]. The total germination of these seeds was over 98%.

The activity of the allelochemicals can vary according to the planting substrate; for this reason, the experiment was carried out on Whatman No. 118 paper and a commercial substrate. The commercial substrate was of universal type, based on 95% peat, 5% green compost, and 1.3 Kg/m^3 fertilizer: 12N + 12P + 17K (Geolia, Aki Bricolage España S.L. B-839857—Spain). The commercial universal soil presented the following characteristics: organic matter per dry matter (60%), electrical conductivity (40 mS/m), apparent dry density (320 g/L), grain size (0–20 mm), and pH 5.5–6.5.

None of the materials (seeds and substrate) was sterilised before the experiment.

4.2. Phytotoxic Activity Test

Chemically pure reagents (propiophenone, 4′-methylacetophenone and 2′,4′-dimethyla cetophenone of >99% purity) were obtained from Sigma-Aldrich (Merck KGaA, Darmstadt, Germany). Different solutions were prepared with Milli-Q water from each component separately. The highest concentration was 1 mM, which is the maximum recommended concentration in allelopathic bioassays [78]. The minimum concentration was 0.1 mM, which was the lowest concentration to show the dissipation of the effect of these compounds. In addition, solutions of 0.5 mM were also prepared. For the mixture of the three phenols, three solutions were prepared with equimolar concentrations of each of the compounds at 1 mM, 0.5 mM, and 0.1 mM. To eliminate the effects of pH, we measured these parameters for each solution. The pH varied between 6 and 6.3 from one solution to another. There were no significant differences in pH among solutions.

The study was carried out in Petri dishes with Whatman paper or 25 g of commercial substrate. Four replicates were conducted for each assay, with 50 seeds of *Lactuca sativa* in each repetition (200 in total). They were watered with 5 mL and 16 mL of the different solutions for the paper and soil plates, respectively. The same amount of Milli-Q water was applied to the control. To prevent evaporation, the plates were sealed with Parafilm and were kept in a germination chamber for 5 days at 22 °C and a photoperiod of 15 h of light and 9 h of darkness.

4.3. Measured Indices to Quantify the Phytotoxic Effect

Germinated seeds were counted daily and total germination (%TG) [35,51] and germination rate (%GR) [69,75] were calculated according to Equations (1) and (2).

$$\%\mathrm{TG} = \frac{\mathrm{N_T} \cdot 100}{\mathrm{N}}, \tag{1}$$

where N_T is the average number of germinated seeds in each treatment and N is the average number of seeds in the control.

$$\mathrm{GR} = (\mathrm{N_1} \cdot 1) + (\mathrm{N_2} - \mathrm{N_1}) \cdot \frac{1}{2} + (\mathrm{N_3} - \mathrm{N_2}) \cdot \frac{1}{3} + \ldots (\mathrm{N_n} - \mathrm{N_{n-1}}) \cdot \frac{1}{n}, \tag{2}$$

where N_1, N_2, N_3, … N_{n-1}, N_n are the proportions of germinated seeds obtained in the first (1), second (2), third (3), … , (n − 1), (n) days (for lettuce it was n = 5 and for onion n = 6). All results are expressed as a percentage relative to the control (%GR).

On the last day of the experiment, 10 seedlings per Petri dish were randomly selected. Their radicle and hypocotyl length were measured [79] and the median was expressed as a percentage relative to the control, thus:

$$\%Radicle\ length = \frac{treatment\ root\ length}{control\ root\ length} \cdot 100 \tag{3}$$

$$\%Hypocotyl\ length = \frac{treatment\ cotyledon\ length}{control\ cotyledon\ length} \cdot 100 \tag{4}$$

4.4. Statistical Analysis

The significance level of the comparisons among treatments was estimated using the Mann-Whitney U test. The differences were considered significant when $p < 0.05$. The interrelationships between germination and seed growth with the concentration of phenolic compounds were determined by Pearson's determination coefficient. The effective concentrations required to induce half-maximal inhibition of growth (IC_{50}) were calculated according to the linear relationship between concentration and percent inhibition or stimulation of plant growth. All statistical analyses were conducted using the statistical software IBM SPSS Statistics version 26.

Author Contributions: Conceptualization, T.S.; methodology, J.G.-M.; validation, T.S.; formal analysis, C.I., I.H.-C., I.G., J.G.-M. and M.E.-C.; investigation, C.I., I.H.-C., I.G., J.G.-M. and M.E.-C.; resources, T.S.; data curation, C.I.; writing—original draft preparation, M.E.-C.; writing—review and editing, T.S.; visualization, I.H.-C. and I.G.; supervision, T.S.; project administration, T.S.; funding acquisition, T.S. All authors have read and agreed to the published version of the manuscript.

Funding: This research was funded by the Regional Government of Extremadura and the European Regional Development Fund, grant number GR21086.

Data Availability Statement: The data presented in this study are available on request from the corresponding author.

Conflicts of Interest: The authors declare no conflict of interest.

References

1. Singh, H.P.; Batish, D.R.; Kohli, R.K. Allelopathic Interactions and Allelochemicals: New Possibilities for Sustainable Weed Management. *Crit. Rev. Plant Sci.* **2003**, *22*, 239–311. [CrossRef]
2. Benyus, J.M. *Biomímesis: Innovaciones Inspiradas Por La Naturaleza*; García, L.A., Ed.; Tusquets Editores: Barcelona, Spain, 2012.
3. Rice, E.L. *Allelopathy*; Academic Press: Orlando, FL, USA, 1984; p. 422.
4. Jouini, A.; Verdeguer, M.; Pinton, S.; Araniti, F.; Palazzolo, E.; Badalucco, L.; Laudicina, V.A. Potential effects of essential oils extracted from mediterranean aromatic plants on target weeds and soil microorganisms. *Plants* **2020**, *9*, 1289. [CrossRef] [PubMed]
5. Mahdavikia, F.; Saharkhiz, M.J. Phytotoxic activity of essential oil and water extract of peppermint (Mentha × piperita L. CV. Mitcham). *J. Appl. Res. Med. Aromat. Plants* **2015**, *2*, 146–153. [CrossRef]
6. Uremis, I.; Arslan, M.; Sangun, M.K. Herbicidal activity of essential oils on the germination of some problem weeds. *Asian J. Chem.* **2009**, *21*, 3199–3210.
7. Blanco, Y. La utilización de la alelopatía y sus efectos en diferentes cultivos agrícolas. *Cultiv. Trop.* **2006**, *27*, 5–16.
8. Kuiters, A.T. Effects of phenolic acids on germination and early growth of herbaceous woodland plants. *J. Chem. Ecol.* **1989**, *15*, 467–479. [CrossRef]
9. Einhellig, F.A. Mode of Allelochemical Action of Phenolic Compounds. In *Allelopathy: Chemistry and Mode of Action of Allelochemicals*; Macías, F.A., Galindo, J.C.G., Molinillo, J.M.G.V., Cutler, H.G., Eds.; CRC Press: Boca Raton, FL, USA, 2003; pp. 217–238. [CrossRef]
10. Inderjit. Plant Phenolics in Allelopathy. *Bot. Rev.* **1996**, *62*, 186–202. [CrossRef]
11. Zubkov, F.I.; Kouznetsov, V.V. Traveling across Life Sciences with Acetophenone—A Simple Ketone That Has Special Multipurpose Missions. *Molecules* **2023**, *28*, 370. [CrossRef]
12. Charoenying, P.; Teerarak, M.; Laosinwattana, C. An allelopathic substance isolated from *Zanthoxylum limonella* Alston fruit. *Sci. Hortic.* **2010**, *125*, 411–416. [CrossRef]
13. Chotsaeng, N.; Laosinwattana, C.; Charoenying, P. Herbicidal Activities of Some Allelochemicals and Their Synergistic Behaviors toward *Amaranthus tricolor* L. *Molecules* **2017**, *22*, 1841. [CrossRef]
14. Zaman, F.; Iwasaki, A.; Suenaga, K.; Kato-Noguchi, H. Allelopathic potential and identification of two allelopathic substances in *Eleocharis atropurpurea*. *Plant Biosyst.-Int. J. Deal. All Asp. Plant Biol.* **2020**, *155*, 510–516. [CrossRef]
15. Márquez-García, F.; García-Alonso, D.; Vázquez, F.M. Notes of Changes in Biodiversity in the Exploited Populations of Cistus Ladanifer, L., (Cistaceae) from SW Iberian Peninsula. *J. Fungi.* **2022**, *2*, 1–7.
16. Agra-Coelho, C.; Rosa, M.L.; Moreira, I. Allelopathic effects of *Cistus ladanifer* L. *III Natl. Symp. Herbol.* **1980**, *1*, 165–178.
17. Pereira, I.P. Allelopathic potential of *Cistus ladanifer* L. and *Cistus salvifolius* L. Master's Thesis, Instituto Superior de Agronomia, Lisboa, Portugal, 1992.
18. Dias, A.S.; Costa, C.T.; Dias, L.S. Allelopathic Plants. XVII. *Cistus ladanifer L. Allelopathy. J.* **2005**, *16*, 1–30.
19. Chaves, N.; Alías, J.C.; Sosa, T. Phytotoxicity of *Cistus ladanifer* L.: Role of allelopathy. *Allelopathy. J.* **2016**, *38*, 113–132.
20. Pérez-Izquierdo, C.; Serrano-Pérez, P.; Rodríguez-Molina, M.C. Chemical composition, antifungal and phytotoxic activities of Cistus ladanifer L. essential oil and hydrolate. *Biocatal. Agric. Biotechnol.* **2022**, *45*, 102527. [CrossRef]
21. Verdeguer, M.; Blázquez, M.A.; Boira, H. Chemical composition and herbicidal activity of the essential oil from a *Cistus ladanifer* L. population from Spain. *Nat. Prod. Res. Former. Nat. Prod. Lett.* **2012**, *26*, 1602–1609. [CrossRef]
22. Frazão, D.F.; Martins-Gomes, C.; Steck, J.L.; Keller, J.; Delgado, F.; Gonçalves, J.C.; Bunzel, M.; Pintado, C.M.B.S.; Díaz, T.S.; Silva, A.M. Labdanum Resin from *Cistus ladanifer* L.: A Natural and Sustainable Ingredient for Skin Care Cosmetics with Relevant Cosmeceutical Bioactivities. *Plants* **2022**, *11*, 1477. [CrossRef]
23. Chaves, N.; Sosa, T.; Alías, J.C.; Escudero, J.C. Identification and effects of interaction phytotoxic compounds from exudate of *Cistus ladanifer* leaves. *J. Chem. Ecol.* **2001**, *27*, 611–621. [CrossRef]
24. Chaves, N.; Sosa, T.; Escudero, J.C. Plant growth inhibiting flavonoids in exudate of *Cistus ladanifer* and in associated soils. *J. Chem. Ecol.* **2001**, *27*, 623–631. [CrossRef]
25. Chaves, N.; Sosa, T.; Alías, J.C.; Escudero, J.C. Germination inhibition of herbs in *Cistus ladanifer* L. soil: Possible involvement of allelochemicals. *Allelopathy. J.* **2003**, *11*, 31–42.
26. Días, A.S.; Dias, L.S.; Pereira, I.P. Activity of water extracts of *Cistus ladanifer* and *Lavandula stoechas* in soil on germination and early growth of wheat and *Phalaris minor*. *Allelopathy. J.* **2004**, *14*, 59–64.
27. Herranz, J.M.; Farrandis, P.; Copete, M.A.; Duro, E.M.; Zalacaín, A. Effect of allelopathic compounds produced by *Cistus ladanifer* on germination of 20 Mediterranean taxa. *Plant Ecol.* **2006**, *184*, 259–272. [CrossRef]
28. Sosa, T.; Valares, C.; Alías, J.C.; Chaves, N. Persistence of flavonoids in *Cistus ladanifer* soils. *Plant Soil.* **2010**, *377*, 51–63. [CrossRef]
29. Ramalho, P.S.; de Freitas, V.A.P.; Macedo, A.; Silva, G.; Silva, A.M.S. Volatile components of *Cistus ladanifer* leaves. *Flavour Fragr. J.* **1999**, *14*, 300–302. [CrossRef]

30. Morales-Soto, A.; Oruna-Concha, M.J.; Stephen Elmore, J.; Barrajón-Catalán, E.; Micol, V.; Roldán, C.; Segura-Carretero, A. Volatile profile of Spanish Cistus plants as sources of antimicrobials for industrial applications. *Ind. Crops Prod.* **2015**, *74*, 425–433. [CrossRef]
31. Weyerstahl, P.; Marschall, H.; Weirauch, M.; Thefeld, K.; Surburg, H. Constituents of comercial labdanum oil. *Flavour Fragance J.* **1998**, *13*, 295–318. [CrossRef]
32. Kato-Noguchi, H.; Seki, T.; Shigemori, H. Allelopathy and allelopathic substance in the moss *Rhynchostegium pallidifolium*. *J. Plant Phys.* **2010**, *167*, 468–471. [CrossRef]
33. Einhellig, F.A.; Schon, M.K.; Rasmussen, J.A. Synergistic effects of four cinnamic acid compounds on grain sorghom. *Plant Growth Regul.* **1983**, *1*, 251–258.
34. Batish, D.R.; Singh, H.P.; Kaur, S.; Kohli, R.K.; Yadav, S.S. Caffeic acid affects early growth, and morphogenetic response of hypocotyl cuttings of mung bean *(Phaseolus aureus)*. *J. Plant Phys.* **2008**, *165*, 297–305. [CrossRef]
35. Requesón, E.; Osuna, D.; del Rosario Santiago, A.; Sosa, T. Evaluation of the Activity of Estragole and 2-Isopropylphenol, Phenolic Compounds Present in *Cistus ladanifer*. *Agronomy* **2022**, *12*, 1139. [CrossRef]
36. Cheng, H.H. Characterization of the mechanisms of allelopathy: Modeling and experimental approaches. In *Allelopathy: Organisms, Processes, and Applications*; Inderjit, Dakshini, K.M.M., Einhellig, F.A., Eds.; American Chemical Society: Washington, WA, USA, 1995; pp. 132–141.
37. National Center for Biotechnology Information. PubChem Compound Summary for CID 8500, 4′-Methylacetophenone. Available online: https://pubchem.ncbi.nlm.nih.gov/compound/8500 (accessed on 8 January 2023).
38. Macías, F.A.; Molinillo, J.M.; Varela, R.M.; Galindo, J.C. Allelopathy a natural alternative for weed control. Pest Management. *Science* **2007**, *63*, 327–348.
39. Einhellig, F.A.; Leather, G.R.; Hobbs, L.L. Use of *Lemna minor L.* as a bioassay in allelopathy. *J. Chem. Ecol.* **1985**, *11*, 65–72. [CrossRef] [PubMed]
40. Li, H.H.; Inoue, M.; Nishimura, H.; Mizutani, J.; Tsuzuki, E. Interactions of trans-cinnamic acid, its related phenolic allelochemicals, and abscisic acid in seedling growth and seed germination of lettuce. *J. Chem. Ecol.* **1993**, *19*, 1775–1787. [CrossRef] [PubMed]
41. Waller, G.R. *Allelochemicals: Role in Agriculture and Forestry. ACS Symposium Series 330*; American Chemical Society: Washington, WA, USA, 1987. [CrossRef]
42. Haugland, E.; Brandsaeter, L.O. Experiments on bioassay sensitivity in the study of allelopathy. *J. Chem. Ecol.* **1996**, *22*, 1845–1859. [CrossRef]
43. Tudi, M.; Daniel Ruan, H.; Wang, L.; Lyu, J.; Sadler, R.; Connell, D.; Chu, C.; Phung, D.T. Agriculture Development, Pesticide Application and Its Impact on the Environment. *Int. J. Environ. Res. Public Health* **2021**, *18*, 1112. [CrossRef]
44. Baucom, R.S. Evolutionary and ecological insights from herbicide-resistant weeds: What have we learned about plant adaptation, and what is left to uncover? *New Phytol.* **2019**, *223*, 68–82. [CrossRef]
45. Chung, I.M.; Miller, D.A. Effect of alfalfa plant and soil extracts on germination and seedling growth. *Agronomy J.* **1995**, *87*, 762–767. [CrossRef]
46. Muller, C.H.; Hanawalt, R.B.; McPherson, J.K. Allelopathic control of herb growth in the fire cycle of Californian chaparral. *Bull. Torrey Bot. Club.* **1968**, *95*, 225–231. [CrossRef]
47. Martínez-Otero, A.; González, L.; Reigosa, M.J. Oxygen electrode for seedling metabolism measurement in allelopathy. *Allelopathy J.* **2005**, *16*, 95–104.
48. Zhou, Y.H.; Yu, J.Q. Allelochemicals and photosynthesis. In *Allelopathy: A Physiological Process with Ecological Implications*; Reigosa, M.J., Pedrol, N., González, L., Eds.; Kluwer Academic Publishers: Amsterdam, The Netherlands, 2006; pp. 127–139. [CrossRef]
49. Lorenzo, P.; Pazos-Malvido, E.; González, L.; Reigosa, M.J. Allelopathic interference of invasive *Acacia dealbata*: Physiological effects. *Allelopath. J.* **2008**, *22*, 64–76.
50. Ishii-Iwamoto, E.L.; Abrahim, D.; Sert, M.A.; Bontato, C.M.; Kelmer-Brancht, A.M.; Bracht, A. Mitochondria as a site of allelochemicals action. In *Allelopathy: A Physiological Process with Ecological Implications*; Reigosa, M.J., Pedrol, N., González, L., Eds.; Kluwer Academic Publishers: Amsterdam, The Netherlands, 2006; Volume 12, pp. 267–284.
51. Tena, C.; del Rosario Santiago, A.; Osuna, D.; Sosa, T. Phytotoxic Activity of p-Cresol, 2-Phenylethanol and 3-Phenyl-1-Propanol, Phenolic Compounds Present in *Cistus ladanifer* L. *Plants* **2021**, *10*, 1136. [CrossRef] [PubMed]
52. Belz, R.G.; Cedergreen, N.; Duke, S.O. Herbicide hormesis—Can it be useful in crop production? *Weed Res.* **2011**, *51*, 321–332. [CrossRef]
53. Kaya, C.; Tuna, A.L.; Yokas, I. The role of plant hormones in plants under salinity stress. In *Salinity and Water Stress: Improving Crop Efficiency*; Ashraf, M., Ozturk, M., Athar, H.R., Eds.; Springer: Berlin/Heidelberg, Germany, 2009; pp. 45–50.
54. Soltys, D.; Krasuska, U.; Bogatek, R.; Gniazdowsk, A. Allelochemicals as Bioherbicides-Present and Perspectives. In *Herbicides-Current Research and Case Studies in Use*; Price, A.J., Kelton, J.A., Eds.; IntechOpen: London, UK, 2013; pp. 517–542. Available online: https://www.intechopen.com/chapters/44466 (accessed on 15 February 2023).
55. Yang, L.; Ruan, X.; Jiang, D.; Zhang, J.; Pan, C.; Wang, Q. Physiological effects of autotoxicity due to DHAP stress on *Picea schrenkiana* regeneration. *PLoS ONE* **2017**, *12*, e0177047. [CrossRef] [PubMed]
56. Oh, M.S.; Yang, J.-Y.; Lee, H.S. Acaricidal Toxicity of 2′-Hydroxy-4′-methylacetophenone Isolated from *Angelicae koreana* Roots and Structure–Activity Relationships of Its Derivatives. *J. Agric. Food Chem.* **2012**, *60*, 3606–3611. [CrossRef]

57. Macías, F.A.; Marín, D.; Oliveros-Bastidas, A.; Molinillo, J.M.G. Optimization of benzoxazinones as natural herbicide models by lipophilicity enhancement. *J. Agric. Food Chem.* **2006**, *54*, 9357–9365. [CrossRef]
58. Macías, F.A.; Marín, D.; Oliveros-Bastidas, A.; Molinillo, J.M.G. Rediscovering the bioactivity and ecological role of 1,4- benzoxazinones. *Nat. Prod. Rep.* **2009**, *26*, 478–489. [CrossRef]
59. Macías, F.A.; Molinillo, J.M.; Galindo, J.C.; Varela, R.M.; Simonet, A.M.; Castellano, D. The use of allelopathic studies in the search for natural herbicides. *J. Crop Prod.* **2001**, *4*, 237–255. [CrossRef]
60. Chinchilla, N.; Marín, D.; Oliveros-Bastidas, A.; Molinillo, J.M.G.; Macías, F.A. Soil biodegradation of a benzoxazinone analog proposed as a natural products-based herbicide. *Plant Soil* **2015**, *393*, 207–214. [CrossRef]
61. Macías, F.A.; De Siqueira, J.M.; Chinchilla, N.; Marín, D.; Varela, R.M.; Molinillo, J.M.G. New herbicide models from benzoxazinones: Aromatic ring functionalization effects. *J. Agric. Food Chem.* **2006**, *54*, 9843–9851. [CrossRef]
62. Macías, F.A.; Chinchilla, N.; Varela, R.M.; Molinillo, J.M.G.; Marín, D.; de Siqueira, J.M. Aromatic-ring-functionalised benzoxazinones in the system *Oryza sativa-Echinochloa crusgalli* as biorational herbicide models. *Pest Manag. Sci.* **2009**, *65*, 1104–1113. [CrossRef] [PubMed]
63. Chotpatiwetchkul, W.; Chotsaeng, N.; Laosinwattana, C.; Charoenying, P. Structure–Activity Relationship Study of Xanthoxyline and Related Small Methyl Ketone Herbicides. *ACS Omega* **2022**, *7*, 29002–29012. [CrossRef] [PubMed]
64. Dayan, F.E.; Duke, S.O. Biological activity of allelochemicals. In *Plant Derived Natural Products: Synthesis, Function, and Application*; Osbourn, A., Lanzotti, V., Eds.; Springer: New York, NY, USA, 2009; pp. 361–384.
65. Bais, H.P.; Vepachedu, R.; Gilroy, S.; Callaway, R.M.; Vivanco, J.M. Allelopathy and Exotic Plant Invasion: From Molecules and Genes to Species Interactions. *Science* **2003**, *301*, 1377–1380. [CrossRef] [PubMed]
66. Cruz-Ortega, A.; Lara-Núñez, A.; Anaya, A.L. Allelochemical stress can trigger oxidative damage in receptor plants mode of action of phytotoxicity. *Plant Signal. Behav.* **2007**, *2*, 269–270. [CrossRef]
67. Pardo-Muras, M.; Puig, C.G.; Pedrol, N. *Cytisus scoparius* and *Ulex europaeus* produce volatile organic compounds with powerful synergistic herbicidal effects. *Molecules* **2019**, *24*, 4539. [CrossRef]
68. Pardo-Muras, M.; Puig, C.G.; Pedrol, N. Complex Synergistic Interactions among Volatile and Phenolic Compounds Underlie the Effectiveness of Allelopathic Residues Added to the Soil for Weed Control. *Plants* **2022**, *11*, 1114. [CrossRef]
69. National Center for Biotechnology Information. PubChem Compound Summary for CID 7148, Propiophenone. Available online: https://pubchem.ncbi.nlm.nih.gov/#query=propiophenone (accessed on 8 January 2023).
70. Doi, M.; Shuto, Y. Oxidation of acetophenone by *Aspergillus* species and their possible contribution to Katsuobushi flavor. *Biosci. Biotechnol. Biochem.* **1996**, *59*, 2324–2325. [CrossRef]
71. Dayan, F.E.; Owens, D.K.; Duke, S.O. Rationale for a natural products approach to herbicide discovery. *Pest Manag. Sci.* **2012**, *68*, 519–528. [CrossRef]
72. Anaya, A.L. Alelopatía. In *Ecología Química*; Anaya, A.L., Ed.; Plaza y Valdés: Mexico, Mexico, 2003; pp. 255–298.
73. Prichoa, F.C.; Leyser, G.; Vladimir-Oliveira, J.; Cansian, R.L. Comparative allelopathic effects of *Cryptocarya moschata* and *Ocotea odorifera* aqueous extracts on *Lactuca sativa*. *Acta Sci.* **2013**, *35*, 197–202. [CrossRef]
74. Chiapusio, G.; Sánchez, M.A.; Reigosa, J.M.; González, L.; Pellissier, F. Do germination indices adequately reflect allelochemical effects on the germination process? *J. Chem. Ecol.* **1997**, *23*, 2445–2454. [CrossRef]
75. Wang, W. Literature review on higher plants for toxicity testing. *Water Air Soil Poll.* **1991**, *59*, 381–400. [CrossRef]
76. International Organization for Standarization (ISO). *Soil Quality-Determination of the Effects of Pollutants on Soil Flora. Part 1. Method for the Measurement of Inhibition of Root Growth*; ISO: Geneva, Switzerland, 1993; p. 9.
77. Organization of Economical Cooperation and Development (OECD). *Terrestrial Plants, Growth Test*; Guideline for Testing of Chemicals 208; OECD: Paris, France, 1984; p. 15.
78. Kuuluvaine, T. Gap disturbance, ground microtopography, and the regeneration dynamics of boreal coniferous forest in Finland: A review. *Ann. Zool. Fenn.* **1994**, *31*, 35–51.
79. Arango, M.C.; Ringuelet, J.; Viña, S. Intervención de los Compuestos Secundarios en las Interacciones Biológicas. In *Productos Naturales Vegetales*; Edulp integra la Red de Editoriales Universitarias Nacionales (REUN): Buenos Aires, Argentina, 2013.

plants

MDPI

Article

Essential Oils and Biological Activities of *Eucalyptus falcata*, *E. sideroxylon* and *E. citriodora* Growing in Tunisia

Ismail Amri [1,2,*], Marwa Khammassi [2], Rayda Ben Ayed [3,4], Sana Khedhri [2], Manel Ben Mansour [5], Oumayma Kochti [1], Ylenia Pieracci [6], Guido Flamini [6], Yassine Mabrouk [1], Samia Gargouri [5], Mohsen Hanana [3] and Lamia Hamrouni [1]

1 Laboratory of Biotechnology and Nuclear Technology, National Center of Nuclear Science and Technology, Sidi Thabet, B.P. 72, Ariana 2020, Tunisia
2 Laboratory of Management and Valorization of Forest Resources, National Institute of Researches on Rural Engineering, Water and Forests, P.B. 10, Ariana 2080, Tunisia
3 Laboratory of Extremophile Plants, Centre of Biotechnology of Borj-Cédria, B.P. 901, Hammam-lif 2050, Tunisia
4 Department of Agronomy and Plant Biotechnology, National Institute of Agronomy of Tunisia (INAT), University of Carthage-Tunis, 43 Avenue Charles Nicolle, El Mahrajène 1082, Tunisia
5 Laboratory of Plant Protection, National Institut of Agronomic Research of Tunisia, P.B. 10, Ariana 2080, Tunisia
6 Dipartimento di Farmacia, via Bonanno 6, University of Pisa, 56126 Pisa, Italy
* Correspondence: amri_amri@live.fr

Abstract: Many plants are able to synthesize essential oils (EOs), which play key roles in defense against weeds, fungi and pests. This study aims to analyze the chemical composition and to highlight the antioxidant, antimicrobial and phytotoxic properties of the EOs from *Eucalyptus falcata*, *E. sideroxylon* and *E. citriodora* growing in Tunisia. EOs were analyzed by gas chromatography coupled to mass spectrometry (GC/MS) and their antioxidant properties were determined by total antioxidant capacity (TAC), DPPH and ABTS assays. The phytotoxic potential was assessed against weeds (*Sinapis arvensis*, *Phalaris canariensis*) and durum wheat crop (*Triticum durum*) and compared to chemical herbicide glyphosate. The antifungal activity was investigated in vitro against eight target fungal strains. All EOs displayed a specific richness in oxygenated monoterpenes (51.3–90%) and oxygenated sesquiterpenes (4.8–29.4%), and 1,8-cineole, citronellal, citronellol, *trans*-pinocarveol, globulol, spathulenol and citronellyl acetate were the main constituents. *Eucalyptus* EOs exhibited remarkable antioxidant activity and *E. citriodora* oil exhibited significant activity when compared with *E. falcata* and *E. sideroxylon* EOs. The phytotoxic potential of the tested oils had different efficacy on seed germination and the growth of seedlings and varied among tested herbs and their chemical composition variability. Their effectiveness was better than that of glyphosate. At the post-emergence stage, symptoms of chlorosis and necrosis were observed. Furthermore, a decrease in chlorophyll and relative water content, electrolyte leakage and high levels of MDA and proline were indicators of the oxidative effects of EOs and their effectiveness as bioherbicides. Moreover, all the EOs exhibited moderate fungitoxic properties against all the tested fungal strains. Therefore, according to the obtained results, *Eucalyptus* EOs could have potential application as natural pesticides.

Keywords: allelochemicals; biopesticides; weeds; fungi; plant interactions

Citation: Amri, I.; Khammassi, M.; Ben Ayed, R.; Khedhri, S.; Mansour, M.B.; Kochti, O.; Pieracci, Y.; Flamini, G.; Mabrouk, Y.; Gargouri, S.; et al. Essential Oils and Biological Activities of *Eucalyptus falcata*, *E. sideroxylon* and *E. citriodora* Growing in Tunisia. *Plants* **2023**, *12*, 816. https://doi.org/10.3390/plants12040816

Academic Editors: Margot Schulz and Vincenzo Tabaglio

Received: 28 December 2022
Revised: 31 January 2023
Accepted: 2 February 2023
Published: 11 February 2023

1. Introduction

Chemical pesticides are among the most applied agrochemicals for pest control. However, their repeated application has resulted in an increased risk of emergence of pest resistance and soil degradation, making the overuse of these synthetic pesticides a new problem for global environment preservation [1,2]. The potential hazardous effect on the environment and on human and animal health associated with the abuse of chemicals has encouraged scientists to search for eco-friendly alternatives for pest control [3,4]. The search for alternatives to these chemical pesticides with natural-based products and biological molecules would provide a sustainable solution to protect both crops and the environment [5]. In this context, allelochemicals are molecules synthesized by plants and known to have several biological activities. These molecules are particularly involved in the defense systems of plants against pests and are also synthesized as part of the allelopathy process [6–8]. It has been reported that several allelochemical molecules have different biological activities. They can act as phytotoxic molecules and thus inhibit the germination and growth of several plant species [8,9]. They can exert different types of effects against insects, both attractive and repellent, and even insecticidal and larvicidal effects [3]. They are also known to inhibit the growth of various microorganisms, such as bacteria, fungi and yeasts [10–12].

In this way, allelochemicals could be of great interest for the development of new eco-friendly bio-pesticides [11,13]. In fact, these molecules have a short half-life and therefore problems regarding the persistence of residues in the environment can be solved and they can be considered safe and eco-friendly for the environment [13].

EOs plays a crucial role in the allelopathic interactions between plants and species of their biotopes, such as weeds, fungi, bacteria and insects [3,5,10]. Terpenes are known as potent allelochemicals with herbicidal and antifungal activity [6]. These molecules can be a good source for the discovery of natural fungicides and herbicides and can limit the harmful environmental effects caused by chemical pesticides [11]. According to the literature, the EOs of various plants have been reported to possess a significant phytotoxic effect against weeds [5,8,11]. Hence, exploiting the allelopathic potential of molecules with herbicidal properties could be a promising approach for weed control [4,7,11,14].

The *Eucalyptus* genus is part of the *Myrtaceae* family, comprising forest trees native to Australia; the *Eucalyptus* genus includes 13 subgenera and more than 700 species [15]. More than 100 species of *Eucalyptus* have been introduced in Tunisia since the 1950s. These species have been used for reforestation programs. They are mainly exploited for wood production, erosion control and the production of EOs. *Eucalyptus* species EOs are a mixture of volatile terpenes, such as 1,8-cineole, globulol, citronellol, spathunelol, limonene and pinenes [5,8,9].

Various molecules extracted from *Eucalyptus* species have been reported to have toxic effects against weeds, insects and microorganisms [5,8,9,15]. Plants' secondary metabolites, known to possess several activities, and their applications in the field make them good alternatives for chemical pesticides without side effects—in particular, the persistence linked to the non-biodegradability of pesticides and resulting in toxicity to the environment, cultivated plants, wild animals, beneficial microorganisms and several insects [13,16].

The essential oils of *E. falcata*, *E. sideroxylon* and *E. citriodora*, which grow in Tunisia, have been the subject of recent studies [15,17–19]; however, studies of their herbicidal potential have not been performed. Likewise, the majority of studies on their antifungal potential are focused on clinical fungi, and few are studies of phyto-pathogenic fungi [17]. The herbicidal activity of *Eucalyptus citriodora* has been reported in India [20], but their mode of action was not described. On the other hand, knowing that chemical ecology is a very important strategy, it is widely exploited for the discovery of metabolites because plants are able to synthesize various molecules with the aim of adaptation and acclimatization to environmental pressure. The synthesis of these molecules continues to evolve according to the pedoclimatic conditions of their biotopes in order to act in a physiological or ecological way on the maintenance of the plant in its ecological environment [13]. This may explain the variability in the production of essential oils depending on the origin of the plant and the usefulness of testing plant material from different origins.

In this sense, the present work aims to determine the chemical compositions of the EOs of *E. falcata*, *E. sideroxylon* and *E. citriodora* growing in Tunisia, to evaluate their antioxidant potential and to test their antifungal effects against eight strains of phyto-pathogenic fungi, as well as to study their herbicidal activity against the germination and seedling growth of weeds and cultivated crops at the germination stage. Moreover, at the post-emergence stage, our objective is to test the effect of EOs applied by spraying on the physiological effects of plants: the relative water content, the relative leakage of electrolytes, the synthesis of chlorophyll, the rate of malondialdehyde resulting from lipid peroxidation and also the proline content, always referring to the activity of a reference synthetic herbicide: glyphosate.

2. Materials and Methods

2.1. Plant Material

Eucalyptus species used for essential oil extraction and weed species tested for herbicide trials, sampling origin, climatic conditions and dates of collection are listed in Table 1. For each *Eucalyptus* species, five samples were harvested from different trees at least 20 m away. *Eucalyptus* samples were then stored in a greenhouse for drying until constant weight.

Table 1. Species, date and site of harvest of the plant material.

Species	Part Collected	Date of Harvest	Harvest Site	Preserved Specimen	Bioclimatic Stage
E. citriodora	Leaves	December 2020	Rimel arboreta, Bizerte	EC2026	Humid inferior with mild winter
Eucalyptus sideroxylon	Leaves	October 2020	Henchir En Naâm arboreta, Siliana	ES2027	Upper semi arid
E. falcata	Leaves	October 2020	Mjez Elbab arboreta, Beja	EF2028	Upper semi-arid with mild winter
S. arvensis	Seeds	July 2019	Sidi Ismail, Beja	SA1948	Subhumid with mild winter
P. canariensis	Seeds			SA1949	
T. durum	Seeds			SA1952	

Plants were identified by Professor Lamia Hamrouni, and the voucher specimens were deposited in the herbarium section of the Institute (INRGREF).

2.2. Extraction of Essential Oils

EOs were obtained by hydrodistillation of dried leaves in a Clevenger-type apparatus. The extraction lasted 4 h for each sample. EOs were collected and dried over anhydrous sodium sulfate and stored in a brown glass bottle at 4 °C until use. Yield was calculated based on dried weight (*w*/*w* %).

2.3. Gas Chromatography and Mass Spectrometry Analysis

Gas chromatography/electron ionization–mass spectrometry (GC/EI–MS) was performed using an Agilent 7890B gas chromatograph (Agilent Technologies Inc., Santa Clara, CA, USA) equipped with an Agilent HP-5MS capillary column (30 m × 0.25 mm; coating thickness 0.25 μm) and an Agilent 5977B single quadrupole mass detector. The analytical conditions were as follows: oven temperature programmed from 60 °C to 240 °C at 3 °C/min; injector temperature 220 °C; transfer-line temperature 240 °C; carrier gas helium at 1 mL/min. The acquisition parameters were as follows: full scan; scan range: 35–300 *m*/*z*; scan time: 1.0 s; threshold: 1 count. The identification of the constituents was based on the comparison of their retention times with the retention times of pure reference samples and comparing their linear retention indices (LRIs) relative to the series of n-alkanes. The mass spectra were compared with those listed in the commercial libraries NIST 14 [21] and Adams (2007) [22] and in a home-made mass-spectral library, built using MS literature combined with data experimentally obtained from pure substances.

2.4. Antioxidant Activity of EOs

2.4.1. Total Antioxidant Capacity

The total antioxidant capacity (TAC) was estimated by an assay based on the reduction of Mo (VI) to Mo (V) by EOs and on the formation of a green phosphate/Mo (V) complex at acidic pH [23]. One hundred μL of each oil was added to 1 mL of reagent solution containing sodium phosphate (28 mM), H_2SO_4 (0.6 M) and ammonium molybdate (4 mM). Then, the mixtures were heated for 90 min at 95 °C, followed by cooling, and the absorbance was read at 695 nm. The antioxidant capacity was expressed as mg gallic acid equivalents per gram of EOs (mg GAE/g EOs). All tested oils were analyzed in triplicate.

2.4.2. DPPH Assay

The free radical scavenging activity of *Eucalyptus* oils was assessed by the DPPH (1,1-diphenyl-2-picrylhydrazyl) radical scavenging method, according to Hanato et al. (1988) [24]. For this, 400 μL of each oil diluted with methanol at different concentrations or pure methanol (control) was mixed with 4 mL of methanolic solution of DPPH (0.1 mM). Then, the samples were vortexed and incubated for 30 min at room temperature in the dark, and the absorbance was measured at 517 nm. The percentage of DPPH scavenging activity was calculated according to the following equation:

$$\% \text{ inhibition} = (\text{A control} - \text{A sample}/\text{A control}) * 100$$

where A sample is the absorbance value of the tested oil and A control is the absorbance value of the control.

The IC_{50} is defined as the concentration of EOs required to scavenge 50% of the free radicals. All tested oils were analyzed in triplicate. Butylated hydroxytoluene (BHT) was used as a positive control.

2.4.3. ABTS Free Radical Scavenging Activity

The ABTS+ assay was assessed following the methods described by Re et al. (1999), with slight modifications [25]. A solution containing ABTS radical cations was prepared by mixing an equal volume of potassium persulfate (2.45 mM) and ABTS (7 mM). The mixture was incubated for 16 h in the dark at room temperature. The absorbance of the ABTS solution was adjusted to 0.70 ± 0.02 at 734 nm. A 400 μL aliquot of various concentrations of *Eucalyptus* EOs or pure methanol for the control was mixed with 4 mL of the ABTS radical solution and allowed to stand at room temperature for 5 min in the dark. Absorbance was then measured at 734 nm. Trolox was used as the standard.

The percentage inhibition of the radical cation $ABTS^+$ was determined using the following formula:

$$\text{Inhibition of ABTS } (\%) = ((\text{Ac} - \text{As})/\text{Ac}) \times 100$$

where Ac is the absorbance of the control and As is the absorbance of the tested oil.

ABTS scavenging activity is expressed as the IC_{50} value (μg/mL). All tested oils were analyzed in triplicate.

2.5. Seed Germination and Seedling Growth Experiments

Seeds of two weeds, *P. canariensis* L. and *S. arvensis* L., and a cultivated species, *T. durum* L., were used in herbicidal activity assays. Before germination tests, seeds were disinfected with 5% sodium hypochlorite, and then rinsed with water. Twenty seeds were placed in Petri dishes lined with a double layer of Whatman No. 1 filter paper. They were then treated with different doses (0, 0.5, 1, 1.5, and 2 μL/mL) of *Eucalyptus* oils in a solution of Tween 20 (0.1%) and incubated for 12 days at 25 °C; glyphosate was used as a positive control and the choice of glyphosate was based on the fact that it is the active molecule of several commercial herbicides used for the control of annual and perennial grasses and broad-leaved weeds in the post-planting/pre-emergence of many crops [7,16].

The tests were carried out in a completely randomized manner with three replicates for each dose. After 12 days, germination percentages and measured root and shoot growth (cm) were calculated.

2.6. Post-Emergence Assays

The post-emergence application of *Eucalyptus* EOs was performed to study its herbicidal effects on 21-day-old plants under controlled conditions in a glass greenhouse. Twenty seeds of the tested plant species were sown in polypropylene pots containing 500 g of mixture of peat, sand and perlite (1/3 of each). Fifteen days after seedling emergence, only five equal-sized plants per pot were kept.

When plants were 21 days old, they were sprayed with the following different solutions at 100 mL/m^2:

- Solution of water/Tween 20 (0.1%) as negative control;
- Solution of 10 µL/mL of *Eucalyptus* EOs dissolved in water/Tween 20 (0.1%);
- Solution of 10 mg/mL of glyphosate dissolved in water/Tween 20 (0.1%) as positive control.

A total of 15 pots were used for each plant test: five treatments and three replicates per treatment. Five days after spraying, the treated plants were used for various tests.

2.6.1. Relative Water Content (RWC)

Relative water content was calculated according to the following equation [26]:

$$RWC = (FW - DW)/FW) \times 100$$

where FW is the fresh weight and DW is the dry weight of treated plants (dried to constant weight at 75 °C).

2.6.2. Determination of Total Chlorophyll Content

Chlorophyll content was determined on fresh leaves five days after plant spraying. Here, 100 mg of leaves were ground in 5 mL acetone and kept for 72 h at 4 °C in the dark. The content of chlorophyll was determined according to the method of Lichtenthaler (1987), by measuring the absorbance at 663 and 646 nm [27].

2.6.3. Relative Electrolyte Leakage

Relative electrolyte leakage was assessed in leaves of treated herbs to study the effect of *Eucalyptus* oils on solute leakage. Leaves were immersed in glass boxes containing distilled water for 60 min, and then the conductivity of the water solution was measured (C1). At this point, the aqueous solution was boiled for 30 min and its conductivity was measured (C2) [28]. The relative electrolyte leakage (REL) was calculated according to the following formula:

$$\%\ REL = (C1/C2) \times 100$$

2.6.4. Determination of Free Proline Content

Proline levels were assessed according to the method adopted by (Bates et al., 1973) [29]. The fresh leaves of tested plants were digested with 6 mL of sulfosalicylic acid (3%) at 100 °C for 30 min, and then centrifuged at 25 °C at 2000× *g* for 5 min.

A volume of 0.5 mL of the obtained extract was mixed with 0.5 mL of distilled water and 2 mL of a mixture containing 0.5 g of ninhydrin, 30 mL of glacial acetic acid, and 20 mL of distilled water. The mixtures were boiled for 1 h. Then, the different samples were cooled and extracted with 6 mL of toluene. The toluene phase was used for absorbance measurements at 520 nm. The proline content was calculated from a standard curve. Proline content was expressed as mg/g fresh weight. All samples were analyzed in triplicate.

2.7. Antifungal Activity

Eight fungal strains (*Bipolaris sorokiniana, Fusarium verticillioides, F. pseudograminearum, F. proliferatum, F. nygamai, F. graminarium, F. avenaceum,* and *F. culmorum*) were obtained from the Laboratory of Plant Protection, Tunisian National Institute of Agronomic Research (INRAT). The antifungal potential of the *Eucalyptus* oils on the mycelial growth of tested fungi was assessed in vitro using the agar dilution method [16]. EOs tested were diluted in a solution (0.1%) of Tween 20 and then mixed in PDA medium to obtain a final concentration of 4 μL/mL. A disk of 5 mm, cut from the periphery of each fungal culture, was placed in the center of the PDA plate and then incubated for 7 days in optimal conditions. The experiments were carried out as three replicates per treatment. Growth inhibition was calculated as the growth percentage of inhibition relative to the control according to the formula:

$$\% \text{ Inhibition } ((C - T)/C) \times 100$$

where C: mean of growth of three replicates (mm) of control plates. T: mean of growth of three replicates (mm) of plates containing *Eucalyptus* oils.

2.8. Statistical Analysis

All of the experiments were carried out in three replicates, with the results represented as mean ± standard deviation. Data were subjected to one-way analysis of variance (ANOVA) using the SPSS 18.0 software package. Differences between means were tested through Student–Newman–Keuls and values with $p \leq 0.05$ were considered significantly different.

3. Results

3.1. Eucalyptus EOs' Yield and Chemical Composition

The hydrodistillation of dried leaves yielded 0.7 ± 0.1%, 1.3 ± 0.1%, and 2.9 ± 0.2% of yellow oils for *E. falcata*, *E. sideroxylon*, and *E. citriodora*, respectively. Oils' components, area percentages, retention indices, formulae, and the chemical classes of components for each oil are reported in Table 2.

Table 2. Chemical compositions of essential oils of *Eucalyptus* species.

Peak	Component	Formula	LRI (Lit.)	LRI	%		
					E. falcata	*E. sideroxylon*	*E. citriodora*
1	α-pinene	$C_{10}H_{16}$	939	941	6	1	
2	*p*-cymene	$C_{10}H_{14}$	1026	1028	2.5	0.9	
3	limonene	$C_{10}H_{16}$	1031	1032	0.7	0.9	0.1
4	1,8-cineole	$C_{10}H_{18}O$	1033	1034	28.4	65.4	0.3
5	bergamal	$C_{11}H_{18}O$	1053	1053			0.3
6	isopentyl isovalerate	$C_{10}H_{20}O$	1103	1104		0.1	
7	*cis*-rose oxide	$C_{11}H_{18}O$	1111	1111			0.1
8	fenchol	$C_{10}H_{18}O$	1110	1112	0.7		
9	α-campholenal	$C_{10}H_{16}O$	1124	1126	0.2		
10	*trans*-rose oxide	$C_{11}H_{18}O$	1126	1127			0.2
11	*trans*-pinocarveol	$C_{10}H_{16}O$	1139	1141	14.2	1.8	
12	isopulegol	$C_{10}H_{18}O$	1146	1146			8.1
13	citronellal	$C_{10}H_{18}O$	1153	1155			48.7
14	*trans*-β-terpineol	$C_{10}H_{18}O$	1162	1160			4.1
15	pinocarvone	$C_{10}H_{14}O$	1162	1164	2.2	0.2	
16	borneol	$C_{10}H_{18}O$	1165	1166	1.1	0.2	
17	*trans*-ocimenol	$C_{10}H_{16}O$	1168	1169		0.2	
18	4-terpineol	$C_{10}H_{18}O$	1177	1179	0.6	0.4	
19	*p*-cymen-8-ol	$C_{10}H_{14}O$	1183	1184	0.4	0.3	
20	*p*-mentha-1(7),8-dien-2-ol	$C_{10}H_{16}O$	1186	1186	0.7	1.5	
21	α-terpineol	$C_{10}H_{18}O$	1189	1191	1.5	1.4	
22	myrtenol	$C_{10}H_{16}O$	1191	1193	0.4		
23	*trans*-carveol	$C_{10}H_{16}O$	1217	1220	0.3	0.3	
24	*cis*-carveol	$C_{10}H_{16}O$	1229	1228		1.6	

Table 2. *Cont.*

Peak	Component	Formula	LRI (Lit.)	LRI	%		
					E. falcata	*E. sideroxylon*	*E. citriodora*
25	citronellol	$C_{10}H_{20}O$	1228	1230			20.2
26	(*Z*)-tagetone	$C_{10}H_{16}O$	1234	1231		0.9	
27	carvone	$C_{10}H_{14}O$	1242	1244		0.2	
28	geraniol	$C_{10}H_{18}O$	1255	1256			0.1
29	citronellyl formate	$C_{11}H_{20}O_2$	1275	1276			0.1
30	carvacrol	$C_{10}H_{14}O$	1298	1298	0.4		
31	*p*-vinylguaiacol	$C_9H_{10}O_2$	1312	1314			0.4
32	3,7-dimethyl-6-octenoic acid	$C_{10}H_{18}O_2$	1314	1315			1.2
33	2-acetoxy-1,8-cineole	$C_{12}H_{20}O_3$	1345	1345	0.2	0.2	
34	α-terpinyl acetate	$C_{12}H_{20}O_2$	1350	1352		0.9	
35	citronellyl acetate	$C_{12}H_{22}O_2$	1354	1354			8.1
36	2-phenylethyl isobutyrate	$C_{12}H_{16}O_2$	1395	1394			0.1
37	(*Z*)-jasmone	$C_{11}H_{16}O$	1394	1395			0.4
38	β-caryophyllene	$C_{15}H_{24}$	1418	1419	0.6	0.3	1
39	isoamyl benzoate	$C_{12}H_{16}O_2$	1439	1439			
40	aromadendrene	$C_{15}H_{24}$	1440	1440	3.2	2.1	
41	α-humulene	$C_{15}H_{24}$	1454	1455			0.2
42	alloaromadendrene	$C_{15}H_{24}$	1460	1462	0.8	0.6	
43	β-selinene	$C_{15}H_{24}$	1486	1487	0.4		
44	α-vetispirene	$C_{15}H_{22}$	1483	1488	0.5		
45	β-phenylethyl isovalerate	$C_{13}H_{18}O_2$	1489	1490		0.3	
46	α-selinene	$C_{15}H_{24}$	1497	1495		0.2	
47	bicyclogermacrene	$C_{15}H_{24}$	1496	1496	1.1	0.2	
48	germacrene B	$C_{15}H_{24}$	1556	1557		1	
49	ledol	$C_{15}H_{26}O$	1565	1566	1	0.7	
50	spathulenol	$C_{15}H_{24}O$	1576	1576	7.2	3.3	1.4
51	caryophyllene oxide	$C_{15}H_{24}O$	1582	1582			2.5
52	globulol	$C_{15}H_{26}O$	1584	1583	9.1	7.4	
53	viridiflorol	$C_{15}H_{26}O$	1590	1591	3.6	1.6	
54	rosifoliol	$C_{15}H_{26}O$	1599	1601	1.5	0.6	
55	10-*epi*-α-eudesmol	$C_{15}H_{26}O$	1617	1620	1.9	0.6	0.5
56	*iso* spathulenol	$C_{15}H_{24}O$	1640	1640	0.7		
57	β-eudesmol	$C_{15}H_{26}O$	1649	1650	4.4	2	
58	α-cadinol	$C_{15}H_{26}O$	1653	1652			0.4
	Monoterpene hydrocarbons %				9.2	2.8	0.1
	Oxygenated monoterpenes %				51.3	75.5	90
	Sesquiterpene hydrocarbons %				6.6	4.4	1.2
	Oxygenated sesquiterpenes %				29.4	16.2	4.8
	Non-terpene derivatives %				0	0.4	2.4
	Total identified %				96.5	99.3	98.5

Analysis of the three oils allowed the identification of 58 compounds distributed across five classes of terpene and non-terpene derivatives. All oils showed a specific richness in oxygenated monoterpenes (51.3–90%) and oxygenated sesquiterpenes (4.8–29.4%). In *E. falcata*, 31 components were identified, representing 96.5% of the total EOs. Oxygenated monoterpenes and oxygenated sesquiterpenes were the two main subclasses in this oil (51.3% and 29.4%, respectively). EOs' main constituents were 1,8-cineole (28.4%) and trans-pinocarveol (14.2%) as oxygenated monoterpenes, and globulol (9.1%) and spathulenol (7.2%) as oxygenated sesquiterpenes. Furthermore, appreciable percentages of the monoterpene hydrocarbon α-pinene (6.0%) were also present. The major and characteristic compounds of the essential oils of *E. falcata* are presented in Figure 1.

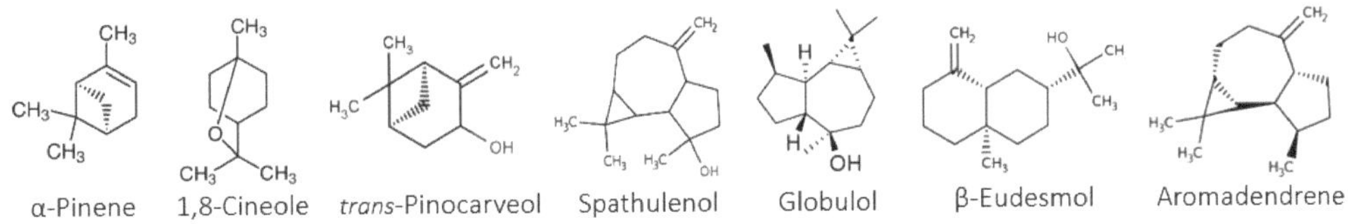

Figure 1. Major compounds of essential oils of *E. falcata*.

In *E. sideroxylon* EOs, 33 components were identified, accounting for 99.3% of the oil. Oxygenated monoterpenes and oxygenated sesquiterpenes were the two main subclasses in this oil (75.5% and 16.2%, respectively). The EOs of this species were characterized by a high level of 1,8-cineole (65.4%) as the major oxygenated monoterpene and globulol (7.4%) as an oxygenated sesquiterpene. The major and characteristic compounds of the essential oils of this species are presented in Figure 2.

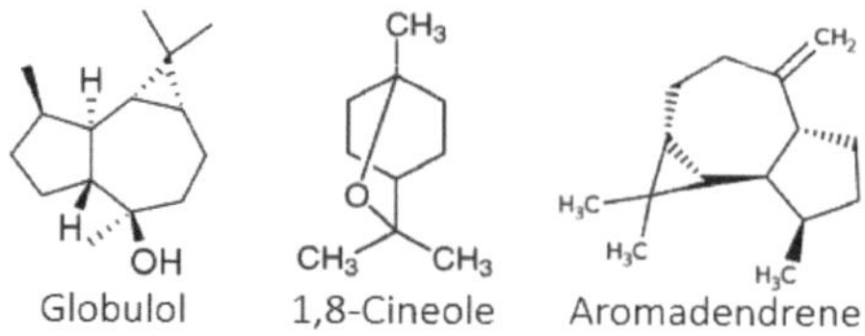

Figure 2. Major compounds of essential oils of *E. sideroxylon*.

In *E. citriodora* EOs, 22 different chemical constituents were identified, accounting for 98.5% of the oil. Globally, *E. citriodora* EOs were characterized by the dominance of oxygenated monoterpenes (90%). The major components were citronellal (48.7%), citronellol (20.2%), aitronellyl acetate (8.1%), and isopulegol (8.1%).

Great diversity in the structure and nature of the compounds of *E. citriodora* EOs has been described. The chemical structure of the major compounds is shown in Figure 3.

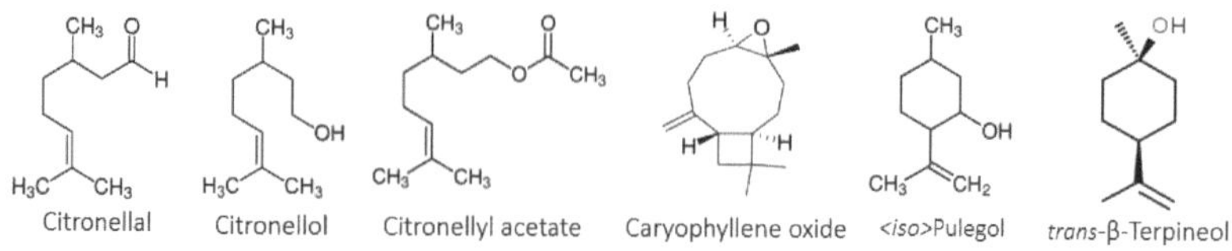

Figure 3. Major compounds of essential oils of *E. citriodora*.

3.2. Antioxidant Activity

To establish a better assessment of the antioxidant potential of the EOs, they were evaluated using three different methods. Table 3 details the results obtained in this study.

Table 3. Antioxidant activity of the essential oils of the three *Eucalyptus* species.

Essential Oils	TAC (mg GAE/g)	DPPH (IC_{50} μg/mL)	ABTS (IC_{50} μg/mL)
E. citriodora	80.21 ± 5.72 a	71.37 ± 3.89 b	53.26 ± 5.57 b
E. sideroxylon	26.68 ± 6.25 b	96.29 ± 5.58 c	84.84 ± 1.52 c
E. falcata	32.59 ± 3.83 b	112.61 ± 3.86 d	90.35 ± 3.27 c
BHT	-	20.53 ± 2.79 a	-
Trolox	-	-	15.75 ± 1.11 a

TAC: total antioxidant capacity; DPPH: 2,2-diphenyl-1-picrylhydrazyl; ABTS: 2,2′-Azino-bis(3-ethylbenzthiazoline-6-sulfonic acid; GAE: gallic acid equivalent; BHT: butylated hydroxytoluene. Results are expressed as mean ± standard deviation of 3 repetitions. Different letters in the same column indicate significant differences ($p \leq 0.05$) through Student–Newman–Keuls.

According to the statistical analysis, significant differences were noted in the TAC of the three tested EOs. The highest TAC value was observed for *E. citriodora* oil (80.21 mg GAE/g), followed by *E. falcata* oil (32.59 mg GAE/g) and *E. sideroxylon* (26.68 mg GAE/g) oils.

E. citriodora EOs showed also the highest antiradical activity with both the DPPH and ABTS assay (IC_{50} = 71.37 and 53.26 μg/mL, respectively).

Hence, the EOs of *E. citriodora* exhibited the highest antioxidant activity in all methods, while the lowest activity was obtained with the EOs of *E. falcata* and *E. sideroxylon*. The observed results could be related to the different compositions of the three oils.

3.3. *Phytotoxic Effects of Eucalyptus EOs*

3.3.1. Anti-Germinative Activity

The phytotoxic effect of the *Eucalyptus* EOs was evaluated against two well-known weeds, *S. arvensis* and *P. canariensis*, and on one cultivated crop, *T. durum*. Germination and growth of all tested herbs decreased in a dose-dependent manner, as shown in Table 4.

Table 4. Inhibitory effect of *Eucalyptus* species EOs on seed germination.

Tested Plant	Dose (µL/mL)	Germination %			
		E. falcata	*E. sideroxylon*	*E. citriodora*	Glyphosate
S. arvensis	0	100 ± 0.0 a	100 ± 0.0 a	100 ± 0.0 a	100 ± 0.0 a
	0.5	46.6 ± 11.54 bC	76.6 ± 5.7 bB	33.3 ± 5.8 bC	96.66 ± 3.33 aA
	1	23.3 ± 5.8 cB	40 ± 10 cA	6.7 ± 5.8 cC	90 ± 5.77 aA
	1.5	16.7 ± 5.8 cB	23.3 ± 5.7 dB	0 ± 0 cC	86.66 ± 3.33 aA
	2	0 ± 0 dC	16.6 ± 5.7 dB	0 ± 0 cC	73.33 ± 3.33 bA
P. canariensis	0	90 ± 10 a	90 ± 10 a	90 ± 10 a	90 ± 10 a
	0.5	76.7 ± 5.8 aB	90 ± 10 aA	66.7 ± 5.8 bB	93.33 ± 3.33 aA
	1	46.7 ± 11.5 bC	100 ± 00 aA	43.3 ± 5.8 cC	80 ± 0.0 aB
	1.5	23.3 ± 5.8 cB	80 ± 10 aA	16.7 ± 5.8 dB	80 ± 0.0 aA
	2	26.7 ± 5.8 cB	30 ± 10 bB	3.3 ± 5.8 eC	66.66 ± 3.33 bA
T. durum	0	96.7 ± 5.7 a	96.7 ± 5.7 a	96.7 ± 5.7 a	96.7 ± 5.7 a
	0.5	86.7 ± 5.8 aB	83.3 ± 5.7 aB	76.7 ± 5.8 bB	100 ± 0.0 aA
	1	63.3 ± 5.8 bB	96.6 ± 5.7 aA	40 ± 10 cC	93.33 ± 3.33 aA
	1.5	30 ± 10 cB	90 ± 10 aA	13.3 ± 5.8 dC	80 ± 0.0 bA
	2	36.7 ± 5.8 cC	63.3 ± 10 bB	0 ± 0 eD	76.66 ± 3.33 bA

Values are means ± standard errors (n = 3). The lowercase letters compare, in the columns, the dose effect for each herb vis-à-vis each oil. The capital letters compare, in the rows, the sensitivity of each herb at the same dose vis-à-vis the samples tested. Means followed by the same letter are not significantly different by the Student–Newman–Keuls test ($p \leq 0.05$).

Data showed that all oils significantly inhibited the seed germination of all tested herbs. The inhibition depended on both the tested species and dose. In fact, the inhibition of seed germination increased with increasing oil concentration.

A significant reduction in the seed germination of all tested herbs was observed at all doses. *E. falcata* oil caused complete germination inhibition in *S. arvensis* at 2 µL/mL. However, at the same dose, seed germination was only reduced for *P. canariensis* and *T. durum* (26.7 and 36.7%), respectively.

The volatile oil of *E. sideroxylon* reduced the germination of *S. arvensis* at the dose of 2.0 µL/mL to 16.6% and that of *P. canariensis* to 30%. On the contrary, *T. durum* appeared to be more resistant at the same dose, observing a higher percentage of germination (63.3%).

The high phytotoxicity of *E. citriodora* oil was remarkable, with a significant inhibitory effect against the germination of all tested herbs at all oil concentrations (0.5, 1.0, 1.5, and 2.0 µL/mL). Indeed, it was also remarkable that the complete inhibition of germination for *S. arvensis* by *E. citriodora* oil was obtained at the dose of 1 µL/mL. However, a partial reduction at the same dose (1.0 µL/mL) was observed in the germination of both *P. canariensis* and *T. durum*, although, at the highest dose (2.0 µL/mL), the seed germination of all tested herbs was totally inhibited.

Similarly, a high herbicidal effect was found with *Eucalyptus* oils against the seedling growth (roots and shoots) of all tested herbs at all doses; see Tables 5 and 6.

Table 5. Inhibitory effect of *Eucalyptus* species essential oils on aerial parts growth of germinated seeds.

Tested Plant	Dose (μL/mL)	Shoots (cm)			
		E. falcata	*E. sideroxylon*	*E. citriodora*	Glyphosate
S. arvensis	0	7.7 ± 0.68 a	7.7 ± 0.68 a	7.7 ± 0.68 a	7.7 ± 0.68 a
	0.5	6.36 ± 1.32 aA	6.3 ± 0.58 bA	1.83 ± 0.74 bB	5.66 ± 0.28 bA
	1	4.1 ± 0.96 bA	4.7 ± 0.37 cA	1.13 ± 1.06 bcB	5.96 ± 0.38 bA
	1.5	2.93 ± 0.49 bA	2.73 ± 0.5 dA	0 ± 0 cB	3.06 ± 0.58 cA
	2	0 ± 0 cC	2.43 ± 0.49 dA	0 ± 0 cC	1.33 ± 0.17 dB
P. canariensis	0	9.76 ± 0.5 a	9.76 ± 0.5 a	9.76 ± 0.5 a	9.76 ± 0.5 a
	0.5	8.4 ± 2.16 abA	11.8 ± 1.58 aA	10.53 ± 0.61 aA	8.46 ± 0.41 bA
	1	9.13 ± 0.3 aA	8.8 ± 0.96 aA	7.5 ± 1.04 bA	4.56 ± 0.34 cB
	1.5	6.06 ± 0.9 bA	6.5 ± 2.02 abA	6.1 ± 0.75 cA	2.23 ± 0.33 dB
	2	3.3 ± 1.6 cA	3.4 ± 4.2 bA	0.4 ± 0.7 dA	0.66 ± 0.16 eA
T. durum	0	9.16 ± 1.2 a	9.16 ± 1.2 b	9.16 ± 1.2 a	9.16 ± 1.2 a
	0.5	8.73 ± 1.27 aB	11.53 ± 0.98 aA	8.56 ± 0.45 aB	5.36 ± 1.12 bC
	1	7.73 ± 0.76 aB	10.1 ± 1.4 abA	6.1 ± 0.2 bB	3.63 ± 0.18 bcC
	1.5	5.56 ± 0.81 bB	8.2 ± 0.36 bcA	4.56 ± 1.15 cB	2.2 ± 0.17 cdC
	2	3.6 ± 1.04 cB	6.5 ± 0.45 cA	0 ± 0 dC	0.9 ± 0.15 dC

Values are means ± standard errors (n = 3). The lowercase letters compare, in the columns, the dose effect for each herb vis-à-vis each oil. The capital letters compare, in the rows, the sensitivity of each herb at the same dose vis-à-vis the samples tested. Means followed by the same letter are not significantly different by the Student–Newman–Keuls test ($p \leq 0.05$).

Table 6. Inhibitory effect of *Eucalyptus* species essential oils on roots growth of germinated seeds.

Tested Plant	Dose (μL/mL)	Roots (cm)			
		E. falcata	*E. sideroxylon*	*E. citriodora*	Glyphosate
S. arvensis	0	11.73 ± 0.7 a	11.73 ± 0.7 a	11.73 ± 0.7 a	11.73 ± 0.7 a
	0.5	10.07 ± 1.4 bA	7.8 ± 0.45 bB	3.77 ± 0.76 bD	5.86 ± 0.24 bC
	1	8.2 ± 0.5 cA	7.03 ± 1.4 bA	1.5 ± 1.32 cB	5.8 ± 0.55 bA
	1.5	5.7 ± 0.6 dA	4.6 ± 0.7 cB	0 ± 0 cD	2.36 ± 0.31 cC
	2	0 ± 0 eC	3.87 ± 0.55 cA	0 ± 0 cC	1.73 ± 0.21 cB
P. canariensis	0	10.3 ± 1.18 a	10.3 ± 1.18 a	10.3 ± 1.18 a	10.3 ± 1.18 a
	0.5	7.8 ± 0.4 abC	9.93 ± 1.33 aBC	11.2 ± 1.31 aA	9.66 ± 0.43 aBC
	1	8.7 ± 1.5 abB	9.46 ± 0.75 aA	7.5 ± 0.47 bBC	6.2 ± 0.25 bC
	1.5	6.5 ± 1.3 bcA	4.5 ± 1.23 bA	6.06 ± 1.0 bA	2.36 ± 0.32 cB
	2	4.8 ± 1.2 cA	2.33 ± 0.85 cB	0.4 ± 0.7 cC	0.63 ± 0.18 dC
T. durum	0	10.2 ± 1.09 a	10.2 ± 1.09 a	10.2 ± 1.09 a	10.2 ± 1.1 a
	0.5	10.4 ± 0.6 aA	7.4 ± 1.11 bB	9.06 ± 1.55 aAB	8.2 ± 0.5 bAB
	1	8.43 ± 0.45 aA	6.5 ± 0.7 bB	5.87 ± 0.76 bB	3.0 ± 0.15 cC
	1.5	5.83 ± 1.51 bA	5.7 ± 0.76 bA	5.73 ± 0.38 bA	2.23 ± 0.14 cdB
	2	3.9 ± 1.02 cB	7.36 ± 0.7 bA	0 ± 0 cD	1.2 ± 0.15 dC

Values are means ± standard errors (n = 3). The lowercase letters compare, in the columns, the dose effect for each herb vis-à-vis each oil. The capital letters compare, in the rows, the sensitivity of each herb at the same dose vis-à-vis the samples tested. Means followed by the same letter are not significantly different by the Student–Newman–Keuls test ($p \leq 0.05$).

The growth of *S. arvensis* seedlings was the most affected by *Eucalyptus* oil application, with high inhibition in both aerial parts and root growth when compared to the control at a low dose (0.5 µL/mL). Conversely, no significant effect at the same dose was observed for *P. canariensis* and *T. durum* (monocot species), which are more resistant to oil application.

3.3.2. Post-Emergence Assays

The post-emergent spray treatment with *Eucalyptus* EOs and glyphosate caused chlorosis and necrosis of the leaves, breaking of the stem, and complete wilting of all the tested weeds (*P. canariensis* and *S. arvensis*) and the cultivated species (*T. durum*) when compared to the negative control. Five days after the application of the EOs, significant effects on chlorophyll synthesis were seen. In fact, a significant decrease in the chlorophyll rate was noted for all the tested herbs. This effect varied according to the applied EOs and the tested species (Figure 4).

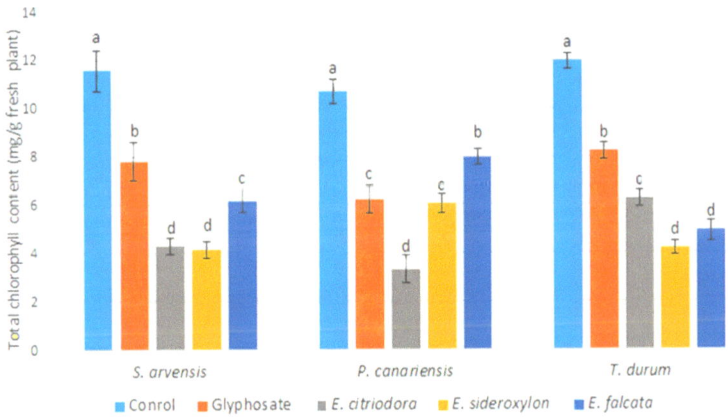

Figure 4. Effects of *Eucalyptus* species essential oils and glyphosate on total chlorophyll content as mg/g fresh tested plants measured five days after application. Error bars and letters on graphs represent standard error means and significant differences at $p \leq 0.05$ through Student–Newman–Keuls.

Generally, all the oils showed similar and even better effects than glyphosate. For *S. arvensis* (dicot weed), a significant decrease in the synthesis of chlorophyll was noted in the order of 48.28, 61.74, and 63.37% for *E. falcata*, *E. citriodora*, and *E. sideroxylon*, respectively. On the contrary, glyphosate showed the lowest rate of inhibition (33.96%).

In the case of *P. canariensis* (monocot weed), all oils showed the significant inhibition of chlorophyll synthesis; *E. citriodora* showed the highest rate of inhibition (67.45%), followed by *E. falcata* (24.24%) and *E. sideroxylon* (44.87%), which had effects similar to glyphosate (42.61%).

For *T. durum*, significant inhibition was noted with all *Eucalyptus* oils and its effects outweighed the chemical herbicide used as a positive control. The inhibition of chlorophyll synthesis may explain the herbicidal potential of the *Eucalyptus* EOs and may explain the effects of chlorosis, necrosis, and stunted growth observed after their application.

In the second phase and in relation to the drying effect observed following the application of *Eucalyptus* oils, the relative water content of the treated herbs was determined. The obtained results are presented in Figure 5.

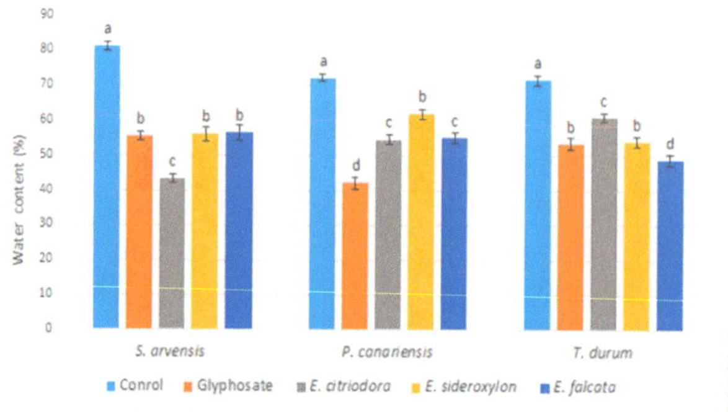

Figure 5. Effects of *Eucalyptus* essential oils and glyphosate on water content % of test plants measured five days after application. Error bars and letters on graphs represent standard error means and significant differences at $p \leq 0.05$ through Student–Newman–Keuls.

According to the statistical analysis, a significant and important decrease in the water content of the different tested herbs can be confirmed. This reduction varied according to the applied oil and the tested herb (mono- or dicotyledons and weeds or cultivated species). The effects observed are similar to those obtained with glyphosate. In fact, the water content was reduced for *S. arvensis* by 28.00, 29.26, and 46.4% by *E. falcata*, *E. sideroxylon*, and *E. citriodora*, respectively. For *P. canariensis*, the observed reductions were in the order of 24.6, 15.16, and 27.36%, respectively. In the case of *T. durum*, a smaller decrease in water content was observed with *E. citriodora* and *E. sideroxylon* (10.47 and 25.49%, respectively), while *E. falcata* caused the most significant decrease (49.57%), which outweighed the effects caused by glyphosate (27.5%).

The obtained results reflect the phytotoxic potential of *Eucalyptus* oils and can explain their mechanism of action, which consists in altering the water balance of the plant.

In light of these observations, the effect of *Eucalyptus* oils on the membrane integrity of the tested herbs was evaluated by measuring the relative electrolyte leakage. It has been observed that exposure to *Eucalyptus* EOs induces significant electrolyte leakage from the leaves of the tested herbs, as assessed by the increase in the conductivity of the bathing medium. Hence, these volatile oils disrupt the membrane integrity and cause solute leakage (Figure 6).

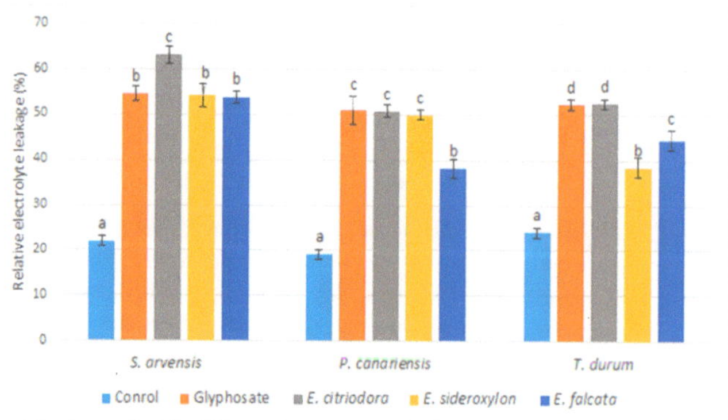

Figure 6. Effects of *Eucalyptus* essential oils and glyphosate on relative electrolyte leakage percentage of tested plants measured five days after spray. The errors bars and letters on graphs represent standard errors means and significant differences at $p \leq 0.05$ through Student–Newman–Keuls.

According to the statistical analysis, very pronounced relative electrolyte leakage was noted following the application of EOs and glyphosate compared to the control.

Electrolyte leakage explains the loss of plants' membrane integrity and therefore the observed phytotoxic effects.

The effect of the three oils on electrolyte leakage varied according to the species of *Eucalyptus* and that of the tested plants. Volatile oils of *E. citriodora* induced the most pronounced effects on all herbs: 54.06, 55.23, and 66.93% for *T. durum*, *P. canariensis*, and *S. arvensis*, respectively.

Again, the dicot weed *S. arvensis* was the most sensitive to the action of all applied oils.

The lipid peroxidation of polyunsaturated fatty acids, of which malondialdehyde is a product, could also provide an explanation for the relative leakage of electrolytes and the loss of membrane integrity. For this reason, we measured the level of MDA in the aerial parts of tested species after the application of the oils and the positive and negative controls. The main results are shown in Figure 7.

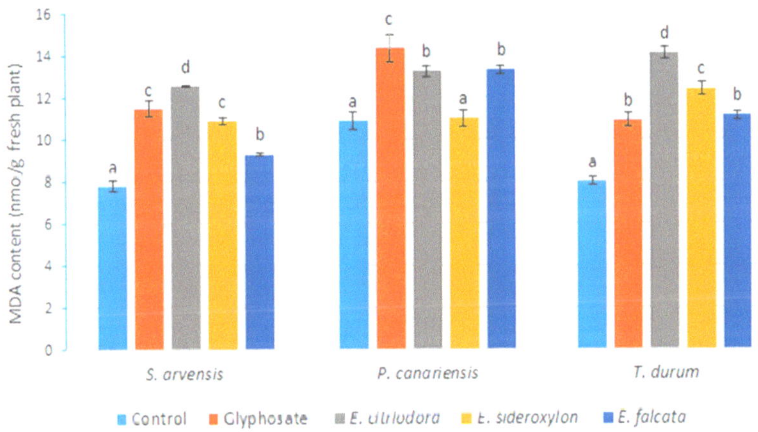

Figure 7. Effects of *Eucalyptus* species essential oils and glyphosate on MDA content as nmol/g fresh tested plants measured five days after application. Error bars and letters on graphs represent standard error means and significant differences at $p \leq 0.05$ through Student–Newman–Keuls.

Based on the statistical analysis of the obtained data, an increase in the MDA level was noted for all oils and for glyphosate compared to the control. The highest MDA levels for all the tested herbs were obtained with *E. citrioodora* oil and glyphosate. These results confirm and can be correlated with the loss of membrane integrity and confirm the phytotoxicity of EO application.

It is also known that allelopathic substances generate oxidative stress in plants. This stress is particularly reflected in the accumulation of proline. In this context, the proline level in the aerial parts of the plants sprayed with the three *Eucalyptus* oils was measured.

The results obtained are summarized in Figure 8, which shows an increase in proline levels in the aerial parts of sprayed weeds and crops.

This accumulation varied according to the oils tested and the herbs used for phytotoxicity tests. Generally, EOs showed similar effects to the commercial herbicide. The effect of *E. citriodora* oil was most pronounced for *P. canariensis* (22.36 mg/g f.w.), three times greater than the control (6.39 mg/g f.w.), and *S. arvensis* (12.41 mg/g f.w.), more than two times greater than the control (5.36 mg/g f.w.). In the case of *T. durum*, all the oils showed a similar effect on the accumulation of proline.

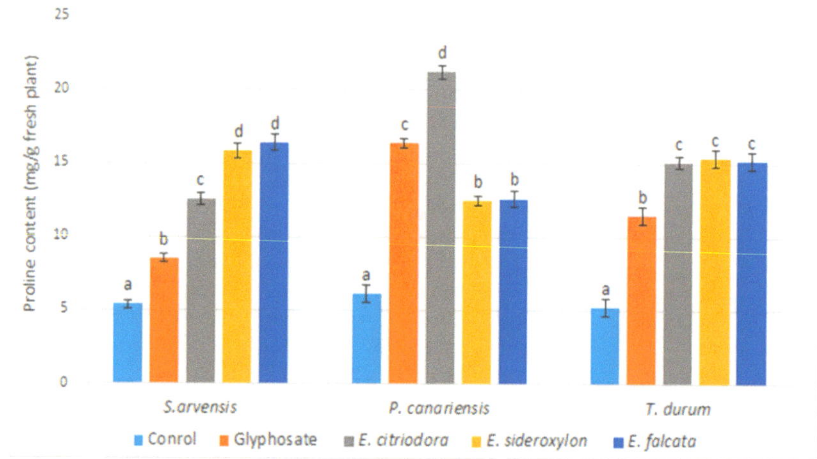

Figure 8. Effects of *Eucalyptus* essential oils and glyphosate on proline content as mg/g fresh tested plants measured five days after application. Error bars and letters on graphs represent standard error means and significant differences at $p \leq 0.05$ through Student–Newman–Keuls.

To evaluate the possible antifungal activity, the three *Eucalyptus* EOs were tested against eight phytopathogenic fungi that attack cereals and fruits; the main results are presented in Table 7.

Table 7. Antifungal activity of *E. citriodora*, *E. sideroxylon*, and *E. falcata* essential oils.

Fungal Strain	Essential Oil Inhibition % at the Dose of 4 µL/mL		
	E. citriodora	***E. sideroxylon***	***E. falcata***
F. culmorum	100 ± 0.0 a	80.01 ± 7.51 b	100 ± 0.0 a
F. pseudograminearum	100 ± 0.0 a	70.49 ± 4.73 b	100 ± 0.0 a
F. graminearum	100 ± 0.0 a	80 ± 6.65 b	81.59 ± 4.27 b
F. proliferatum	100 ± 0.0 a	51.07 ± 6.96 b	87.06 ± 7.41 a
F. avenaceum	100 ± 0.0 a	62.55 ± 5.01 c	74.77 ± 2.39 b
F.nygamai	100 ± 0.0 a	74.55 ± 1.11 c	90.00 ± 4.71 b
F. verticillioides	100 ± 0.0 a	78.11 ± 5.23 b	73.52 ± 4.15 b
B. sorokiniana	77.64 ± 6.65 a	48.70 ± 8.26 a	66.66 ± 11.78 a

Results are expressed as mean ± standard errors. Different letters in the same row indicate significant differences ($p \leq 0.05$).

The data in Table 7 show that the oils, at the dose of 4 µL/mL, significantly inhibited the growth of all tested fungi. According to the statistical analysis, the oils exhibited varying degrees of inhibition. *B. sorokiniana* was the most resistant to all the oils, while *Fusarium* species were the most sensitive. *E. citriodora* EOs showed the best antifungal properties, causing total inhibition for all strains, except for *B. sorokiniana*. The other two EOs caused partial inhibition on nearly all tested strains.

4. Discussion

The study of the chemical composition of the essential oils of the three species of *Eucalyptus* revealed the great importance and richness of various compounds. According to the literature, among the few studies on *E. falcata* EOs, three studies were performed on plant material from Tunisia [17–19]. The chemical composition of the EOs of *E. falcata* is in agreement with that reported in the present study. In fact, the authors stated that the major components were 1,8-cineole (30.7%), trans-pinocarveol (26%), globulol (7%), and α-pinene (6%) [17]. However, the main differences with the previous studies consist in the absence of spathunelol, the presence of globulol, and the high levels of trans-pinocarveol [17,18].

Similar results were obtained in previous reports for *E. sideroxylon* [15,19]. According to these studies, the EOs of *E. sideroxylon* were characterized by their high content of 1,8-cineole (69–81%), which is in agreement with the present study.

Several investigations on *E. citriodora* EOs have been reported worldwide. In fact, in Colombian *E. citriodora* oils [30,31], in India [20], in Taiwan [32], and in the Congo [33], the major components were citronellal (40–72%), citronellol (6–22%), and iso-pulegol (3–13%), which is in agreement with our obtained data. However, citronellyl acetate was detected for the first time in this study. On the other hand, our study is in contrast with the results of a Tunisian report [2,34] wherein the authors evidenced a different chemotype characterized by a specific richness in 1,8-cineole (54%) and α-pinene (23%) and the total absence of citronellal, citronellol, and iso-pulegone. Apart from this, the data of the current study are in good agreement with the literature. The small differences in the percentages may be due to several factors, such as the genetic background, biotic and abiotic factors, season of collection, and extraction methods [35].

The evaluation of the antioxidant potential of the three oils showed interesting activities that can contribute to the valorization of essential oils of forest species, particularly of the *Eucalyptus* genus. These results are in agreement with the literature.

Indeed, several studies have investigated the antioxidant properties of different *Eucalyptus* species, such as *E. oleosa*, *E. grandis* × *E. urophylla*, *E. gracilis*, *E. citriodora*, *E. salubris*, and *E. salmonophloia* [36–39].

Furthermore, the EOs of *E. citriodora* exhibited higher antioxidant potential than those of *E. falcata* and *E. sideroxylon*. This appears to be related to the high level of oxygenated terpenes in *E. citriodora* oil [36,37].

Similarly, the EOs of *E. citriodora*, collected in India, contain oxygenated monoterpenes such as citronellal (60.66%), exhibiting strong antioxidant activity, in agreement with the current study [38]. In the same report, the authors evaluated and showed the good antioxidant activity of α-citronellal, β-citronellol, and isopulegol, the major compounds of the oil. Furthermore, the synergism between oil components could be related to the remarkable antioxidant capacity of the tested oils [40].

The study of the allelopathic properties of the EOs obtained from the three species of *Eucalyptus* revealed remarkable herbicidal potential, which even exceeds the activity of the synthetic herbicide glyphosate.

The germination- and growth-inhibiting effects of the *Eucalyptus* oils tested in the present study appear to be linked to the presence of several chemical components, particularly monoterpenes and sesquiterpenes, which are the main constituents and are known for their herbicidal potential [41]. Indeed, EOs of species belonging to the *Myrtaceae* family are known for their richness in allelochemicals [5,8,9,16,42]. In particular, in previous studies, we showed the phytotoxic potential of volatile oils and crude extracts of *Eucalyptus erythrocorys* against weeds [16,42].

Recently, the herbicidal potential of 22 *Eucalyptus* species growing in Tunisia has been reported. These studies reveal enormous allelopathic potential linked to the EOs of *Eucalyptus* species; this could be in agreement with the present study [5,8,9,16].

Observing the chemical composition of the *Eucalyptus* oils tested in the present work (Table 2), we can assume that the activity is due to different components, such as 1,8-cineole (0.3–65.4% in *Eucalyptus* oils), citronellal (48.7% in *E. citriodora* oil), citronellol (20.2% in *E. citriodora* oil), and α-pinene (ranging between 0 and 6% in *Eucalyptus* oils), already known for their significant phytotoxic effects [20,43,44].

Several studies report that terpenes, both hydrocarbons and oxygenated, are phytotoxic molecules. Additionally, oxygenated monoterpenes (51.3, 75.5, and 90%, respectively, in *E. falcata*, *E. sideroxylon*, and *E. citriodora*) have been reported to have stronger herbicidal properties than the corresponding hydrocarbon derivatives [20,43,44]; this can explain the high herbicidal activity of *E. citriodora* compared to *E. falcata* and *E. sideroxylon*.

To explain the mode of action of terpenes on the inhibition of germination, various physiological and biochemical mechanisms have been reported.

In fact, such monoterpenes can affect physiological functions, such as cell division, viability, and growth inhibition by the alteration of membrane integrity [45,46].

Other studies have reported that EOs reduce germination and growth by inhibiting the incorporation of nitrogen into amino acid synthesis, leading to the accumulation of ammonia, and by impairing photosynthesis and photorespiration [47].

In the same way, Abrahim et al., 2003 demonstrated that the monoterpene hydrocarbon α-pinene reduces the growth of maize by inhibiting electron transfer and uncoupling oxidative phosphorylation, resulting in the alteration of energy metabolism and the blocking of ATP synthesis in the mitochondria [48]. Similarly, some sesquiterpenes have been reported to have phytotoxic effects by causing oxidative stress, the inhibition of photosynthesis, and the induction of microtubular alterations [47]. Previous reports have shown that EOs induce oxidative stress to weeds and increase peroxidase and superoxide dismutase activities [49–51]. Other studies have shown that EOs and their constituents have remarkable herbicidal potential, the effects of which result in anatomical and physiological effects, such as the formation of lipid globules in the cytoplasm, inhibition of mitochondrial development, and alteration of the integrity of the membrane of the nucleus and mitochondria [52,53].

For post-emergence trials and the application of EOs by spraying, few studies have been undertaken and the exact modes of action still remain poorly known. On the other hand, data obtained in this study are in agreement with those reported in the literature. Several studies have reported that various constituents of EOs, when applied at the post-emergence stage, are able to inhibit the growth of weeds [54], hypothesizing oxidative stress and alteration of the water content [47]. The evidence that emerged in the present study showed a correlation between the decrease in the water content and the arrest of the growth of the plants treated with *Eucalyptus* EOs.

Following the application of *Eucalyptus* EOs, a marked decrease in chlorophyll synthesis was noted, suggesting a correlation between the visible effects on weeds (i.e., chlorosis, necrosis, and photosynthetic functions) and their growth. According to previous studies, several EOs and their pure components have been described to inhibit photosynthesis and chlorophyll synthesis [55]. Indeed, some oxygenated monoterpenes have been reported to induce a photosynthetic decrease by reducing the chlorophyll content in various weeds [56].

In addition, citronellol and 1,8-cineole, major compounds of the *Eucalyptus* species in the current study, reduced the chlorophyll content of *Ageratum conyzoides* by 60 and 66%, respectively [57,58]. Earlier studies concluded that the decrease in chlorophyll content appeared to be related to the inhibition of chlorophyll synthesis and degradation of photosynthetic pigments [59,60].

Among the physiological effects of *Eucalyptus* EOs on membrane integrity, the excessive leakage of electrolytes was noted, which can be explained by the loss of integrity of membrane cells. On the other hand, membrane integrity is essential for vital functions and plays several roles, including that of a barrier, selective permeability to nutrients, and, indeed, control of the electrolyte balance [61]. Various secondary metabolites have been reported to alter the integrity of the cell membrane in plants, particularly EOs and their pure constituents. A previous study has shown that linalool caused increased membrane permeability, further demonstrating that EOs can cause important damage to membrane permeability [62]. According to the data obtained, the application of *Eucalyptus* EOs induced, in the tested herbs, an increase in the rate of MDA and proline compared to the control, in agreement with the previous observations. Indeed, it has been reported that the application of EOs during the post-emergence stage increases the permeability of the membrane by disrupting its integrity because of lipid peroxidation [47,63–65]. Additionally, some EOs and monoterpenes have been reported to induce oxidative stress. α-pinene caused lipid peroxidation when applied to *Cassia occidentalis*, resulting in increased solute leakage [25]. Furthermore, the increase in MDA levels was related to the lipid peroxidation of the membrane and alteration of its permeability.

Finally, the current study has shown that *Eucalyptus* oils have significant in vitro antifungal potential against the growth of phytopathogenic fungi.

The differences in the resistance of different fungal strains to *Eucalyptus* volatile oil seem to be related to the different capacities of the oil components to disrupt the cell membrane [66].

Eucalyptus oils are known to have remarkable antimicrobial activity that is related to their richness in oxygenated compounds [67].

Indeed, we have previously reported the antifungal potential of *Eucalyptus erhytrocorys* EOs and its crude extracts against phytopathogenic fungi [16,42], which is in agreement with the current study. Similarly, the inhibitory activity of the volatile oil obtained from *Eucalyptus camaldulensis* against the growth of fungi such as *Thanatephorus cucumeris*, *Aspergillus niger*, *F. oxysporum*, and *Rhizopus oryzae* is also known [67].

In agreement with the current study, Ivanov et al. (2021) demonstrated the fungicidal activity of 1,8-cineole and camphor (major compounds in *Eucalyptus* volatile oils) against the growth of different *Candida* species [68]. In another research work, it was reported that terpinen-4-ol and α-terpineol, terpinolene, and 1,8-cineole, constituents of *Eucalyptus* oils, are fungitoxic against *Botrytis cinerea* [69]. However, with regard to the antimicrobial mechanism, it has been reported that the apolar nature of terpenes may favor their penetration into the lipid bilayer of the fungal membrane, inducing its disruption. Many terpenes have been reported to exert their antifungal activity by increasing the permeability of fungal cells and membrane fluidity [70].

5. Conclusions

To conclude, our data showed that *Eucalyptus* EOs are a complex mixture of various compounds. The chemical profile of the tested EOs was characterized by the high content of oxygenated monoterpenes. Great diversity was noted between the three species with the identification of different major compounds; however, the three species belong to the same genus. This is an interspecific diversity, with the identification of 58 compounds. Similarly, the variability of the results obtained in this present study with those in the literature could be considered within the context of ecological variability, which makes this work necessary to appreciate the results obtained with these three species of *Eucalyptus* growing in Tunisia. The essential oils tested for their biological activities have shown efficacy for the control of weeds and phyto-pathogenic fungi, and their antioxidant potential has also been described. These biological activities offer several areas of application for these oils, whether in agronomy or in the chemical and pharmaceutical industries. These molecules are of natural identity; they are biodegradable molecules and are compatible with the environment. Their applications could solve several problems, including the problem of resistance and the undesirable environmental effects. Plants synthesize these substances as allelochemicals; their exploitation can provide an enhancement in forest species and also allow the exploitation of the allelopathic potential of aromatic and medicinal plants. Consequently, the adoption of such bio-molecules as novel compounds in herbicidal formulations could facilitate not only sustainable weed control but also sustainable agriculture. Further studies are required to evaluate their application under field conditions. Such results encourage the use of *Eucalyptus* EOs as natural products in agriculture and for their agro-industrial potential.

Author Contributions: Conceptualization, I.A., M.H. and L.H.; Data curation, R.B.A., S.K., O.K. and Y.P.; Formal analysis, M.K. and M.B.M.; Funding acquisition, L.H.; Investigation, I.A. and M.K.; Methodology, R.B.A., G.F. and L.H.; Project administration, L.H.; Resources, I.A.; Supervision, S.G.; Validation, I.A., M.H. and L.H.; Writing—original draft, I.A., M.K., G.F. and Y.M.; Writing—review and editing, I.A. All authors have read and agreed to the published version of the manuscript.

Funding: This research received no external funding.

Data Availability Statement: All data are available in the manuscript file.

Conflicts of Interest: The authors declare no conflict of interest.

References

1. Oyekunle, J.A.O.; Adekunle, A.S.; Adewole, A.M. Determination of Organochlorine Pesticide Residues in Some Evaporated Milk Samples in Nigeria Using Gas Chromatography-Mass Spectrometry. *Chem. Afr.* **2021**, *4*, 349–366. [CrossRef]
2. Fahed, C.; Hayar, S. Conventional vs. Organic Cucumber Production in Lebanon: Risk Assessment of the Recommended Agrochemicals on Consumer Health and the Environment. *Chem. Afr.* **2021**, *4*, 463–476. [CrossRef]
3. Amri, I.; Lamia, H.; Mohsen, H.; Bassem, J.; Kaouthar, L. Essential oils as biological alternatives to protect date palm (*Phoenix dactylifera* L.) against *Ectomyelois ceratoniae* Zeller (*Lepidoptera*: *Pyralidae*). *Chil. J. Agric. Res.* **2014**, *74*, 273–279. [CrossRef]
4. Oliveira, A.F.; Costa Junior, L.M.; Lima, A.S.; Silva, C.R.; Ribeiro, M.N.S.; Mesquista, J.W.C.; Rocha, C.Q.; Tangerina, M.M.P.; Vilegas, W. Anthelmintic Activity of Plant Extracts from Brazilian Savanna. *Vet. Parasitol.* **2017**, *236*, 121–127. [CrossRef] [PubMed]
5. Khammassi, M.; Polito, F.; Amri, I.; Khedhri, S.; Hamrouni, L.; Nazzaro, F.; Fratianni, F.; De Feo, V. Chemical Composition and Phytotoxic, Antibacterial and Antibiofilm Activity of the Essential Oils of *Eucalyptus occidentalis*, *E. striaticalyx* and *E. stricklandii*. *Molecules* **2022**, *27*, 5820. [CrossRef]
6. Malaspina, P.; Papaianni, M.; Ranesi, M.; Polito, F.; Danna, C.; Aicardi, P.; Cornara, L.; Woo, S.L.; De Feo, V. Eucalyptus cinerea and E. nicholii by-Products as Source of Bioactive Compounds for Agricultural Applications. *Plants* **2022**, *11*, 2777. [CrossRef]
7. Ismail, A.; Mancini, E.; De Martino, L.; Hamrouni, L.; Hanana, M.; Jamoussi, B.; Gargouri, S.; Scognamiglio, M.; De Feo, V. Chemical Composition and Biological Activities of Tunisian *Cupressus arizonica* Greene Essential Oils. *Chem. Biodivers.* **2014**, *11*, 150–160. [CrossRef]
8. Kouki, H.; Polito, F.; De Martino, L.; Mabrouk, Y.; Hamrouni, L.; Amri, I.; Fratianni, F.; De Feo, V.; Nazzaro, F. Chemistry and bioactivities of six Tunisian Eucalyptus species. *Pharmaceuticals* **2022**, *15*, 1265. [CrossRef]
9. Polito, F.; Kouki, H.; Khedhri, S.; Hamrouni, L.; Mabrouk, Y.; Amri, I.; Nazzaro, F.; Fratianni, F.; De Feo, V. Chemical Composition and Phytotoxic and Antibiofilm Activity of the Essential Oils of *Eucalyptus bicostata*, *E. gigantea*, *E. intertexta*, *E. obliqua*, *E. pauciflora* and *E. tereticornis*. *Plants* **2022**, *11*, 3017. [CrossRef]
10. Ullah, N.; Haq, I.U.; Safdar, N.; Mirza, B. Physiological and biochemical mechanisms of allelopathy mediated by the allelochemical extracts of *Phytolacca latbenia* (Moq.) H. Walter. *Toxicol. Ind. Health* **2015**, *31*, 931–937. [CrossRef]
11. Werrie, P.Y.; Durenne, B.; Delaplace, P.; Fauconnier, M.L. Phytotoxicity of Essential Oils: Opportunities and Constraints for the Development of Biopesticides. A Review. *Foods* **2020**, *9*, 1291. [CrossRef]
12. Najar, B.; Mecacci, G.; Nardi, V.; Cervelli, C.; Nardoni, S.; Mancianti, F.; Ebani, V.V.; Giannecchini, S.; Pistelli, L. Volatiles and Antifungal-Antibacterial-Antiviral Activity of South African Salvia Spp. Essential Oils Cul-tivated in Uniform Conditions. *Molecules* **2021**, *26*, 2826. [CrossRef]
13. Shen, S.; Xu, G.; Roy Clemen, D.; Jin, G.; Liu, S.; Zhang, F.; Yang, Y.; Chen, A.; Kato-Noguc, H. Effects of Invasive Plant *Mikania micrantha* on Plant Community and Diversity in Farming Systems. *Asian J. Plant Sci.* **2014**, *14*, 27–33. [CrossRef]
14. Alves-Silva, J.M.; Cocco, E.; Piras, A.; Gonçalves, M.J.; Silva, A.; Falconieri, D.; Porcedda, S.; Cruz, M.T.; Maxia, A.; Salgueiro, L. Unveiling the Chemical Composition and Biological Properties of Salvia cacaliifolia Benth. Essential Oil. *Plants* **2023**, *12*, 359. [CrossRef]
15. Elaissi, A.; Marzouki, H.; Medini, H.; Larbi Khouja, M.; Farhat, F.; Lynene, F.; Harzallah-Skhiri, F.; Chemli, R. Variation in volatile leaf oils of 13 *Eucalyptus* species harvested from Souinet Arboreta (Tunisia). *Chem. Biodivers.* **2010**, *7*, 909–921. [CrossRef]
16. Khedhri, S.; Polito, F.; Caputo, L.; Manna, F.; Khammassi, M.; Hamrouni, L.; Amri, I.; Nazzaro, F.; De Feo, V.; Fratianni, F. Chemical Composition, Phytotoxic and Antibiofilm Activity of Seven *Eucalyptus* Species from Tunisia. *Molecules* **2022**, *27*, 8227. [CrossRef]
17. Elaissi, A.; Rouis, Z.; Salem, N.A.B.; Mabrouk, S.; ben Salem, Y.; Salah, K.B.H.; Aouni, M.; Farhat, F.; Chemli, R.; Harzallah-Skhiri, F.; et al. Chemical composition of 8 *eucalyptus* species' essential oils and the evaluation of their antibacterial, antifungal and antiviral activities. *BMC Complement. Altern. Med.* **2012**, *12*, 81. [CrossRef]
18. Elaissi, A.; Medini, H.; Khouja, M.L.; Monique, S.; Lyenne, F.; Farhat, F.; Chemli, R.; Harzallah-Skhiri, F. Variation in volatile leaf oils of eleven *Eucalyptus* species harvested from Korbous arboreta (Tunisia). *Chem. Biodivers.* **2010**, *7*, 1841–1854. [CrossRef]
19. Elaissi, A.; Medini, H.; Simmonds, M.; Lynen, F.; Farhat, F.; Chemli, R.; Harzallah-Skhiri, F.; Khouja, M.L. Variation in Volatile Leaf Oils of Seven *Eucalyptus* Species Harvested from Zerniza Arboreta (Tunisia). *Chem. Biodivers.* **2011**, *8*, 362–372. [CrossRef]
20. Batish, D.R.; Singh, H.P.; Setia, N.; Kaur, S.; Kohli, R.K. Chemical Composition and Phytotoxicity of Volatile Essential Oil from Intact and Fallen Leaves of *Eucalyptus citriodora*. *Z. Für Nat. C* **2006**, *61*, 465–471. [CrossRef]
21. National Institute of Standards and Technology. *NIST/EPA/NIH Mass Spectral Library*; The NIST Mass Spectrometry Data Center: Gaithersburg, MD, USA, 2014.
22. Adams, R.P. *Identification of Essential Oil Components by Gas Chromatography/Mass Spectrometry*; Allured Publishing Co.: Carol Stream, IL, USA, 2007.
23. Prieto, P.; Pineda, M.; Aguilar, M. Spectrophotometric quantitation of antioxidant capacity through the formation of a phosphomolybdenum complex: Specific application to the determination of vitamin E1. *Anal. Biochem.* **1999**, *269*, 337–341. [CrossRef]

24. Hanato, T.; Kagawa, H.; Yasuhara, T.; Okuda, T. Two new flavonoids and other constituents in licorice root: Their relative astringency and radical scavenging effect. *Chem. Pharm. Bull.* **1988**, *36*, 1090–1097.
25. Re, R.; Pellegrini, N.; Proteggente, A.; Pannala, A.; Yang, M.; Rice-Evans, C. Antioxidant activity applying an improved ABTS radical cation decolorization assay. *Free. Radic. Biol. Med.* **1999**, *26*, 1231–1237. [CrossRef]
26. Downey, L.A.; Miller, J.W. Rapid measurements of relative turgidity in maize (*Zea mays* L.). *New Phytol.* **1971**, *70*, 555–560. [CrossRef]
27. Lichtenthaler, K.H. Chlorophylls and carotenoids: Pigments of photosynthetic biomembranes. *Methods Enzymol.* **1987**, *148*, 350–382. [CrossRef]
28. Singh, H.P.; Batish, D.R.; Kaur, S.; Arora, K.; Kohli, R.K. α-pinene inhibits growth and induces oxidative stress in roots. *Ann. Bot.* **2006**, *98*, 1261–1269. [CrossRef]
29. Bates, L.S.; Waldren, R.P.; Teare, I.D. Rapid determination of free proline for water-stress studies. *Plant Soil* **1973**, *39*, 205–207. [CrossRef]
30. Olivero-Verbel, J.; Nerio, L.S.; Stashenko, E.E. Bioactivity against *Tribolium castaneum* Herbst (*Coleoptera*: *Tenebrionidae*) of *Cymbopogon citratus* and *Eucalyptus citriodora* essential oils grown in Colombia. *Pest Manag. Sci. Former. Pestic. Sci.* **2010**, *66*, 664–668. [CrossRef]
31. Olivero-Verbel, J.; Tirado-Ballestas, I.; Caballero-Gallardo, K.; Stashenko, E.E. Essential oils applied to the food act as repellents toward *Tribolium castaneum*. *J. Stored Prod. Res.* **2013**, *55*, 145–147. [CrossRef]
32. Su, Y.C.; Ho, C.L.; Eugene, I.C.; Chang, T. Antifungal activities and chemical compositions of essential oils from leaves of four *Eucalypts*. *Taiwan J. For. Sci.* **2006**, *21*, 49–61.
33. Cimanga, K.; Kambu, K.; Tona, L.; Apers, S.; De Bruyne, T.; Hermans, N.; Totté, J.; Pieters, L.; Vlietinck, A.J. Correlation between chemical composition and antibacterial activity of essential oils of some aromatic medicinal plants growing in the Democratic Republic of Congo. *J. Ethnopharmacol.* **2002**, *79*, 213–220. [CrossRef]
34. Elaissi, A.; Salah, K.H.; Mabrouk, S.; Larbi, K.M.; Chemli, R.; Harzallah-Skhiri, F. Antibacterial activity and chemical composition of 20 *Eucalyptus* species' essential oils. *Food Chem.* **2011**, *129*, 1427–1434. [CrossRef]
35. Abd-ElGawad, A.M.; Bonanomi, G.; Al-Rashed, S.A.; Elshamy, A.I. Persicaria lapathifolia Essential Oil: Chemical Constituents, Antioxidant Activity, and Allelopathic Effect on the Weed *Echinochloa colona*. *Plants* **2021**, *10*, 1798. [CrossRef]
36. Gatto, L.J.; Fabri, N.T.; Souza, A.M.; Fonseca, N.S.T.; Furusho, A.; Miguel, O.G.; Dias, J.F.G.; Zanin, S.M.W.; Miguel, M.D. Chemical composition, phytotoxic potential, biological activities and antioxidant properties of *Myrcia hatschbachii* D. Legrand essential oil. *Braz. J. Pharm. Sci.* **2020**, *56*, 18402. [CrossRef]
37. Ben Marzoug, H.N.; Bouajila, J.; Ennajar, M.; Lebrihi, A.; Mathieu, F.; Couderc, F.; Abderraba, M.; Romdhane, M. *Eucalyptus* (*gracilis*, *oleosa*, *salubris*, and *salmonophloia*) Essential Oils: Their Chemical Composition and Antioxidant and Antimicrobial Activities. *J. Med. Food* **2010**, *13*, 1005–1012. [CrossRef]
38. Singh, H.P.; Kaur, S.; Negi, K.; Kumari, S.; Saini, V.; Batish, D.R.; Kohli, R.K. Assessment of in vitro antioxidant activity of essential oil of *Eucalyptus citriodora* (lemon-scented Eucalypt; *Myrtaceae*) and its major constituents. *LWT Food Sci. Technol.* **2012**, *48*, 237–241. [CrossRef]
39. Zhou, L.; Li, J.; Kong, Q.; Luo, S.; Wang, J.; Feng, S.; Yuan, M.; Chen, T.; Yuan, S.; Ding, C. Chemical composition, antioxidant, antimicrobial, and phytotoxic potential of *Eucalyptus grandis* × *E. urophylla* leaves essential oils. *Molecules* **2021**, *26*, 1450. [CrossRef]
40. Nafis, A.; Kasrati, A.; Jamali, C.A.; Mezrioui, N.; Setzer, W.; Abbad, A.; Hassani, L. Antioxidant activity and evidence for synergism of *Cannabis sativa* (L.) essential oil with antimicrobial standards. *Ind. Crops Prod.* **2019**, *137*, 396–400. [CrossRef]
41. Amri, I.; Hanana, M.; Jamoussi, B.; Hamrouni, L. Chemical composition of *Thuja orientalis* L. essential oils and study of their allelopathic potential on germination and seedling growth of weeds. *Arch. Phytopathol. Plant Prot.* **2015**, *48*, 18–27. [CrossRef]
42. Ben Ghnaya, A.; Hamrouni, L.; Amri, I.; Ahoues, H.; Hanana, M.; Romane, A. Study of allelopathic effects of *Eucalyptus erythrocorys* L. crude extracts against germination and seedling growth of weeds and wheat. *Nat. Prod. Res.* **2016**, *30*, 2058–2064. [CrossRef]
43. De Martino, L.; Mancini, E.; Rolim de Almeida, L.F.; De Feo, V. The antigerminative activity of twenty-seven monoterpenes. *Molecules* **2010**, *15*, 6630–6637. [CrossRef]
44. Chowhan, N.; Singh, H.P.; Kohl, R.K. Phytotoxic effects of β-pinene on early growth and associated biochemical changes in rice. *Acta Physiol. Plant.* **2011**, *33*, 2369–2376. [CrossRef]
45. Mahdavikia, F.; Saharkhiz, M.J.; Karami, A. Defensive response of radish seedlings to the oxidative stress arising from phenolic compounds in the extract of peppermint (*Mentha* × *piperita* L.). *Sci. Hortic.* **2017**, *214*, 133–140. [CrossRef]
46. Fagodia, S.K.; Singh, H.P.; Batish, D.R.; Kohli, R.K. Phytotoxicity and cytotoxicity of *Citrus aurantiifolia* essential oil and its major constituents: Limonene and citral. *Ind. Crops Prod.* **2017**, *108*, 708–715. [CrossRef]
47. Araniti, F.; Costas-Gil, A.; Cabeiras-Freijanes, L.; Lupini, A.; Sunseri, F.; Reigosa, M.J.; Abenavoli, M.R.; Sánchez-Moreiras, A.M. Rosmarinic acid induces programmed cell death in *Arabidopsis* seedlings through reactive oxygen species and mitochondrial dysfunction. *PLoS ONE* **2018**, *13*, e0208802. [CrossRef]
48. Abrahim, D.; Francischini, A.C.; Pergo, E.M.; Kelmer-Bracht, A.M.; Ishii-Iwamoto, E.L. Effects of α-pinene on the mitochondrial respiration of maize seedlings. *Plant Physiol. Biochem.* **2003**, *41*, 985–991. [CrossRef]

49. Bogatek, R.; Gniazdowska, A.; Stępień, J.; Kupidłowska, E. Sunflower allelochemicals mode of action in germinating mustard seeds. In Proceedings of the 4th World Congress on Allelopathy, "Establishing the Scientific Base", Wagga Wagga, NSW, Australia, 21–26 August 2005; pp. 365–369. Available online: https://www.cabdirect.org/cabdirect/abstract/20073229801 (accessed on 11 July 2022).
50. Hayat, S.; Ahmad, H.; Ali, M.; Ren, K.; Cheng, Z. Aqueous garlic extract stimulates growth and antioxidant enzymes activity of tomato (*Solanum lycopersicum*). *Sci. Hortic.* **2018**, *240*, 139–146. [CrossRef]
51. Ji, X.; Li, J.; Meng, Z.; Dong, S.; Zhang, S.; Qiao, K. Inhibitory effect of allicin against *Meloidogyne incognita* and *Botrytis cinerea* in tomato. *Sci. Hortic.* **2019**, *253*, 203–208. [CrossRef]
52. Koitabashi, R.; Suzuki, T.; Kawazu, T.; Sakai, A.; Kuroiwa, H.; Kuroiwa, T. 1,8-Cineole inhibits root growth and DNA synthesis in the root apical meristem of *Brassica campestris* L. *J. Plant Res.* **1997**, *110*, 1–6. [CrossRef]
53. Zunino, M.P.; Zygadlo, J.A. Effect of monoterpenes on lipid oxidation in maize. *Planta* **2004**, *219*, 303–309. [CrossRef]
54. Sánchez-Muñoz, B.A.; Aguilar, M.I.; King-Díaz, B.; Rivero, J.F.; Lotina-Hennsen, B. The Sesquiterpenes β-Caryophyllene and Caryophyllene Oxide Isolated from *Senecio salignus* Act as Phytogrowth and Photosynthesis Inhibitors. *Molecules* **2012**, *17*, 1437–1447. [CrossRef] [PubMed]
55. Pouresmaeil, M.; Nojadeh, M.S.; Movafeghi, A.; Maggi, F. Exploring the bio-control efficacy of *Artemisia fragrans* essential oil on the perennial weed *Convolvulus arvensis*: Inhibitory effects on the photosynthetic machinery and induction of oxidative stress. *Ind. Crops Prod.* **2020**, *155*, 112785. [CrossRef]
56. Vaid, S.; Batish, D.; Singh, H.; Kohli, R.; Phytotoxic effect of eugenol towards two weedy species. Bioscan 2010, 5: 339–341. Available online: https://www.semanticscholar.org/paper/PHYTOTOXIC-EFFECT-OF-EUGENOL-TOWARDS-TWO-WEEDY-Vaid-Batish/541b664b28a6e3e90cf08e467328286f91bd4024 (accessed on 25 September 2022).
57. Singh, H.P.; Batish, D.R.; Kohli, R.K. Allelopathic effect of two volatile monoterpenes against bill goat weed (*Ageratum conyzoides* L.). *Crop Prot.* **2002**, *21*, 347–350. [CrossRef]
58. Grana, E.; Daaaz-Tielas, C.M.; Sanchez-Moreiras, A.; Reigosa, M. Mode of Action of Monoterpenes in Plant-Plant Interactions. *Curr. Bioact. Compd.* **2012**, *8*, 80–89. [CrossRef]
59. Romagni, J.G.; Allen, S.N.; Dayan, F.E. Allelopathic effects of volatile cineoles on two weedy plant species. *J. Chem. Ecol.* **2000**, *26*, 303–313. [CrossRef]
60. Poonpaiboonpipat, T.; Pangnakorn, U.; Suvunnamek, U.; Teerarak, M.; Charoenying, P.; Laosinwattana, C. Phytotoxic effects of essential oil from *Cymbopogon citratus* and its physiological mechanisms on barnyardgrass (*Echinochloa crus-galli*). *Ind. Crops Prod.* **2013**, *41*, 403–407. [CrossRef]
61. McNeil, P.L.; Steinhardt, R.A. Loss, Restoration, and Maintenance of Plasma Membrane Integrity. *J. Cell Biol.* **1997**, *137*, 1–4. [CrossRef]
62. Varona, S.; Martín, Á.; Cocero, M.J. Formulation of a natural biocide based on lavandin essential oil by emulsification using modified starches. *Chem. Eng. Process* **2009**, *48*, 1121–1128. [CrossRef]
63. Dmitrović, S.; Perišić, M.; Stojić, A.; Živković, S.; Boljević, J.; Nestorović, Ž.J.; Aničić, N.; Ristić, M.; Mišić, D. Essential oils of two *Nepeta* species inhibit growth and induce oxidative stress in ragweed (*Ambrosia artemisiifolia* L.) shoots in vitro. *Acta Physiol. Plant.* **2015**, *37*, 64. [CrossRef]
64. Mahdavikia, F.; Saharkhiz, M.J. Secondary metabolites of peppermint change the morphophysiological and biochemical characteristics of tomato. *Biocatal. Agric. Biotechnol.* **2016**, *7*, 127–133. [CrossRef]
65. Hazrati, H.; Saharkhiz, M.J.; Niakousari, M.; Moein, M. Natural herbicide activity of *Satureja hortensis* L. essential oil nanoemulsion on the seed germination and morphophysiological features of two important weed species. *Ecotoxicol. Environ. Saf.* **2017**, *142*, 423–430. [CrossRef] [PubMed]
66. Obame, L.C.; Edou, P.; Bassolé, I.H.N.; Koudou, J.; Agnaniet, H.; Eba, F. Chemical composition, antioxidant and antimicrobial properties of the essential oil of *Dacryodes edulis* (G. Don) H. J. Lam from Gabon. *Afr. J. Microbiol. Res.* **2008**, *2*, 148–152.
67. Siramon, P.; Ohtani, Y.; Ichiura, H. Chemical Composition and Antifungal Property of *Eucalyptus camaldulensis* Leaf Oils from Thailand. *Rec. Nat. Prod.* **2013**, *7*, 49–53.
68. Ivanov, M.; Kannan, A.; Stojković, D.S.; Glamočlija, J.; Calhelha, R.C.; Ferreira, I.C.F.R.; Sanglard, D.; Soković, M. Camphor and Eucalyptol—Anticandidal Spectrum, Antivirulence Effect, Efflux Pumps Interference and Cytotoxicity. *Int. J. Mol. Sci.* **2021**, *22*, 483. [CrossRef]
69. Yu, D.; Wang, J.; Shao, X.; Xu, F.; Wang, H. Antifungal modes of action of tea tree oil and its two characteristic components against *Botrytis cinerea*. *J. Appl. Microbiol.* **2015**, *119*, 1253–1262. [CrossRef]
70. Cristani, M.; D'Arrigo, M.; Mandalari, G.; Castelli, F.; Sarpietro, M.G.; Micieli, D.; Venuti, V.; Bisignano, G.; Saija, A.; Trombetta, D. Interaction of Four Monoterpenes Contained in Essential Oils with Model Membranes: Implications for Their Antibacterial Activity. *J. Agric. Food Chem.* **2007**, *55*, 6300–6308. [CrossRef]

plants

Article

Phytotoxicity of *Quillaja lancifolia* Leaf Saponins and Their Bioherbicide Potential

Maria E. M. Marques [1], Ana C. de Carvalho [1], Anna C. A. Yendo [2], Yve V. S. Magedans [1], Eliane Zachert [1] and Arthur G. Fett-Neto [1,*]

[1] Plant Physiology Laboratory, Center for Biotechnology and Department of Botany, Federal University of Rio Grande do Sul, Porto Alegre 91501-970, RS, Brazil
[2] Arborea Biotechnology, Center for Biotechnology Start Up Incubator (IECBiot), UFRGS, Porto Alegre 91501-970, RS, Brazil
* Correspondence: fett.neto@ufrgs.br

Abstract: Weeds are major threats to the integrity of agricultural and natural environments due to their invasive and competing potential. Bioherbicides are substances based on natural compounds that are biodegradable and often have low residual effects. Plant species able to produce and release phytotoxic compounds may represent effective bioherbicide sources. Leaves of *Quillaja lancifolia* D.Don (formerly *Q. brasiliensis* (A.St.-Hil. & Tul.) Mart.) produce water-soluble specialized metabolites of the saponin class that could be evaluated for phytotoxic activity and potential as natural herbicides. This study was conducted to examine the impacts of *Q. lancifolia* total saponins aqueous extract (AE) at 4 and 10% (*w*/*v*) and of two combined reverse-phase chromatography purified saponin fractions (QB) at 1 and 2% (*w*/*v*) on morpho-physiological parameters of *Lactuca sativa* (lettuce) and *Echinochloa crus-galli* (barnyardgrass) in pre- and post-emergence bioassays. QB was only tested in pre-emergence assays. In pre-emergence bioassays, the germination rate and germination kinetics were determined. Post-emergence evaluations included effects on seedling morphology, root and shoot length, dry mass, and chlorophyll content. Osmotic potential and pH analyses ruled out roles for these factors in the observed responses. AE had a high inhibitory impact on the germination of both lettuce and barnyardgrass. QB at 1% and 2% (*w*/*v*) significantly decreased the growth of lettuce seedlings germinated in its presence by more than 10 fold. Phytotoxic effects on the post-emergence growth of lettuce, especially at the highest concentration tested of AE (10% *w*/*v*), was also observed. The presence of quillaic acid-based triterpene saponins in AE and QB was confirmed using different analytical methods. Therefore, both saponin-enriched fraction and aqueous extracts of *Q. lancifolia* inhibited tested plant growth and development. The water solubility of saponins and the availability of a sustainable source of these molecules from the leaves of cultivated young *Q. lancifolia* plants make them attractive candidates for use as bioherbicides.

Keywords: soap tree; triterpene saponin; secondary metabolism; germination; seedling

Citation: Marques, M.E.M.; de Carvalho, A.C.; Yendo, A.C.A.; Magedans, Y.V.S.; Zachert, E.; Fett-Neto, A.G. Phytotoxicity of *Quillaja lancifolia* Leaf Saponins and Their Bioherbicide Potential. *Plants* **2023**, *12*, 663. https://doi.org/10.3390/plants12030663

Academic Editors: Margot Schulz and Vincenzo Tabaglio

Received: 15 December 2022
Revised: 27 January 2023
Accepted: 30 January 2023
Published: 2 February 2023

1. Introduction

The combination of efficient agricultural land use with biodiversity conservation is a challenge [1]. It is estimated that agriculture will need to provide almost 50% more food by 2050 [2]. To sustain this level of production and knowing that 20–40% of global crop yields are lost each year to pests and diseases, pesticide use in agriculture has increased in recent years. Due to its extensive planting area, Brazil is one of the three largest consumers of pesticides in the world [2,3].

"Pests" are considered invasive organisms or pathogens of crops or forest species [4] and include animal pests (insects, mites, nematodes, rodents, slugs, snails, birds), plant pathogens (viruses, bacteria, fungi, chromists), and weeds (i.e., competitive plants) [5]. It has been estimated that weeds represent the most damaging category of pests in agriculture [5,6]. The Weed Science Society of America defines a weed as a plant that causes economic losses or ecological damage, creates health problems for humans or animals, or is undesirable where it is growing [7]. Weeds negatively affect crop yields due to a number of features, including: (a) capacity to grow in disturbed habitats, (b) high environmental plasticity, (c) ability to physically hinder or smother crop growth, (d) capacity to compete with crops for light, nutrients, moisture, and space, (e) rapid emergence, seedling growth, and quick maturation, (f) production of a large numbers of seeds per plant, (g) ability to release natural substances that inhibit crop growth (allelopathy), and (h) host pests or pathogens that may contaminate crops [8,9].

Although pesticides prevent, destroy, and repel most pests growing on crop plants, their continued use can have a significant impact on the environment and human health by contaminating water bodies, air, soil, and food. Moreover, unintentional negative effects on local biodiversity are often observed [10,11]. Thus, there is an increasing demand for pesticides that are safe for the environment and may replace or complement synthetic management methods [12]. This is an important concern in organic farming, given its expansion and the need for environmentally safe management tools [9,10]. Weed management techniques can vary (manual, mechanical, chemical, and biological control) but depend mostly on the use of synthetic herbicides (pre-emergent and post-emergent) [11,13]. Hence, it is important to search for new and effective alternatives for weed management.

Substances obtained from living organisms, e.g., the natural metabolites produced during growth and development, can be ingredients of bioherbicides. Although the use of these metabolites can be efficient and beneficial, few conventional herbicides are derived from natural compounds [14,15]. Accordingly, the use of plant species able to produce and release phytotoxic compounds may represent an effective tool for use alone or in association with other plant protection methods for weed management [12,16].

Saponins are specialized metabolites with complex chemical structures and high variability that are involved in the environmental adaptation of plants [17,18]. These molecules are widely distributed among plants and have several significant industrial and pharmacological applications [19]. Some studies have reported phytotoxic or allelopathic properties for saponins [20–26], which suggests that untested saponins could be evaluated as potential bioherbicides [14].

Quillaja lancifolia D.Don (formerly *Q. brasiliensis* (A.St.-Hil. & Tul.) Mart.) (Quillajaceae), known as the soap tree, is distributed throughout Southern Brazil [27]. The species is well known for the abundant presence of saponins and, therefore, a large spectrum of biological activities, including immunoadjuvant properties [28,29], antifungal, and antiherbivore effects [30]. The structure of *Q. lancifolia* saponins is remarkably similar to that of *Quillaja saponaria* Molina bark saponins [31]. The latter is a related Chilean species that has been widely used as an adjuvant in vaccine formulations and is one of the main sources of industrial saponins present in plants. The commercially used saponin fraction obtained from barks of *Q. saponaria* is known as Quil-A® [19]. Studies have reported that *Q. saponaria* shows strong aphicidal, deterrent, nematocidal, molluscicidal, and antifungal activity, indicating that the plant could potentially be used as a biopesticide [32–36]. Some of these features have also been observed for leaf saponins of *Q. lancifolia* [30]. If *Q. lancifolia* saponins prove to be phytotoxic, there is also potential use of these natural products as bioherbicides. In addition, saponins may be used in combination with other molecules at lower concentrations, thereby preventing resistance development in plant enemies. Regardless of its potential use as a biopesticide or bioherbicide, *Q. lancifolia* leaf saponins are a more sustainable and easily renewable alternative compared to *Q. saponaria* bark saponins since they do not require phloem stripping or the use of adult native forests [37]. Miniclonal gardens of young *Q. lancifolia* provide a readily available year-round source of leaf biomass [19].

Herein, the phytotoxic activities of leaf aqueous extract and purified saponin fractions of *Q. lancifolia* on different morpho-physiological parameters of lettuce and barnyard grass were examined. Parameters included germination, growth, chlorophyll concentration, and the presence of morphological anomalies. Exposure of soil-like substrate to *Q. lancifolia* extracts, followed by washing with water, was used to assess retention on the substrate and potential residual effects of bioactive metabolites. Chemical analyses of aqueous extracts and purified saponin fractions were also carried out.

2. Results

2.1. *Phytotoxicity Assay*

2.1.1. Osmotic Potential, pH of Extracts, and Presence of Saponins

PEG solutions up to 0.02 M did not significantly affect the emergence of lettuce and barnyardgrass seeds. Germination reached between 80 and 100% in both cases, being statistically equivalent to respective water controls. Since the osmotic potential of the 0.02 M solution was equivalent to that of the 10% *Q. lancifolia* extract, it was determined that, up to this concentration, the effects on seeds were not due to any osmotic limitation (Table S1). Similarly, the pH of the extract was in the 4–6 range, which was not capable of inhibiting germination of the test species.

Triterpene saponins' presence in aqueous extracts and purified fraction QB-90 were confirmed unequivocally by the analytical tests carried out. The TLC data indicated the presence of saponins matching the Rf of Quil-A, a commercial fraction obtained from *Q. saponaria* bark. Although, as expected, other compounds (e.g., phenolics) were present in aqueous extracts, a prominent band matching Quil-A was evident (Figure S1). Further analyses by HPLC revealed peaks with saponin spectral features within the elution times expected for these metabolites (Figure S2). GC-MS analysis of QB-90 after acid hydrolysis yielded quillaic acid as the main aglycone (Figure S3), confirming the triterpene saponin core identity.

2.1.2. Pre-Emergence Bioassay

Both AE 4% and 10% caused significant inhibitory effects on seed germination. In lettuce, no germination was recorded after 4 days, while the negative control had 97% germinability (Figure 1A). Barnyardgrass had 16% germination in 4% AE and 75% in the negative control; the 10% extract completely inhibited germination (Figure 2A). Both 4 and 10% AE reduced not only the percentage of germination but also the speed of the process in the two species (Table S2). Purified saponins (QB fraction) did not inhibit lettuce germination significantly (Figure 1B). On the other hand, the growth of seedlings germinated in the presence of QB was severely impaired (Figure 3).

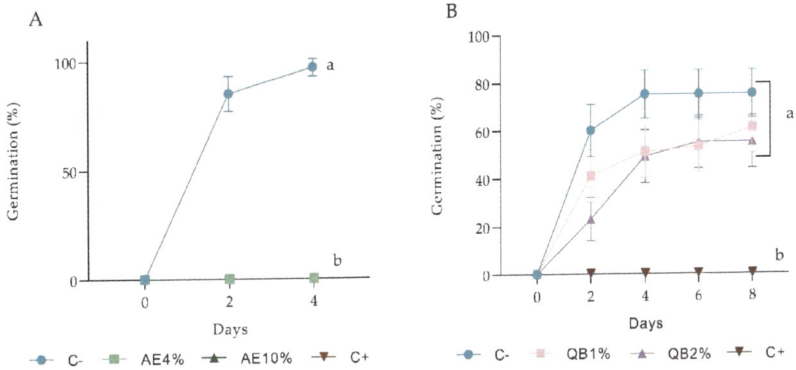

Figure 1. Pre-emergence bioassay with lettuce. Effects of (**A**) aqueous extracts (AE) (at 4 and 10%) and (**B**) saponin fraction (QB) (at 1 and 2%) on the number of germinated seeds of *L. sativa* at the end of 4 days and 8 days, respectively. C- represents the negative control (distilled water), and C+ represents the positive control (NaCl). Bars show 95% confidence intervals (Wilson score interval). Treatments not sharing the same letter at the final sampling time differ in germinability ($p \leq 0.05$).

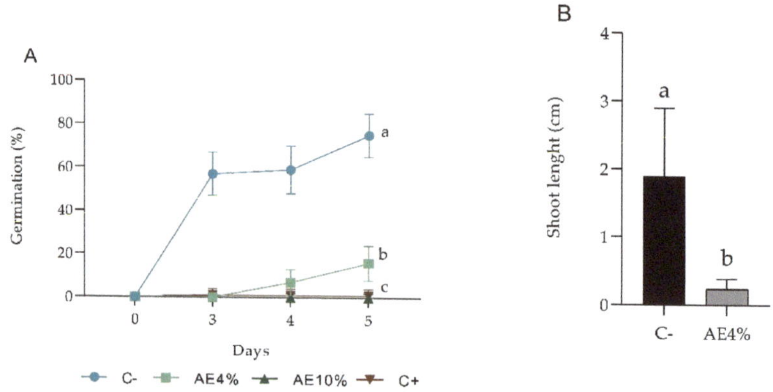

Figure 2. Pre-emergence bioassay in barnyardgrass. (**A**) Seed germination in distilled water (C-), aqueous extract (AE) at 4%, AE at 10% and NaCl (C+) and (**B**) seedling growth (shoot length) at the end of 5 days in treatments that showed germination. In (**A**), bars show 95% confidence intervals (Wilson score interval). Treatments not sharing the same letter at the final sampling time differ in germinability ($p \leq 0.05$). In (**B**), bars represent the mean ± SD. Treatments are significantly different by *t*-test ($p \leq 0.05$).

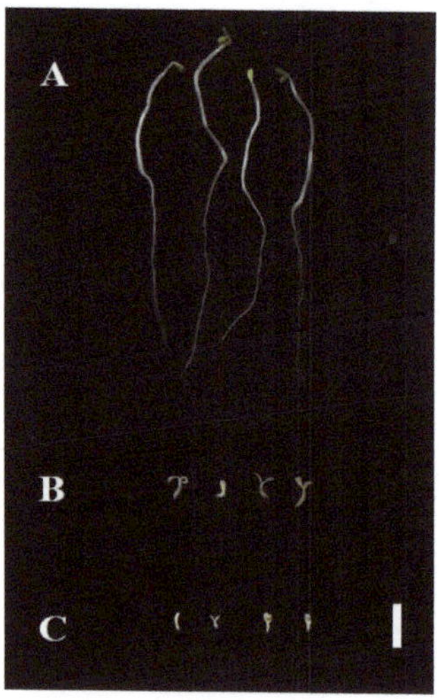

Figure 3. Pre-emergence bioassay with lettuce. Seedings at the end of 8 days of application of (**A**) distilled water, (**B**) purified saponin (QB) at 1%, and (**C**) QB 2%. Bar = 1 cm. Statistical data on whole seedling length (cm): distilled water (7.65 ± 0.93 a); QB at 1% (0.60 ± 0.14 b) and QB at 2% (0.38 ± 0.05 b). Numbers in brackets are means ± SD. Means sharing the same letter do not differ by one-way ANOVA and Tukey test ($p \leq 0.05$).

2.1.3. Post-Emergence Bioassay

Seven days after the application on lettuce seedlings, exposure to both 4 and 10% AE yielded significant differences from the negative control in every parameter analyzed, except for shoot length and dry weight in the lower concentration (4%) (Figure 4). The chlorophyll content of AE 10% seedlings decreased as much as that of their positive control treatment (NaCl) counterparts (Figure 4C). Plants in AE 4% showed abnormalities and brown spots on the leaves (Figure 5). Moreover, most of the seedlings of the AE 4% were notably fragile. The loss of seedling viability in AE 10% (24%) was comparable to the one recorded in the positive control (16%) (Table S3).

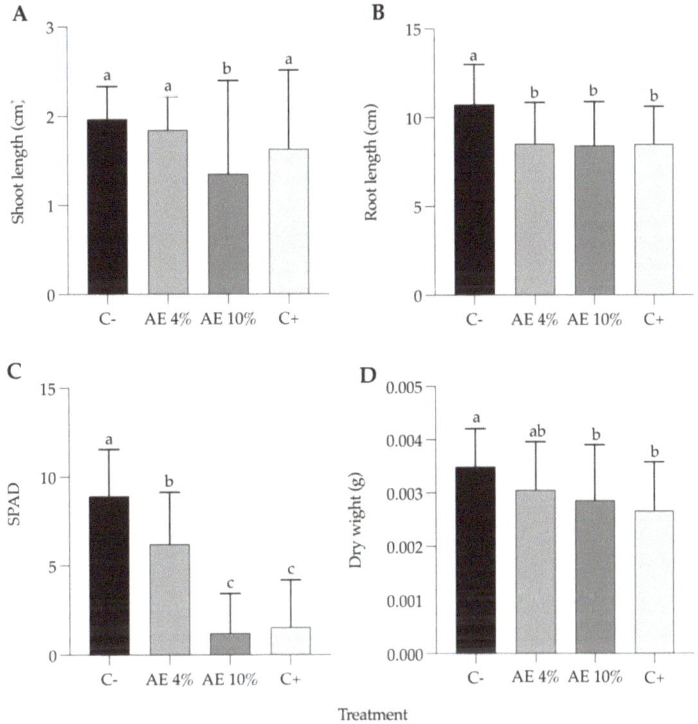

Figure 4. Post-emergence bioassay with lettuce. Parameters (7 days after treatment): (**A**) shoot length; (**B**) root length; (**C**) chlorophyll content (SPAD readings); and (**D**) seedling dry weight. Abbreviations: C- (distilled water), AE 4% (aqueous extract at 4%), AE 10% (aqueous extract at 10%), and C+ (NaCl). Bars represent the mean + SD. Different letters indicate significant differences by Kruskal–Wallis test and Dunn's test ($p \leq 0.05$).

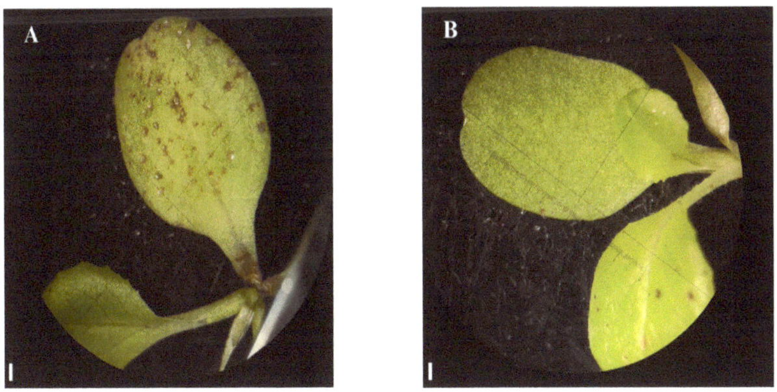

Figure 5. *Lactuca sativa* leaves at day 7 in the post-emergence bioassay treated with (**A**) aqueous extract (AE) at 4% and (**B**) negative control (distilled water). Note the brown spots on the leaves of (**A**). Bar = 1 mm.

In barnyardgrass, no significant difference was observed in root length, shoot length, chlorophyll content, and seedling dry weight between the treatments and the controls (Figure S4). However, the plants from the 10% AE appeared wilty (Figure S5).

In the additional post-emergence bioassay growing lettuce in Petri dishes, a significant reduction was observed in root and shoot elongation of seedlings in both AE treatments

(4% and 10%) and the positive control (NaCl) (Figure S6). No significant difference was observed in the seedling dry mass (Figure S6C). The chlorophyll content was not measured in this experiment because the plants were too small.

2.1.4. Substrate Leaching Bioassay

The soil-like substrate leaching bioassay carried out with lettuce evaluated germination rates and plant development. Regarding seedling growth, the 10% extract resulted in a significant reduction of shoot length, root length (Figures S7 and 6A,B) and dry mass (Figure 6D). In addition, plants treated with the extracts had a dose–response decrease in chlorophyll content (Figure 6C). Interestingly, the germination rate was higher with AE 10% and AE 4% (56% and 51%, respectively), while the negative control had only 28% seedling emergence (Figure S8). The dry weight was also higher in the 4% extract compared to the control (Figure 6D).

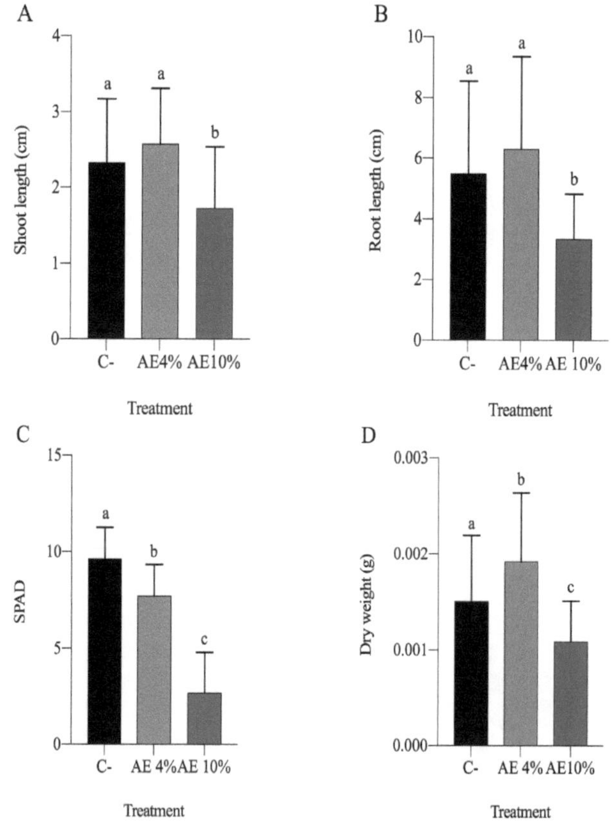

Figure 6. Substrate leaching bioassay with lettuce. Parameters (12 days of application of treatments): (**A**) shoot length, (**B**) root length, (**C**) chlorophyll content, and (**D**) seedling dry weight. Abbreviations: C- (distilled water), AE 4% (aqueous extract at 4%), AE 10% (aqueous extract at 10%). Bars represent the mean + SD. Different letters indicate a significant difference by one-way ANOVA and Tukey test ($p \leq 0.05$).

Barnyardgrass plants under extract treatments showed significant growth inhibition, in contrast with those of the negative control (Figure 7). AE 4% and AE 10% revealed a concentration-dependent inhibitory effect on seedlings in terms of shoot length and dry weight (Figure 8A,C).

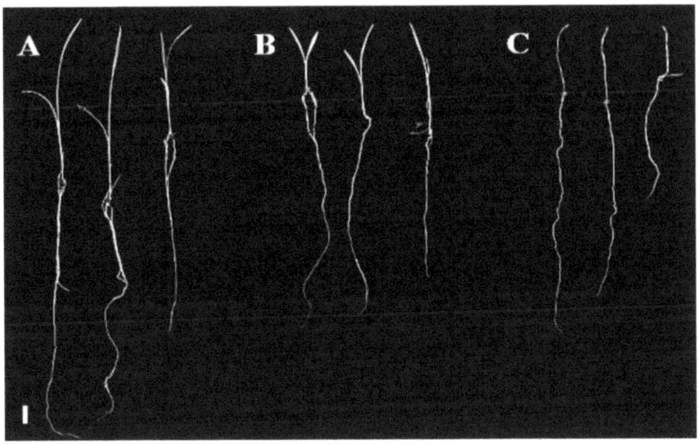

Figure 7. Seedlings of barnyardgrass at the end of 8 days in the leaching bioassay. (**A**) Negative control (plants treated with distilled water); (**B**) plants with aqueous extract (AE) at 4% and (**C**) plants with AE at 10%. Bar = 1 cm.

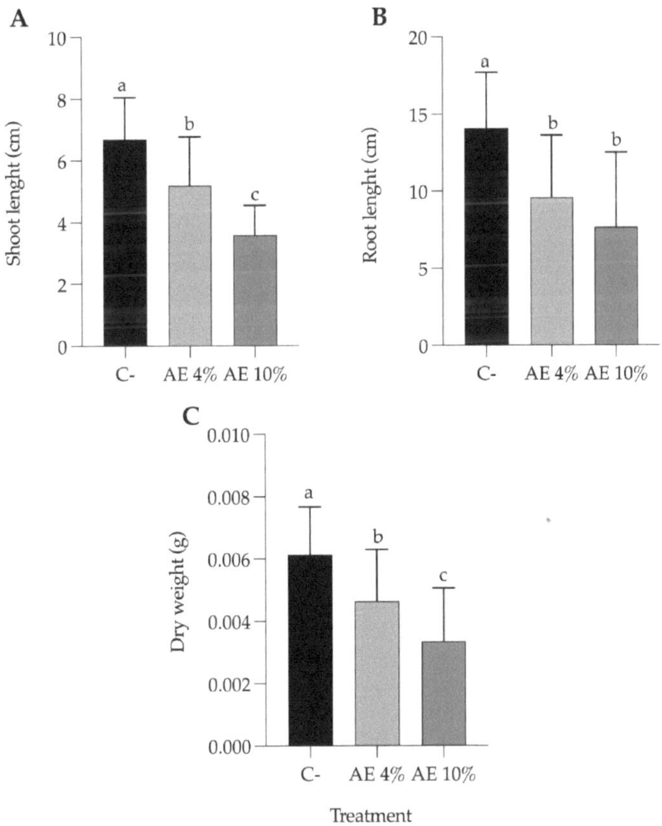

Figure 8. Substrate leaching bioassay with barnyardgrass. Parameters after 8 days of leaching (and 4 water applications): (**A**) shoot length, (**B**) root length, and (**C**) seedling dry weight. Abbreviations: distilled water (C-), aqueous extract at 4% (AE 4%), and aqueous extract at 10% (AE 10%). Bars represent the mean + SD. Different letters indicate significant differences by one-way ANOVA and Tukey test ($p \leq 0.05$).

3. Discussion

Possible significant impacts of nonspecific factors on the phytotoxic activities of *Q. lancifolia* aqueous extracts and purified saponins at the concentrations used were ruled out. Therefore, the validity of the subsequently recorded data was confirmed. Osmotic and pH values were within non-inhibitory ranges for both tested species [38–40].

Pre-emergence bioassays using 4% and 10% aqueous extracts of *Q. lancifolia* against both lettuce and barnyardgrass revealed a significant reduction of germination rates and root and shoot length of the emerged seedlings. In *L. sativa*, the extracts completely inhibited germination. Extracts and QB fraction significantly inhibited one or more parameters, such as germination percentage and shoot and root length. Considering that even at low concentrations of AE, germination was suppressed, it appears that the extracts showed phytotoxic potential that could be at least partly attributed to the saponins present in the plant. Indeed, analytical methods consistently showed the presence of triterpene saponins with typical quillaic acid aglycone [19,31]. The possible participation of saponins in the phytotoxic effects is also supported by the fact that seeds germinated in the presence of the purified saponin fraction QB-yielded seedlings severely reduced in size. Further studies of these compounds could lead to possible use as an effective pre-emergence bioherbicide.

Phytotoxic effects on the post-emergence growth of lettuce were also observed, especially at AE 10%. Under the AE 4% treatment, numerous leaves exhibited abnormalities that could explain the significantly reduced chlorophyll content recorded. Although the effects on post-emergence were not as intense as those in the germination experiments, *Q. lancifolia* extracts also had phytotoxic effects on seedling growth. In the leaching bioassay, however, there was a significant decrease in the inhibitory effect of the extracts on lettuce, particularly at 4% AE. This could lead to possible use as a post-emergence bioherbicide, alone or in combination, effective at pre- and post-germination with a reduced residual effect.

However, the lack of significant changes shown in the post-emergence bioassays with barnyardgrass can indicate that the extracts are not as effective in repressing the seedling growth of *E. crus-galli* as they are inhibiting its germination. This possibly reflects a higher resistance to biotic stresses in weeds [8]. On the other hand, the leaching bioassay with barnyardgrass showed a significant reduction in the root and shoot length as well as dry weight of plants treated with AE. Perhaps, for this species, longer exposure to the extract residue could affect plants negatively.

Ideally, a bioherbicide residue on the soil surface should be able to undergo leaching and/or degradation, thus avoiding the exposure of subsequent crops to potentially phytotoxic compounds [11]. The leaching assay with lettuce is also considered a pre-emergence experiment since it used seeds grown on substrates exposed to the extracts. Since mixtures of sand and vermiculite are only soil-like substrates, the data obtained with them cannot be directly extrapolated to natural soil conditions. Nonetheless, the use of these substrates provides useful information on the overall behavior of phytotoxic agents in soil. The higher dry biomass of seedlings in AE 4% relative to the water control may indicate a nutritional effect of the extracts caused, for example, by the presence of sugars in saponins [41]. The application of the 10% extract, nonetheless, resulted in lower growth and chlorophyll content, perhaps due to a negative impact on photosynthesis [42].

Antiherbivore and antifungal effects of *Q. lancifolia* saponins were observed in the 1 to 2% concentration range [30]. For saponins to constitute a potential biopesticide, the absence or limited phytotoxic effects at concentrations in which they can inhibit herbivores and pathogens are necessary. Further studies focusing on the compatible use of these molecules in biopesticide and bioherbicide formulations are needed.

In most cases, as has been reported for other saponins, the activity of these triterpenes depends essentially on their effects on membranes [43]. As a result of their amphipathic nature, saponins can interfere with the membrane structure, affecting its integrity and functionality [44]. It seems likely that the growth inhibitory effects recorded in test plants are at least partly caused by saponins present in extracts and their fractions on cell and organelle membranes. However, other compounds, as well as synergistic interactions among metabolites, can also be involved, particularly in AE treatments.

Taken together, the data suggest that AE and QB of *Q. lancifolia* may be considered for use in organic agriculture both as pre- and post-emergence bioherbicide components. In a preliminary assessment, it seems that these agents may be effective for broad leaf and grass weeds. Given the surfactant properties of saponins and based on pilot tests, their application using ordinary sprayers is not difficult. Although it is not possible to estimate more precisely at this point, doses per hectare and the costs of treatment are not likely to be prohibitive. Support for this possibility is based on the renewable and continuous source of saponin supply from the leaves of cultivated plants, as well as on the relatively low active concentrations required. In addition, the combined use of these agents as adjuvants with other products may further contribute to economic feasibility.

4. Conclusions

In summary, *Quillaja lancifolia* aqueous extracts and purified saponin fractions showed strong phytotoxic activity against *E. crus-galli* and *L. sativa* (with greater injury levels on the latter) and weed-suppressing abilities. These findings warrant further studies to evaluate the possible uses of these agents in formulations of bioherbicides. Key features of *Q. lancifolia* saponins, such as their water solubility and sustainable supply from leaves obtained in cultivated clonal gardens, are important for potential deployment as bioherbicides. Future work is required to better understand the mechanisms of action of these phytotoxins, notably their interaction with biological membranes and their components.

5. Materials and Methods

5.1. Plant Material and Extracts

5.1.1. Plant Source and Extract Preparation

Approximately 5 kg of fresh *Q. lancifolia* leaves were collected from adult plants (about 10 m tall of unknown age) growing in the rural zone of the city of Canguçu, RS, Brazil (31°23′42″ S–52°40′32″ W) in the early fall of 2018 (28 March 2018) (harvest authorization by the Brazilian Board of Management of Genetic Resources—CGEN under number 010540/2011-3). Harvest was done on a sunny day with temperature variations between 17 and 24 °C. Monthly precipitation was typical for the area, around 120 mm. The climate of the site of collection is warm temperate Cfb (Köppen classification).

A voucher specimen was deposited at the ICN Herbarium of the Federal University of Rio Grande do Sul (142,953). Air-dried powdered leaves were extracted in distilled water (1:10, *w*/*v*) for 8 h, filtered, and partitioned with ethyl acetate, followed by lyophilization of the aqueous phase, yielding aqueous extract (AE). AE was submitted to reverse-phase chromatography and gradient of water and methanol to obtain fraction QB-90, as previously described [28]. QB-80 was obtained using the same protocol. QB-80 and QB-90 were also analyzed by thin-layer chromatography (TLC) to further confirm fraction isolation. The two purified saponin fractions were combined for the bioassays constituting QB.

5.1.2. Extract and Saponin Chemical Characterization

Characterization of AE and QB was initially performed by TLC. Silica gel aluminum plates (F254, Supelco, Darmstadt, Germany) were used as the stationary phase. The mobile phase was n-butanol:acetic acid:water (5:1:4, *v*/*v*). Plates were revealed using the anisaldehyde-sulfuric acid protocol [28]. Additional confirmation of saponin presence was obtained by high-performance liquid chromatography (HPLC) analysis using a Shimadzu chromatographer equipped with a PDA-UV detector. Saponins were tracked at 214 nm. The HPLC method was based on Wallace et al. [45] with minor modifications. Finally, Gas Chromatography-Mass Spectrometry (GC-MS) analysis of acid-hydrolyzed saponins was carried out with an oven set to 120 °C for 1 min, then increasing at 6 °C/min to 320 °C. The final hold time was 11.6 min. Spectral data were generated with electron impact ionization at 70 eV (positive mode). Data were acquired in full scans ranging from m/z 50 to 550.

The Agilent MassHunter workstation (version B.07 service pack 2) was used for peak integration of total and extracted ion chromatograms and extraction of mass spectra [46].

5.2. Phytotoxicity Assay

5.2.1. Target Species

The experiments on the phytotoxic effect of the extract were carried out with diaspores (herein referred to as seeds) of *Lactuca sativa* L. (a common model in allelopathy studies), as well as seedlings of the same species; and with seeds and seedlings of *Echinochloa crus-galli* L.P.Beauv (an economically relevant weed species, particularly in rice fields). For these assays, commercially available lettuce seeds were used (Isla®, batch 116377-001, Porto Alegre, Brazil), and barnyardgrass seeds (batches grown at the agronomy campus of UFRGS and harvested in late summer/early fall) were kindly provided by the Department of Crop Plants—Faculty of Agronomy, UFRGS.

5.2.2. Extract Osmotic Potential and pH Measurements

Both the osmotic potential and pH of the extracts were determined in triplicate. The osmotic potential of *Q. lancifolia* extracts was measured in different concentrations (0.1%, 1%, 2%, 4%, 10%, 20%, 40%) with a refractometer-based curve prepared with sucrose. To determine the influence of the osmotic potential of the extracts on both lettuce and barnyardgrass germination bioassays, additional tests were performed with solutions of polyethylene glycol (PEG-6000) [38] at the following concentrations: 0.01 M, 0.02 M, and 0.03 M. This experiment was carried out using the same germination procedure (i.e., pre-emergence) bioassays (Table S1 and Section 5.2.3). The osmotic potential of the different concentrations of PEG was also measured with a refractometer, and it was possible to correlate the solution of PEG at 0.02 M with the 10% *Q. lancifolia* aqueous extract. Extract pH values were determined with a pH meter.

5.2.3. Pre-Emergence Bioassay

The pre-emergence bioassay analyzed the germination of *L. sativa* (lettuce) and *E. crus-galli* (barnyardgrass). Petri dishes (9 cm in diameter) lined with qualitative filter paper were used, and assays were conducted in a growth chamber (25 °C, 12 h/12 h dark/light, 40 μmol photons·m^{-2}·s^{-1}).

For each one of the species, the Petri dishes were separated into experimental groups and the control, each containing 25 seeds per plate, with 3 plates per experimental group. Seeds were treated with 5 mL in each Petri dish of the AE of *Q. lancifolia* (at 4 and 10% *w*/*v*), distilled water (negative control), and NaCl 0.5 M (positive control). In the case of lettuce, QB at 1% and 2% (*w*/*v*) were also evaluated.

Germination parameters were measured daily (germination time) or on the 5th day of incubation (germination percentage, root and shoot length). The presence of morphological abnormalities was recorded. Germination velocity indexes (GVI) were calculated [47]:

$$GVI = N1/D1 + N2/D2 + \ldots + Nn/Dn$$

where: GVI = germination velocity index; N = number of seeds germinated within the day interval of counting; D = days after sowing until data collection.

An additional experiment using purified saponin fraction QB was performed with lettuce under the same growth conditions described above. Germination and seedling length were analyzed, with treatments consisting of 1 and 2% of QB and controls, in this case ending on the 7th day of incubation.

5.2.4. Bioassay for Post-Emergence

To obtain lettuce seedlings for the effect on initial growth, seeds were put in separate pot trays (each slot with a volume of 15 mL, one plant per slot) containing autoclaved substrate (washed sand: vermiculite—ratio 1:1; *v*/*v*) and kept for 12 days under controlled light and temperature (photoperiod of 16 h per day; 25 ± 2 °C, irradiance 50 μmol·m^{-2}·s^{-1}).

On the 12th day, plants from germinated seeds were exposed to AE (at 4 and 10% *w*/*v*) and controls (2 mL per slot containing a single plant and sprinkling to dew point). Barnyardgrass seeds were kept for 13 days on the same substrate growth conditions described above, and on the 14th day, seedlings were exposed to the treatments (AE at 4 and 10% *w*/*v*).

Seven days after treatment application, growth parameters were measured (radicle length, shoot length, and whole seedling dry mass). Root and shoot length were recorded with a ruler, and oven dry mass at 60 °C was obtained in an analytical balance. In addition, for lettuce, the chlorophyll content was measured with a SPAD chlorophyllometer (Minolta, Osaka, Japan). With barnyardgrass, due to restrictions of leaf size on the SPAD reading probe, a conventional spectrophotometric evaluation of acetone chlorophyll extracts was carried out [17]. Each individual plant was considered an experimental unit. The number of replicates per treatment and per sampling time was 50 for root length, shoot length, whole seedling dry weight, and SPAD readings. For spectrophotometric chlorophyll measurements, a pooled set of 5 plants was the experimental unit and 4 replicates were used per treatment and sampling time.

In another set of experiments, *Q. lancifolia* extracts and control treatments were applied to post-emergent 12-day-old lettuce seedlings grown in Petri dishes. After 12 days in growth chamber conditions as described above, 5 mL of extracts and controls were applied to each dish containing 25 plants and replicated 3 times. Plants were measured after 7 days of application (root length, shoot length, and whole seedling dry mass).

5.2.5. Substrate Leaching Bioassay

To evaluate whether the extract could be leached from a soil-like substrate, two different bioassays were performed. With *L. sativa*, 50 individual slots of pot trays (each slot with a volume of 15 mL) containing autoclaved soil-like substrate (washed sand: vermiculite—ratio 1:1; *v*/*v*) were inoculated with 2 mL of AE at concentrations of 4 and 10%. In addition, the area comprised of 50 slots was superficially sprayed with 30 mL of the respective extract. Then, each slot was subjected to leaching with 10 mL of distilled water every 2 days until completing 8 days (i.e., 4 washings). In the day of the last water application, 2 lettuce seeds were sown per slot to evaluate germination for 12 days. Root length, shoot length, and seedling dry mass were also recorded at the end of this period. The experiment with barnyardgrass evaluated the leaching potential of seedlings grown on the same untreated substrate previously described for 12 days, with application of the extracts and controls at the end of the 13th day. After 7 days of application, 10 mL of distilled water were applied in each separate slot to leachate the extract away every 2 days until completing 8 days (i.e., 4 washings). The same growth parameters above were measured after the last water application.

5.2.6. Statistical Analyses

Statistical analysis was done with GraphPad Prism 8.1. Results were pre-evaluated for normal distribution by the Shapiro–Wilk or Levene test and then analyzed by one-way ANOVA followed by Tukey test (normal distribution) or Kruskal–Wallis followed by Dunn's test (non-parametric distribution), all at $p \leq 0.05$. For the two treatment comparisons, a *t*-test was applied (after normality verification and at $p \leq 0.05$). Data were expressed as mean ± standard deviation (SD). Germination data were analyzed using 95% confidence intervals (Wilson score interval) [48].

Supplementary Materials: The following supporting information can be downloaded at: https://www.mdpi.com/article/10.3390/plants12030663/s1, Figure S1: Thin layer chromatography of *Quillaja* preparations. AE—aqueous extract prepared from *Q. lancifolia* leaves; Quil-A—commercial preparation obtained from *Q. saponaria* barks; QB-90—purified from AE after column chromatography. Triterpenoids stained brown with anisaldehide-sulfuric. Figure S2: HPLC chromatogram (214 nm) of QB-90 (4 mg/mL) from the leaf extract of *Quillaja lancifolia*. The main triterpene saponins elute in the range of 15 to 25 min. The known saponin QS-21 elutes between 20.5 and 20.7 min. Figure S3: GC-MS analysis of QB-90 after acid hydrolysis. Breakdown of saponins produced quillaic acid as the main aglycone (quillaic acid chemical structure shown on the right). The presence of quillaic acid was confirmed with an authentic standard. IS—internal standard (cholestane). Figure S4. Post-emergence bioassay with barnyardgrass. Parameters (7 days after treatment): (A) shoot length, (B) root length, (C) chlorophyll concentration, and (D) seedling dry weight. Abbreviations: C- (distilled water), AE 4% (aqueous extract at 4%), AE 10% (aqueous extract at 10%) and C+ (NaCl). Bars represent the mean + SD. Different letters indicate significant difference by Kruskal-Wallis and Dunn's test (A and D) or one-way ANOVA and Tukey test (B and C) ($p \leq 0.05$). Figure S5. Post-emergence bioassay with barnyardgrass. Seedlings at the end of 7 days treated with (A) distilled water—C- and (B) aqueous extract at 10%—AE 10%. Bar = 1 cm. Figure S6. Post-emergence bioassay with lettuce carried out in petri dishes. Parameters (7 days after treatment): (A) shoot length, (B) root length, and (C) seedling dry weight. Abbreviations: C- (distilled water), AE 4% (aqueous extract at 4%), AE 10% (aqueous extract at 10%) and C+ (NaCl). Bars represent the mean + SD. Different letters indicate significant difference by Kruskal-Wallis and Dunn's test ($p \leq 0.05$). Figure S7. Substrate leaching bioassay with lettuce. Seedlings at the end of 12 days in substrate treated with (A) distilled water—C-, (B) aqueous extract at 4%—AE 4% and (C) aqueous extract at 10%—AE 10%. Bar = 1 cm. Figure S8. Substrate leaching bioassay. Germination rates of lettuce seeds in soil-like substrate with the previous application of the following treatments: distilled water (C-), aqueous extract at 4% (AE 4%) and aqueous extract at 10% (AE 10%), during 12 days of experiment. Bars show 95% confidence intervals (Wilson score interval). Treatments not sharing the same letter at the final sampling time differ in germinability ($p \leq 0.05$). Reference [46] is cited in the supplementary materials.

Author Contributions: Conceptualization, M.E.M.M., A.C.A.Y. and A.G.F.-N.; methodology, M.E.M.M., A.C.d.C., A.C.A.Y., Y.V.S.M. and E.Z., investigation, M.E.M.M., A.C.d.C. and A.C.A.Y.; resources, A.G.F.-N.; writing—original draft preparation, M.E.M.M.; writing—review and editing, A.G.F.-N.; supervision, A.G.F.-N.; project administration, A.G.F.-N.; funding acquisition, A.G.F.-N. All authors have read and agreed to the published version of the manuscript.

Funding: This research was funded by the Conselho Nacional de Desenvolvimento Científico e Tecnológico (CNPq—Brazil), grant number 310775/2021-3), Rio Grande do Sul State Foundation for Research Support (Fapergs—Brazil), grant number 2019/15477-3, and the Coordenação de Aperfeiçoamento de Pessoal de Nível Superior—(CAPES—Brazil)—Finance Code 001.

Institutional Review Board Statement: Not applicable.

Informed Consent Statement: Not applicable.

Data Availability Statement: The original contributions presented in the study are included in the article/Supplementary Material, and further inquiries can be directed to the corresponding author.

Conflicts of Interest: The authors declare no conflict of interest.

References

1. Rehman, A.; Farooq, M.; Lee, D.J.; Siddique, K.H.M. Sustainable agricultural practices for food security and ecosystem services. *Environ. Sci. Pollut. Res.* **2022**, *29*, 84076–84095. [CrossRef] [PubMed]
2. Food and Agriculture Organization of the United Nations. FAOSTAT-Pesticides Use. 2018. Available online: http://www.fao.org/faostat/en/#data/RP/visualize (accessed on 25 May 2019).
3. Pignati, W.A.; de Lima, F.A.N.S.; de Lara, S.S.; Correa, M.L.M.; Barbosa, J.R.; da Leão, L.H.C.; Pignatti, M.G. Distribuição espacial do uso de agrotóxicos no Brasil: Uma ferramenta para a Vigilância em Saúde (Spacial distribution of agrotoxics use in Brazil: A tool for Health Surveillance). *Cien. Saude Colet.* **2017**, *22*, 3281–3293. [CrossRef]
4. Bebber, D.P.; Holmes, T.; Gurr, S.J. The global spread of crop pests and pathogens. *Glob. Ecol. Biogeogr.* **2014**, *23*, 1398–1407. [CrossRef]
5. Oerke, E.C. Crop losses to pests. *J. Agric. Sci.* **2006**, *144*, 31–43. [CrossRef]
6. Stewart, C.N. Becoming weeds. *Nat. Genet.* **2017**, *49*, 654–655. [CrossRef] [PubMed]
7. Weed Science Society of America. Weeds. 2016. Available online: http://wssa.net/wssa/weed/ (accessed on 25 May 2019).
8. Zimdahl, R.L. *Fundamentals of Weed Science*, 5th ed.; Academic Press: London, UK, 2018; 735p.
9. Schonbeck, M. *Principles of Sustainable Weed Management in Organic Cropping Systems, Workshop for Farmers and Agricultural Professionals on Sustainable Weed Management*, 3rd ed.; Clemson University: Clemson, SC, USA, 2011; 20p.
10. Brühl, C.A.; Zaller, J.G. Biodiversity decline as a consequence of an inappropriate environmental risk assessment of pesticides. *Front. Environ. Sci.* **2019**, *7*, 177. [CrossRef]
11. Abbas, T.; Zahir, Z.A.; Naveed, M.; Kremer, R.J. Limitations of existing weed control practices necessitate development of alternative techniques based on biological approaches. *Adv. Agron.* **2018**, *147*, 239–280. [CrossRef]

12. Cordeau, S.; Triolet, M.; Wayman, S.; Steinberg, C.; Guillemin, J.P. Bioherbicides: Dead in the water? A review of the existing products for integrated weed management. *Crop Prot.* **2016**, *87*, 44–49. [CrossRef]
13. Chauhan, B.S.; Mahajan, G. *Recent Advances in Weed Management*, 1st ed.; Springer: New York, NY, USA, 2014; 411p. [CrossRef]
14. Dayan, F.E.; Duke, S.O. Natural compounds as next-generation herbicides. *Plant Physiol.* **2014**, *166*, 1090–1105. [CrossRef]
15. Duke, S.O.; Pan, Z.; Bajsa-Hirschel, J.; Boyette, C.D. The potential future roles of natural compounds and microbial bioherbicides in weed management in crops. *Adv. Weed Sci.* **2022**, *40*, e020210054. [CrossRef]
16. Puig, C.G.; Reigosa, M.J.; Valentão, P.; Andrade, P.B.; Pedrol, N. Unravelling the bioherbicide potential of *Eucalyptus globulus* Labill: Biochemistry and effects of its aqueous extract. *PLoS ONE* **2018**, *13*, e0192872. [CrossRef] [PubMed]
17. De Costa, F.; Fleck, J.; Yendo, A.C.A.; Gosmann, G.; Fett-Neto, A.G. Accumulation of a bioactive triterpene saponin fraction of *Quillaja brasiliensis* leaves is associated with abiotic and biotic stresses. *Plant Physiol. Biochem.* **2013**, *66*, 56–62. [CrossRef] [PubMed]
18. Yu, B.; Patterson, N.; Zaharia, L.I. Saponin biosynthesis in pulses. *Plants* **2022**, *11*, 3505. [CrossRef] [PubMed]
19. Magedans, Y.V.S.; Phillips, M.A.; Fett-Neto, A.G. Production of plant bioactive triterpenoid saponins: From metabolites to genes and back. *Phytochem. Rev.* **2021**, *20*, 461–482. [CrossRef]
20. Faizal, A.; Geelen, D. Saponins and their role in biological processes in plants. *Phytochem. Rev.* **2013**, *12*, 877–893. [CrossRef]
21. Hoagland, R.E.; Zablotowicz, R.M.; Reddy, K.N. Studies of the phytotoxicity of saponins on weed and crop plants. In *Saponins Used in Food and Agriculture*; Waller, G.R., Yamasaki, K., Eds.; Plenum Press: New York, NY, USA, 1996; pp. 57–73.
22. Jelassi, A.; Ayeb-Zakhama, A.E.; Nejma, A.B.; Chaari, A.; Harzallah-Skhiri, F.; Jannet, H.B. Phytochemical composition and allelopathic potential of three Tunisian *Acacia* species. *Ind. Crops Prod.* **2016**, *83*, 339–345. [CrossRef]
23. Pérez, A.J.; Simonet, A.M.; Pecio, Ł.; Kowalczyk, M.; Calle, J.M.; Macías, F.A.; Oleszec, W.; Stochmal, A. Triterpenoid saponins from the aerial parts of *Trifolium argutum* Sol. and their phytotoxic evaluation. *Phytochem. Lett.* **2015**, *13*, 165–170. [CrossRef]
24. Ribeiro, R.C.; de Carvalho, M.G.; de Moraes, M.L.L.; Rossiello, R.O.P.; de Oliveira, D.R.; de Amorim, R.M.Q.; Barbieri, E., Jr. Chemical screening of *Urochloa humidicola*: Methods for characterizing secondary metabolites and allelopathic activity on forage legumes. *Am. J. Plant Sci.* **2018**, *9*, 1260–1278. [CrossRef]
25. Stavropoulou, M.I.; Angelis, A.; Aligiannis, N.; Kalpoutzakis, E.; Mitakou, S.; Duke, S.O.; Fokialakis, N. Phytotoxic triterpene saponins from *Bellis longifolia*, an endemic plant of Crete. *Phytochemistry* **2017**, *144*, 71–77. [CrossRef]
26. Wyman-Simpson, C.L.; Waller, G.R.; Jurzysta, M.; McPherson, J.K.; Young, C.C. Biological activity and chemical isolation of root saponins of six cultivars of alfalfa (*Medicago sativa* L.). *Plant Soil.* **1991**, *135*, 83–94. [CrossRef]
27. Reitz, R.; Reis, A.; Klein, R.M. *Flora Ilustrada Catarinense—Rosáceas (Illustrated Flora of the State of Santa Catarina—Rosaceae)*; Herbário Barbosa Rodrigues: Itajaí, Brazil, 1996; pp. 113–116.
28. Fleck, J.D.; Kauffmann, C.; Spilki, F.; Lencina, C.L.; Roehe, P.M.; Gosmann, G. Adjuvant activity of *Quillaja brasiliensis* saponins on the immune responses to bovine herpesvirus type 1 in mice. *Vaccine* **2006**, *24*, 7129–7134. [CrossRef] [PubMed]
29. Kumar, A.; Sharma, A.; Tirpude, N.V.; Padwad, Y.; Hallan, V.; Kumar, S. Plant-derived immuno-adjuvants in vaccines formulation: A promising avenue for improving vaccines efficacy against SARS-CoV-2 virus. *Pharmacol. Rep.* **2022**, *74*, 1238–1254. [CrossRef] [PubMed]
30. Yendo, A.C.A. Saponinas de *Quillaja brasiliensis*: Purificação e atividade biológica. (Saponins of *Quillaja brasiliensis*: Purification and biological activity). Ph.D. Thesis, Graduate Program in Botany—Universidade Federal do Rio Grande do Sul (UFRGS), Porto Alegre, Brazil, 2016.
31. Kauffmann, C.; Machado, A.M.; Fleck, J.D.; Provensi, G.; Pires, V.S.; Guillaume, D.; Sonnet, P.; Reginatto, F.H.; Shenkel, E.P.; Gosmann, G. Constituents from leaves of *Quillaja brasiliensis*. *Nat. Prod. Res.* **2004**, *18*, 153–157. [CrossRef]
32. González-Cruz, D.; Martín, R.S. Molluscicidal effects of saponin-rich plant extracts on the grey field slug. *Cien. Investig. Agr.* **2013**, *40*, 341–349. [CrossRef]
33. De Geyter, E.; Smagghe, G.; Rahbe, Y.; Geelen, D. Triterpene saponins of *Quillaja saponaria* show strong aphicidal and deterrent activity against the pea aphid *Acyrthosiphon pisum*. *Pest Manag. Sci.* **2011**, *68*, 164–169. [CrossRef] [PubMed]
34. Moya, E.; San, R.M.; Gamboa, M.; Hidalgo, G.E.A. Evaluation of a *Quillaja saponaria* saponin extract for control of powdery mildew of wheat and squash evaluación de un extracto de saponinas de *Quillaja saponaria* para el control de oídios de trigo y zapallo. *Agro Sur* **2010**, *38*, 87–96. [CrossRef]
35. González-Castillo, J.A.; Quezada-D'Angelo, T.P.; Silva-Aguayo, G.I.; Moya, E.A. Effect of saponins of *Quillaja saponaria* extracts in combination with *Pseudomonas protegens* to control *Gaeumannomyces graminis* var. tritici in wheat. *Chil. J. Agric Res.* **2018**, *78*, 378–390. [CrossRef]
36. Giannakou, I.O. Efficacy of a formulated product containing *Quillaja saponaria* plant extracts for the control of root-knot nematodes. *Eur. J. Plant Pathol.* **2011**, *130*, 587. [CrossRef]
37. Martín, R.; Briones, R. Industrial uses and sustainable supply of *Quillaja saponaria* (Rosaceae) saponins. *Econ. Bot.* **1999**, *53*, 302–311. [CrossRef]
38. Meneses, C.H.S.G.; Lima, L.H.G.M.; Lima, M.M.A.; Pereira, W.E.; Bruno, R.L.A.; Vidal, M.S. Potencial hídrico induzido por Polietilenoglicol-6000 na viabilidade de sementes de algodão (Water potential induced by Polyethylene glycol-6000 on cotton seed viability). *Embrapa Agrobiol.* **2007**, *19*, 1–22.
39. Sadeghloo, A.; Asghari, J.; Ghaderi-Far, F. Seed germination and seedling emergence of velvetleaf (*Abutilon theophrasti*) and Barnyardgrass (*Echinochloa crus-galli*). *Planta Daninha* **2013**, *31*, 259–266. [CrossRef]

40. Rice, E.L. *Allelopathy*, 2nd ed.; Academic Press Inc.: New York, NY, USA, 1984; 422p.
41. Yendo, A.C.A.; De Costa, F.; Da Costa, C.T.; Colling, L.C.; Gosmann, G.; Fett-Neto, A.G. Biosynthesis of plant triterpenoid saponins: Genes, enzymes, and their regulation. *Mini Rev. Org. Chem.* **2014**, *11*, 292–306. [CrossRef]
42. Poonpaiboonpipat, T.; Pangnakorn, U.; Suvunnamek, U.; Teerarak, M.; Charoenying, P.; Laosinwattana, C. Phytotoxic effects of essential oil from *Cymbopogon citratus* and its physiological mechanisms on barnyardgrass (*Echinochloa crus-galli*). *Ind. Crop. Prod.* **2013**, *41*, 403–407. [CrossRef]
43. De Costa, F.; Yendo, A.C.A.; Fleck, J.D.; Gosmann, G.; Fett-Neto, A.G. Immunoadjuvant and anti-inflammatory plant saponins: Characteristics and biotechnological approaches towards sustainable production. *Mini Rev. Med. Chem.* **2011**, *11*, 857–880. [CrossRef]
44. De Groot, C.; Müller-Goymann, C.C. Saponin interactions with model membrane systems—Langmuir monolayer studies, hemolysis and formation of ISCOMs. *Planta Med.* **2016**, *82*, 1496–1512. [CrossRef]
45. Wallace, F.; Bennadji, Z.; Ferreira, F.; Olivaro, C. Analysis of an immunoadjuvant saponin fraction from *Quillaja brasiliensis* leaves by electrospray ionization ion trap multiple-stage mass spectrometry. *Phytochem. Lett.* **2017**, *20*, 228–233. [CrossRef]
46. Magedans, Y.V.S.; Phillips, M.A. Soapbark Triterpenes: *Quillaja brasiliensis* cell culture sapogenin and free sterol analysis by GCMS. *Methods Mol. Biol.* **2022**, *2469*, 119–128. [CrossRef]
47. Popinigis, F. *Fisiologia da Semente (Seed Physiology)*; Brazilian Ministry of Agriculture: Brasíla, Brazil; Inter-American Development Bank (BID): Washington, DC, USA, 1977; pp. 275–277.
48. Newcombe, R.G. Two-sided confidence intervals for the single proportion: Comparison of seven methods. *Statist. Med.* **1998**, *17*, 857–872. [CrossRef]

plants

MDPI

Article

Allelopathic Substances of *Osmanthus* spp. for Developing Sustainable Agriculture

Hisashi Kato-Noguchi [1,*], Yuri Hamada [1], Misuzu Kojima [1], Sanae Kumagai [1], Arihiro Iwasaki [2] and Kiyotake Suenaga [2]

[1] Department of Applied Biological Science, Faculty of Agriculture, Kagawa University, Miki, Kagawa 761-0795, Japan
[2] Department of Chemistry, Faculty of Science and Technology, Keio University, 3-14-1 Hiyoshi, Kohoku, Yokohama 223-8522, Japan
* Correspondence: kato.hisashi@kagawa-u.ac.jp

Abstract: *Osmanthus fragrans* Lour. has been cultivated for more than 2500 years because of the fragrance and color of the flowers. The flowers and roots have been used in tea, liquors, foods, and traditional Chinese medicine. The species contains more than 180 compounds including terpenoids, phenylpropanoids, polyphenols, flavonoids, and sterols. However, there has been limited information available on the allelopathic properties and allelopathic substances of *O. fragrans*. We investigated the allelopathy and allelopathic substances of *O. fragrans* and *Osmanthus heterophyllus* (G.Don) P.S. Green, as well as *Osmanthus* × *fortunei* Carrière, which is the hybrid species between *O. fragrans* and *O. heterophyllus*. The leaf extracts of *O. fragrans*, *O. heterophyllus*, and *O.* × *fortunei* suppressed the growth of cress (*Lepidium sativum* L.), alfalfa (*Medicago sativa* L.), *Lolium multiflorum* Lam., and *Vulpia myuros* (L.) C.C.Gmel with the extract concentration dependently. The extract of the hybrid species *O.* × *fortune* was the most active among the extracts. The main allelopathic substances of *O.* × *fortunei* and *O. fragrans* were isolated and identified as (+)-pinoresinol and 10-acetoxyligustroside, respectively. (+)-Pinoresinol was also found in the fallen leaves of *O.* × *fortunei*. Both compounds showed an allelopathic activity on the growth of cress and *L. multiflorum*. On the other hand, several allelopathic substances including (+)-pinoresinol may be involved in the allelopathy of *O. heterophyllus*. *O. fragrans*, *O. heterophyllus*, and *O.* × *fortunei* are evergreen trees. but their senescent leaves fall and cover the soil under the trees. It is possible that those allelopathic substances are liberated through the decomposition process of the leaves into their rhizosphere soil, and that they accumulate in the soil and provide a competitive advantage to the species through the inhibition of the growth of the neighboring competing plants. Therefore, the leaves of these *Osmanthus* species are allelopathic and potentially useful for weed management options in some agriculture settings to reduce commercial herbicide dependency for the developing sustainable agriculture systems.

Keywords: allelochemical; 10-acetoxyligustroside; decomposition; fallen leaf; growth inhibition; *Osmanthus*; (+)-pinoresinol

Citation: Kato-Noguchi, H.; Hamada, Y.; Kojima, M.; Kumagai, S.; Iwasaki, A.; Suenaga, K. Allelopathic Substances of *Osmanthus* spp. for Developing Sustainable Agriculture. *Plants* **2023**, *12*, 376. https://doi.org/10.3390/plants12020376

Academic Editors: Margot Schulz and Vincenzo Tabaglio

Received: 7 December 2022
Revised: 10 January 2023
Accepted: 10 January 2023
Published: 13 January 2023

1. Introduction

Osmanthus fragrans Lour., belonging to the Oleaceae family, is an evergreen tree that is 3–5 m high, and has been cultivated for more than 2500 years in China and some other countries because of the fragrance and color of the flowers [1–3]. The flowers are used in traditional tea, liquors, and foods. The essential oil is extracted for perfume. Its roots are used in traditional Chinese medicine such as in treatments for stomach ache, toothache, rheumatism, and muscle pain [4]. Pharmacological and phytochemical investigations of the chemical constituents in the plants and its essential oil suggest that the species contains more than 180 compounds, including terpenoids, phenylpropanoids, polyphenols, flavonoids, and sterols. Some of those compounds, such as ionone, flavonoids, iridoids, and polyphenols, have shown pharmacological functions through an anti-hyperglycemic activity, anti-inflammatory activity, anti-melanogenesis activity, antioxidant activity, and antitumor activity [3,5–10]. There has been limited information available on the allelopathic activity of the leaf extracts of *O. fragrans* in a publication using it as Chinese medicine [11]. However, no information is available on the compounds involved in the allelopathy of *O. fragrans*.

Allelopathy is the interaction between donor plants and receiver plants through certain secondary metabolites defined as allelochemicals, which are synthesized in some plant parts and released into the vicinity of the donner plants, including the rhizosphere soil, through the volatilization and rainfall leachates from the plants, root exudation, and decomposition process of plant residues, including fallen leaves. Allelopathy is able to provide a competitive advantage to the donner plants for resources such as nutrients, water, and light, through the inhibition of the germination and growth of the neighboring competitive plants [12–15]. The plant parts and the extracts of several plant species have shown an excellent weed control ability as mulch and soil additives due to their allelopathic characteristic [16–18]. Therefore, the allelopathy of plants is potentially useful for weed management options in several agriculture settings in order to reduce commercial herbicide dependency for the development of sustainable agriculture [19,20].

The *Osmanthus* genus consists of about 30 species, mostly native to Eastman Asia. *Osmanthus* × *fortunei* Carrière is a hybrid species between *O. fragrans* and *Osmanthus heterophyllus* (G.Don) P.S.Green. The leaves of *O. heterophyllus* show a serrate leaf margin and *O. fragran* leaves show no serration, and the hybrid species *O.* × *fortunei* has a combination of both of these characteristics (Figure 1). These plant species are evergreen trees but their senescent leaves fall, and the fallen leaves accumulate on the soil surfaces under the trees (Figure 2). Tree plants are able to release allelopathic substances into their rhizosphere soil over many years through the decomposition process of their fallen leaves. Those liberated compounds from the leaves may accumulate in the soil, and suppress the germination and growth of other plant species [21–23]. Therefore, some allelopathic substances can be released into the rhizosphere soil during the decomposition process of the fallen leaves of *O. fragrans*, *O. heterophyllus*, and *O.* × *fortune*, and act as allelopathic agents [13–15]. The objective of this study was the investigation of the allelopathic activity of the leaves of *O. fragrans*, *O. heterophyllus*, and *O.* × *fortune*, and the determination of the main allelopathic substances in their leaves for the development of sustainable agriculture.

Figure 1. The leaves of *O. fragrans*, *O.* × *fortunei*, and *O. heterophyllus*.

Figure 2. The accumulation of the fallen leaves on the soil under the tree of *O.* × *fortune*.

2. Results

2.1. Allelopathic Activity of the Leaves of O. fragrans, O. heterophyllus and O. × fortunei

The extracts of *O. fragrans* leaves significantly inhibited the hypocotyl and root growth of the dicotyledonous plants, namely cress and alfalfa at concentrations greater than 30–100 mg leaf equivalent extract per mL. The extract also significantly inhibited the coleoptile and root growth of the monocotyledonous weed plants, namely *V. myuros* and *L. multiflorom* at concentrations greater than 10–30 mg leaf equivalent extract per mL (Figure 3). The extracts of *O. heterophyllus* and *O.* × *fortunei* also showed an inhibitory activity on the growth of cress, alfalfa, *V. myuros*, and *L. multiflorom* (Figures 4 and 5). Increasing the extract concentration of *O. fragrans*, *O. heterophyllus*, and *O.* × *fortunei* resulted in an increase in the growth suppression of the hypocotyls/coleoptiles and roots of both dicotyledonous and monocotyledonous plant species (Figures 3–5).

Table 1 shows the IC_{50} values of the extracts of *O. fragrans*, *O.* × *fortune*, and *O. heterophyllus*, which can cause 50% growth inhibition on the hypocotyls/coleoptiles and roots of the test plant species, and indicate that smaller values show a higher growth inhibitory activity than large values. All IC_{50} values of the *O.* × *fortune* extracts, except the IC_{50} value for *V. myuros* roots, were smaller than those of the *O. fragrans* and *O. heterophyllus* extracts.

Table 1. IC_{50} values (mg leaf equivalent extract per mL) of the extracts of the leaves of *O. fragrans*, *O.* × *fortunei*, and *O. heterophyllus* on the growth of hypocotyls/coleoptiles and the roots of bioassay plant species. IC_{50} values were determined with GraphPad Prism. Different letters in the same column indicate significant differences (Tukey's HSD test, $p \leq 0.05$).

	Osmanthus fragrans		*Osmanthus* × *fortunei*		*Osmanthus heterophyllus*	
	Hypocotyl/ Coleoptile	**Root**	**Hypocotyl/ Coleoptile**	**Root**	**Hypocotyl/ Coleoptile**	**Root**
Cress	63.8 ± 4.2 b	63.7 ± 4.7 b	24.8 ± 1.9 c	36.5 ± 1.9 c	41.7 ± 4.1 bc	59.1 ± 4.7 b
Alfalfa	107 ± 8.1 a	153 ± 9.7 a	53.1 ± 4.2 bc	68.5 ± 7.4 b	76.5 ± 6.7 b	114 ± 9.9 a
Lolium multiflorum	82.6 ± 4.9 ab	30.4 ± 2.5 c	21.5 ± 1.6 c	12.4 ± 0.9 d	29.5 ± 3.1 c	13.8 ± 0.9 d
Vulpia myuros	25.9 ± 2.1 c	8.8 ± 0.9 d	21.4 ± 1.7 c	11.5 ± 0.9 d	36.8 ± 3.2 c	13.8 ± 1.1 d

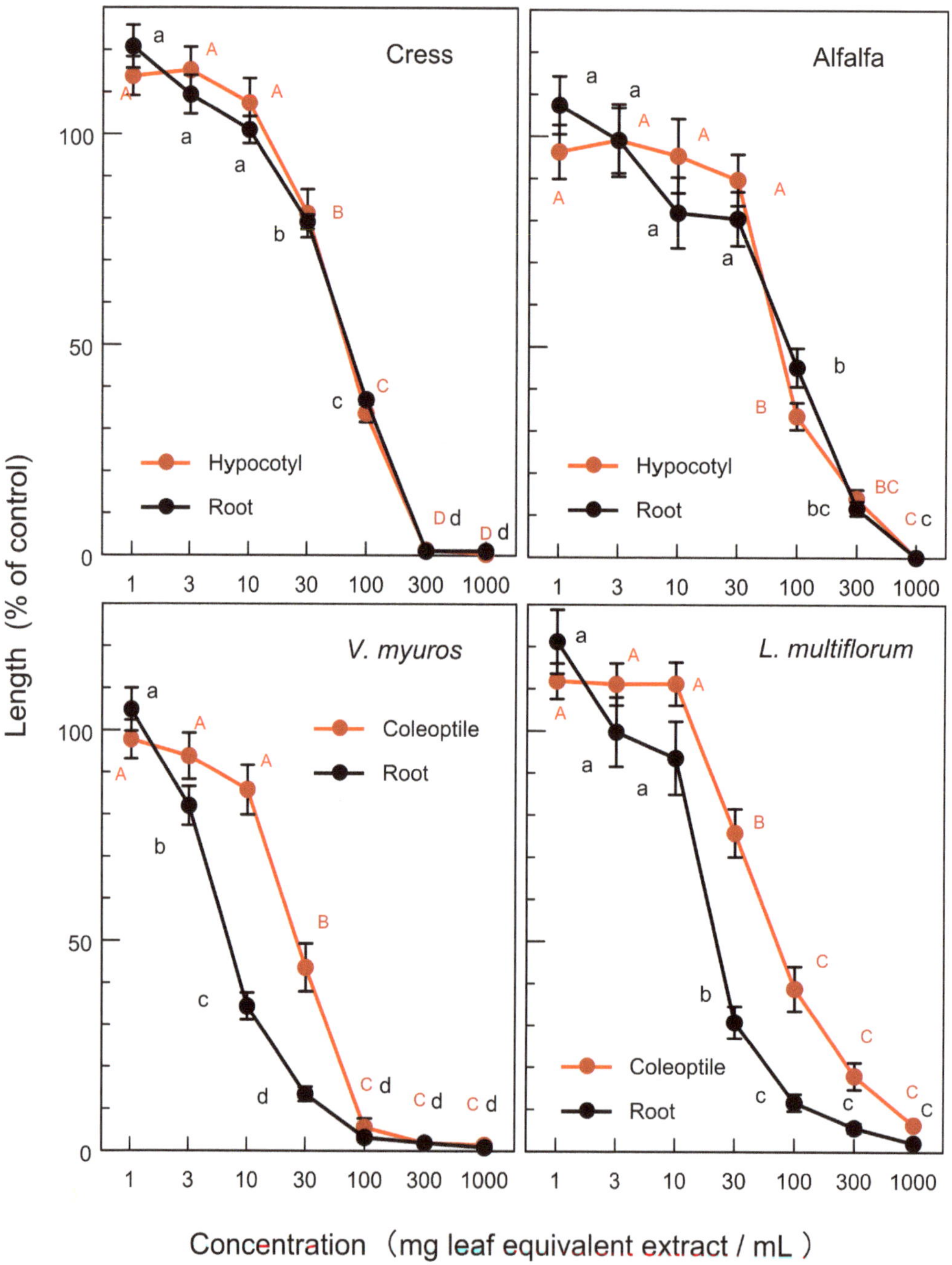

Figure 3. Effects of the extracts of *O. fragrans* leaves on the growth of the hypocotyls and roots of cress and alfalfa, and the growth of the coleoptiles and roots of *L. multiflorum* and *V. myuros*. Concentration (mg leaf equivalent extract/mL) indicates the concentration of the tested sample corresponding to the extracts obtained from 1, 3, 10, 30, 100, 300, and 1000 mg leaves per mL. Different letters on the symbols in the same panels indicate significant differences (Tukey's HSD test, $p \leq 0.05$).

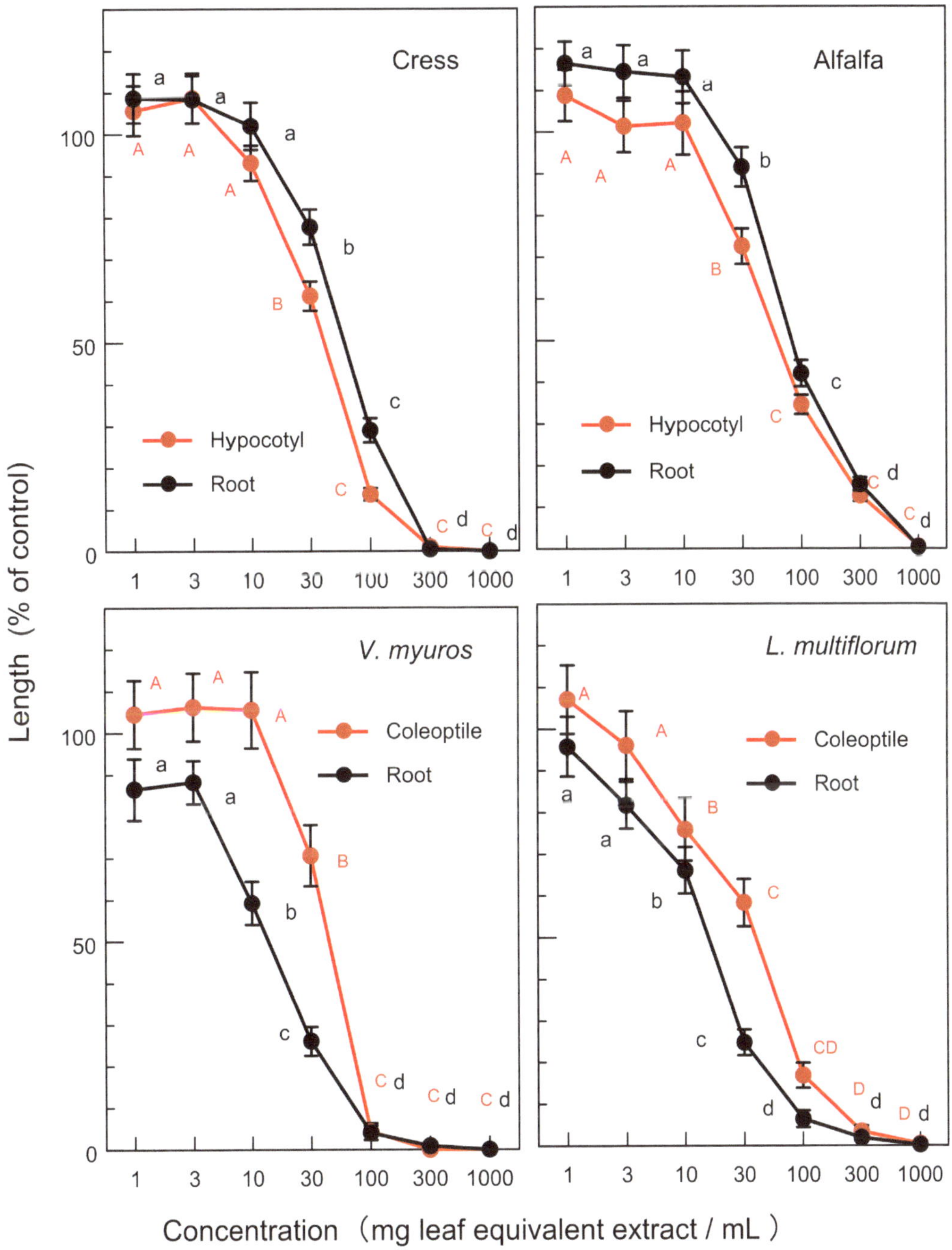

Figure 4. Effects of the extracts of *O. heterophyllus* leaves on the growth of the hypocotyls and roots of cress and alfalfa, and the growth of the coleoptiles and roots of *L. multiflorum* and *V. myuros*. Concentration (mg leaf equivalent extract/mL) indicates the concentration of the tested sample corresponding to the extracts obtained from 1, 3, 10, 30, 100, 300, and 1000 mg leaves per mL. Different letters on the symbols in the same panels indicate significant differences (Tukey's HSD test, $p \leq 0.05$).

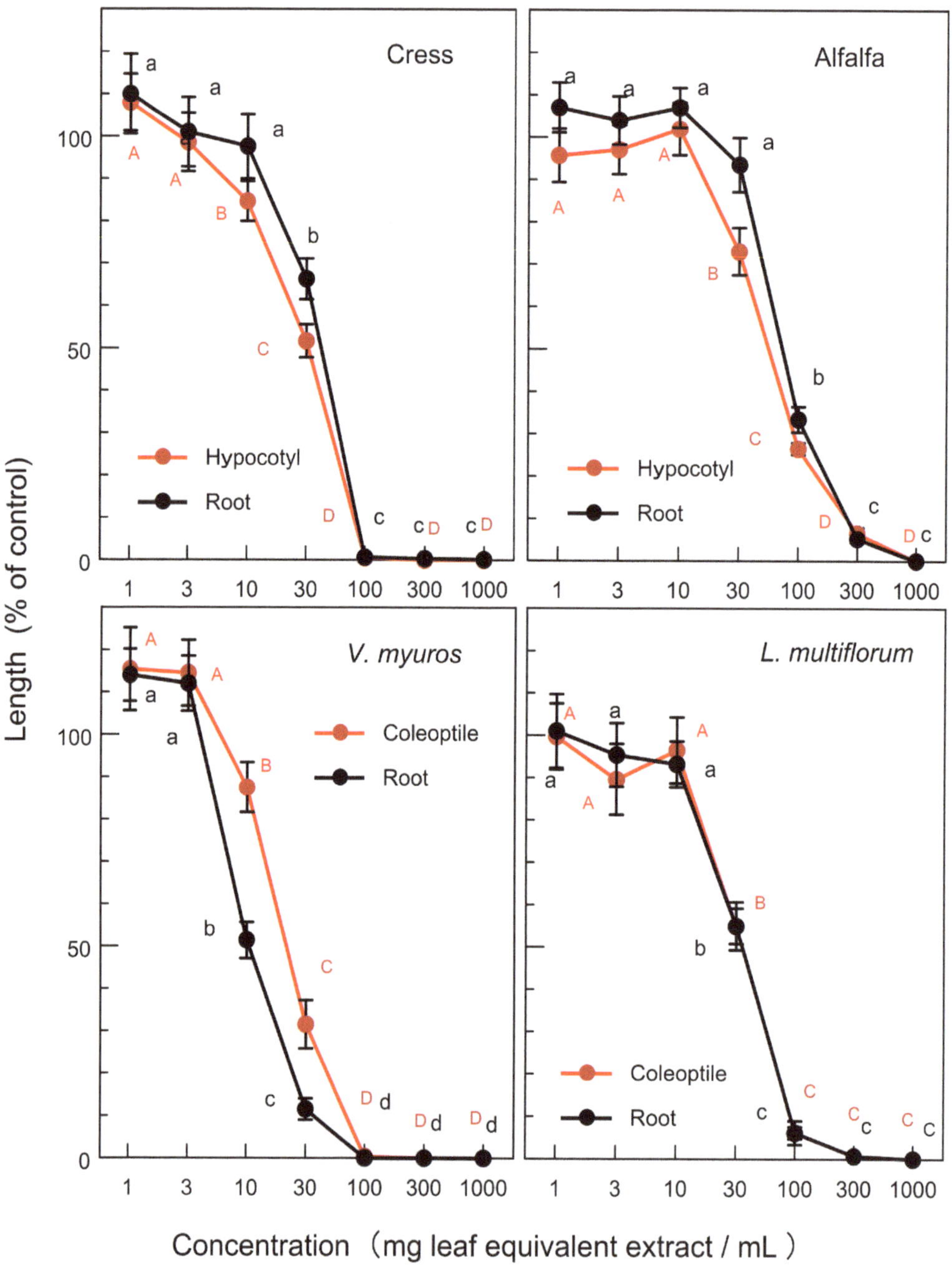

Figure 5. Effects of the extracts of *O.* × *fortunei* leaves on the growth of the hypocotyls and roots of cress and alfalfa, and the growth of the coleoptiles and roots of *L. multiflorum* and *V. myuros*. Concentration (mg leaf equivalent extract/mL) indicates the concentration of tested sample corresponding to the extracts obtained from 1, 3, 10, 30, 100, 300, and 1000 mg leaves per mL. Different letters on the symbols in the same panels indicate significant differences (Tukey's HSD test, $p \leq 0.05$).

2.2. Identification of Compound 1 and Its Activity

The extract of *O.* × *fortunei* leaves was partitioned with ethyl acetate. The activity of the obtained ethyl acetate phase was much greater than that of the aqueous phase (Table 2). Therefore, the ethyl acetate phase was separated using the chromatography of the silica gel, Sephadex LH-20, and ODS cartridge. The allelopathic activity of all of the separated fractions was determined by a cress bioassay after each chromatographic separation, as described above, and the most active fraction was carried out following chromatographic separation. Finally, active compound **1** was isolated by HPLC.

Table 2. Effects of the aqueous and ethyl acetate phase obtained after the partition of the extracts of *O. fragrans*, *O. heterophyllus*, and *O.* × *fortunei* on the growth of the hypocotyl and roots of the cress (% of control). The activity was determined using the test samples obtained from 100 mg of leaf equivalent extract/mL. Different letters in the same column indicate significant differences (Tukey's HSD test, $p \leq 0.05$).

	Osmanthus fragrans		*Osmanthus* × *fortunei*		*Osmanthus heterophyllus*	
	Hypocotyl	**Root**	**Hypocotyl**	**Root**	**Hypocotyl**	**Root**
Aqueous phase	76.8 ± 7.5 a	101 ± 9.8 a	76.2 ± 4.3 a	62.4 ± 4.8 a	87.1 ± 4.0 a	70.9 ± 4.6 b
Ethyl acetate phase	46.8 ± 4.4 b	27.1 ± 2.7 b	22.4 ± 3.6 b	19.9 ± 3.9 b	31.9 ± 3.7 b	29.7 ± 2.9 b

The molecular formula of compound **1** is $C_{20}H_{22}O_6$, as suggested by HRESIMS at m/z 359.1478 $[M + H]^+$ (calcd for $C_{20}H_{23}O_6$ 359.1495). The ^{1}H NMR (400 MHz, CD_3OD) spectrum of compound **1** showed δ_H 6.95 (d, J = 1.9 Hz, 2 H, H-2/2′), 6.81 (dd, J = 8.4, 1.9 Hz, 2 H, H-6/6′), 6.77 (d, J = 8.4 Hz, 2 H, H-5/5′), 4.71 (d, J = 4.5 Hz, 2 H, H-7/7′), 4.23 (dd, J = 9.5, 7.0 Hz, 2 H, H-9a/9a′), 3.86 (s, 6 H, H-10/10′), 3.84 (dd, J = 9.5, 3.9 Hz, 2 H, H-9b/9b′), and 3.14 (m, 2 H, H-8/8′). The optical rotation of the compound was $[\alpha]_D^{27}$ = +84.5 (c = 0.215, CH_3OH). The chemical structure of compound **1** was determined as (+)-pinoresinol (Figure 6) by the comparison of those spectra data with the published data [24]. (+)-Pinoresinol (1) significantly inhibited the growth of hypocotyls/coleoptiles and the roots of the cress and *L. multiflorum* at concentrations greater than 0.3 mM. The inhibition increased with the increasing concentration of (+)-pinoresinol (Figure 7). The IC_{50} values of (+)-pinoresinol of the growth hypocotyls/coleoptiles and the roots of cress and *L. multiflorum* are shown in Table 3.

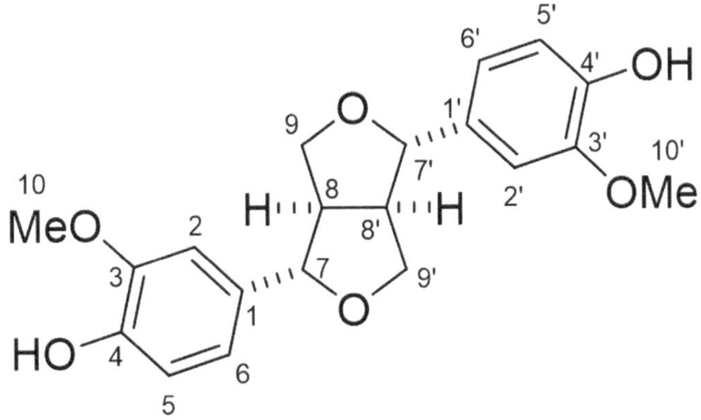

Figure 6. Chemical structure of (+)-pinoresinol.

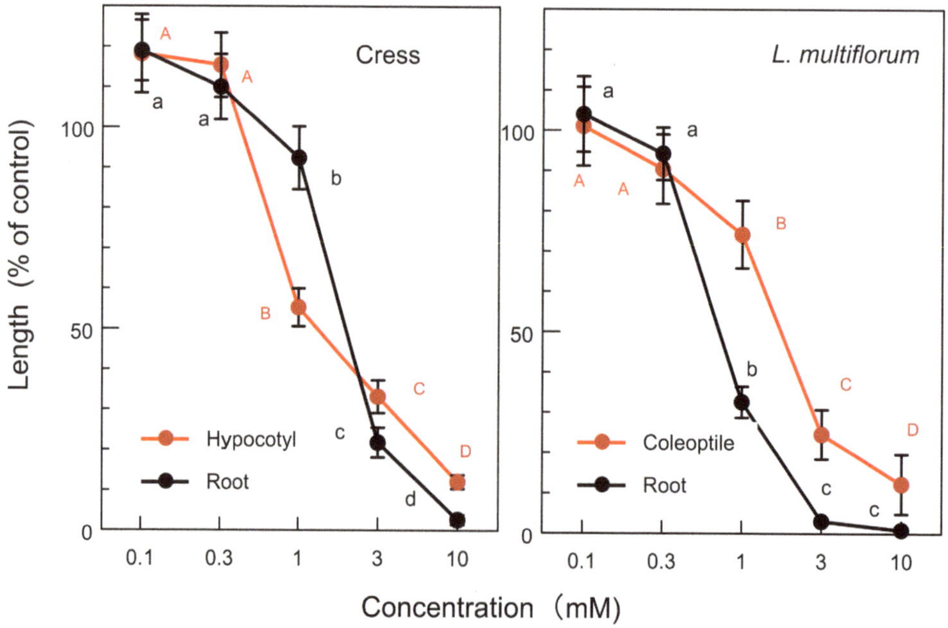

Figure 7. Effects of (+)-pinoresinol on the hypocotyl/coleoptile and root growth of cress and *L. multiflorum*. Means ± SE from three independent experiments with 10 seedlings for each determination are shown. Different letters on the symbols in the same panels indicate significant differences (Tukey's HSD test, $p \leq 0.05$).

Table 3. IC_{50} values (mM) of (+)-pinoresinol and 10-acetoxyligustroside of the growth of the hypocotyls/coleoptiles and roots of the cress and *L. multiflorum*. IC_{50} values were determined with GraphPad Prism. Different letters in the same column indicate significant differences (Tukey's HSD test, $p \leq 0.05$).

	(+)-Pinoresinol		**10-Acetoxyligustroside**	
	Hypocotyl/ Coleoptile	**Root**	**Hypocotyl/ Coleoptile**	**Root**
Cress	1.9 ± 0.2 a	2.5 ± 0.3 b	4.7 ± 0.3 b	4.4 ± 0.3 a
Lolium multiflorum	2.1 ± 0.3 a	0.7 ± 0.1 d	3.1 ± 0.3 b	2.2 ± 0.2 b

2.3. Concentration of (+)-Pinoresinol in the Leaves of O. × fortune, O. fragrans, and O. heterophyllus

The concentration of (+)-pinoresinol (1) in the mature leaves of *O.* × *fortune* was 4.1 mg/g and the concentration of its fallen leaves was 2.1 mg/g. However, the mature leaves *O. fragrans* and *O. heterophyllus* showed a small amount of (+)-pinoresinol (Table 4).

Table 4. The concentration of (+)-pinoresinol in the mature and fallen leaves of *O.* × *fortunei*, and in the mature leaves of *O. fragrans* and *O. heterophyllus*. Different letters indicate significant differences (Tukey's HSD test, $p \leq 0.05$).

Osmanthus × fortunei	(+)-Pinoresinol (mg/g)
Mature leaf	4.1 ± 0.5 a
Fallen leaf	2.1 ± 0.2 b
Osmanthus fragrans	0.22 ± 0.03 c
Osmanthus heterophyllus	0.12 ± 0.02 c

2.4. Identification of Compound 2 and Its Activity

The extract of *O. fragrans* leaves was partitioned with ethyl acetate. The activity of the obtained ethyl acetate phase was much greater than that of the aqueous phase (Table 2). The ethyl acetate phase was separated using the chromatography of the silica gel, Sephadex LH-20, and C_{18} cartridge. The allelopathic activity of all separated fractions was determined by a cress bioassay after each separation step, and the most active fraction was applied to the next separation, and an active compound **2** was finally isolated by HPLC.

The molecular formula of compound **2** is $C_{27}H_{34}O_{1}$, as suggested by HRESIMS at *m/z* 605.1836 $[M + Na]^+$ (calculated for $C_{27}H_{34}O_{14}Na$, 605.1846). The ^1H NMR (400 MHz, CD_3OD) spectrum of the compound **2** showed δ_H 7.53 (s, 1 H, H3), 7.05 (d, *J* = 8.6 Hz, 2 H, H2′, H6′), 6.71 (d, *J* = 8.6 Hz, 2 H, H3′, H5′), 6.08 (brdd, *J* = 7.7, 5.9 Hz, 1 H, H8), 5.98 (brs, 1 H, H1), 4.82 (d, *J* = 7.7 Hz, 1 H, H1″), 4.76 (dd, *J* = 13.1, 7.7 Hz, 1 H, H10), 4.56 (ddd, *J* = 13.3, 5.9, 1.8 Hz, 1 H, H10), 4.24 (ddd, *J* = 10.8, 7.3, 7.3 Hz, 1 H, H8′), 4.14 (ddd, *J* = 10.8, 7.3, 7.3 Hz, 1 H, H8′), 4.00 (dd, *J* = 10.0, 4.1 Hz, 1 H, H5), 3.90 (dd, *J* = 11.8, 1.4 Hz, 1 H, H6″), 3.72 (s, 3 H, H14), 3.67 (dd, *J* = 11.8, 5.9 Hz, 1 H, H6″), 3.41 (dd, *J* = 8.6, 9.2 Hz, H3″), 3.35 (m, 1 H, H5″), 3.34 (m, 1 H, H2″) 3.33 (m, 1 H, H4″), 2.83 (t, *J* = 7.3 Hz, 1 H, H7′), 2.77 (dd, *J* = 15.4, 4.1 Hz, 1 H, H6), 2.50 (dd, *J* = 15.4, 9.5 Hz, 1 H, H6), and 2.02 (s, 3 H, H13). The ^{13}C NMR (100 MHz, CD_3OD) spectrum of the compound showed δ_C 172.8 (C7), 172.5 (C12), 168.3 (C11), 157.1 (C4′), 154.9 (C3), 133.9 (C9), 131.0 (C2′, C6′), 130.0 (C1′), 124.4 (C8), 116.3 (C3′, C5′), 109.1 (C4), 100.8 (C1″), 94.2 (C1), 78.5 (C3″), 77.9 (C5″), 74.7 (C2″), 71.4 (C4″), 66.9 (C8′), 62.7 (C6″), 61.8 (C10), 52.0 (C14), 41.0 (C6), 35.2 (C7′), 32.4 (C5), and 20.8 (C13). The optical rotation of the compound was $[\alpha]_D^{27}$-163 (*c* 0.32, CH_3OH). Those spectra indicate that the chemical structure of compound **2** is 10-acetoxyligustroside (Figure 8), in comparison with those reported in the literature [25]. The compound significantly inhibited the growth of the hypocotyls/coleoptiles and roots of cress and *L. multiflorum* at concentrations greater than 1 mM (Figure 9). IC_{50} values of 10-acetoxyligustroside of the growth of the hypocotyls/coleoptiles and roots of cress and *L. multiflorum* are shown in Table 3.

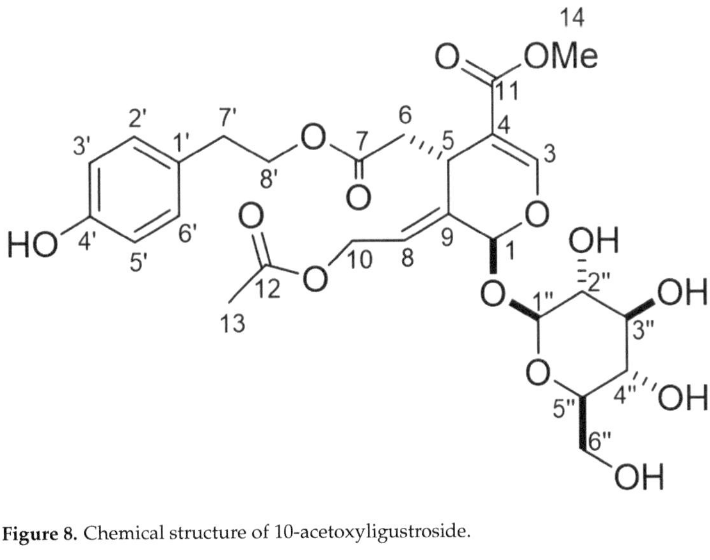

Figure 8. Chemical structure of 10-acetoxyligustroside.

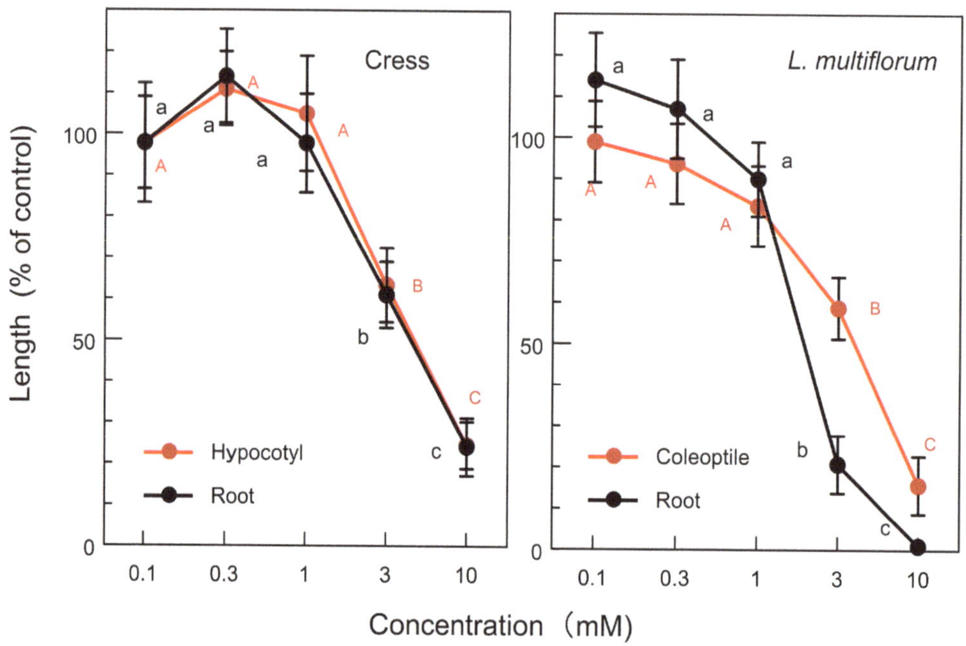

Figure 9. Effects of 10-acetoxyligustroside on the hypocotyl/coleoptile and root growth of cress and *L. multiflorum*. Means ± SE from three independent experiments with 10 seedlings for each determination are shown. Different letters on the symbols in the same panels indicate significant differences (Tukey's HSD test, $p \leq 0.05$).

3. Discussion

The leaf extracts of *O. fragrans*, *O. heterophyllus*, and *O.* × *fortunei* showed allelopathic activity against the growth of dicotyledonous plants (cress and alfalfa) and monocotyledonous weed plants (*L. multiflorum* and *V. myuros*) (Figures 3–5). As allelopathic substances are synthesized and stored in some plant parts until their release into the environments [12–15], the inhibitory activity of these leaves suggest that these plant species may produce and accumulate the allelopathic substances in their leaves. All of the IC_{50} values of *O.* × *fortunei* on the growth of the hypocotyls/coleoptiles and roots of four test plant species were the smallest compared with those of *O. fragrans* and *O. heterophyllus* except for the IC_{50} value on the root growth of *V. myuros*. IC_{50} values on the root growth of *V. myuros* and *O.* × *fortunei* are not significantly different (Table 1). Thus, the leaf extract of a hybrid species of *O.* × *fortunei* was the most active among the extracts of *O. fragrans*, *O. heterophyllus*, and *O.* × *fortunei*. The accumulation of the allelopathic substances in the leaves of *O.* × *fortunei* could be greater than that of the other two species. *Fallopia japonica* (Houtt.) Ronse Decraene and *Fallopia sachalinensis* (F. Schmidt) Ronse Decraene are highly competitive invasive plant species [26,27]. *F. japonica* and *F. sachalinensis* hybridize unnaturally by asexual reproduction and create a hybrid species, *Fallopia* × *bohemica* (Chrteket Chrtková) J.P. Bailey [28]. Hybridization of the both species gives genetic diversity to *F.* × *bohemica* [29,30], and the invasiveness of the hybrid species *F.* × *bohemica* is high [31,32].

The allelopathic activity is usually evaluated on the basis of the specific activity of the compound such as ppm and IC_{50}. However, when we evaluated the contribution of the allelochemicals to the certain plant species, the concentration of the allelochemicals in the plants needs to be considered. For example, allelochemicals, which were contained in large amounts with a moderate activity, contributed more than those that contained a small amount with a very strong activity as the allelopathic agents of the plants. The concept of the total activity included both factors; the specific activity of a compound and its concentration in the plant materials, which indicates the contribution of the compounds to the allelopathy of the plants [33]. When we compare the activity of the extracts and separated fractions obtained from same weight of the plant materials, we could evaluate the total activity of the extracts and fractions, and we chose the most active extracts and fractions as described by Dr. Fujil's group [33]. Therefore, we evaluated the activity of the separated fractions using the concept of total activity, and selected the most active separated fraction.

The most active extract of *O.* × *fortunei* leaves among the extracts of the three species was partitioned with ethyl acetate, and the ethyl acetate phase was separated as described in Section 4.4. After each chromatographic separation, the total activity of all of the separated fractions was determined by a cress bioassay, and the most active fraction was carried out next in chromatographic separation, and an active compound was isolated. The total activity of the fractions was equivalent by the plant material basis (10 and 100 mg plant materials) [33]. During the isolation process of the active compound, the cress seeds were used as the test plant because the sensitivity of the cress to the extracts was not low or high among the four test plant species (Table 1). Based on its spectrum data compared with the published data [24], the chemical structure of the isolated compound **1** was identified as (+)-pinoresinol (Figure 6). (+)-Pinoresinol may be the main contributor to the allelopathic activity of the extract of *O.* × *fortunei* leaves because the compound was isolated from the most active fraction (total activity) of every separation step.

(+)-Pinoresinol is a tetrahydrofuran lignan and is synthesized from two coniferyl alcohol monomers [34–36]. The compound inhibited the growth of cress and *L. multiflorum* (Figure 7), and the IC_{50} values of the compound on the growth of the hypocotyls/coleoptiles and roots of cress and *L. multiflorum* were 0.7–2.5 mM (Table 3). The compound was also reported to show a growth inhibitory activity on the wheat coleoptiles [37], as well as an antifungal activity, antioxidative activity, and hepatoprotective activity in mammalian cells [38–40].

The concentration of (+)-pinoresinol in the fallen leaves of *O.* × *fortune* was less than that in its manure leaves, which indicates that degradation of the compound may occur during the senescence of the leaves or through some other factors. However, the fallen leaves of *O.* × *fortune* still contained 51% (+)-pinoresinol of its mature leaves (Table 4). *O.* × *fortune* is evergreen, but their senescent leaves cover the soil under the trees (Figure 2). The fallen leaves were gradually decomposed and decayed into the soil in the same way as other tree leaves [41–43]. It is possible that (+)-pinoresinol in the fallen leaves may be liberated through the decomposition process of the leaves and accumulate in their rhizosphere soil over many years. The accumulated (+)-pinoresinol may be able to suppress the germination and growth of other plant species as an allelopathic agent.

The concentration of (+)-pinoresinol in the mature leaves *O. fragrans* and *O. heterophyllus* was 0.22 mg/g and 0.12 mg/g, respectively (Table 4). It was also reported that 13 mg of (+)-pinoresinol-β-D-glucopyranoside was isolated from 2.6 kg of *O. heterophyllus* leaves [44], which is equivalent to 0.005 mg/g. The concentration of (+)-pinoresinol in the mature leaves of *O.* × *fortune* was 18.6-fold and 34.2-hfold greater than that of *O. fragrans* and *O. heterophyllus*, respectively. *O. fragrans* and *O. heterophyllus* are the parents of the hybrid species *O.* × *fortune*. Thus, the hybridization of both species may increase the biosynthesis of (+)-pinoresinol.

Based on the IC_{50} values, the allelopathic activity of *O. fragrans* was 38–94% and 40–130% that of *O.* × *fortune* for the growth of the hypocolyls/coleoptiles and roots, respectively, of the four test plant species, and the allelopathic activity of *O. heterophyllus* was 59–72% and 60–89% that of *O.* × *fortune* for their growth of hypocolyls/coleoptiles and roots, respectively (Table 1). However, the concentration of (+)-pinoresinol in the mature leaves of *O. fragrans* and *O. heterophyllus* was only 5% and 3%, respectively, of that in the mature leaves of *O.* × *fortune* (Table 4). Thus, (+)-pinoresinol may not be a main allelopathic substance in the leaves of *O. fragrans* and *O. heterophyllus*.

The extract of *O. fragrans* leaves after being separated is described in Section 4.5, and an allelopathic substance was isolated and identified as 10-acetoxyligustroside (Figure 8) based on its spectrum data compared with the published data [25]. 10-Acetoxyligustroside is an iridoid glycoside and is synthesized from iridodial [45,46]. The compound inhibited the growth of cress and *L. multiflorum* (Figure 9), and the IC_{50} values of the compound on the hypocotyls/coleoptiles and roots of the cress and *L. multiflorum* were 2.2–4.7 mM (Table 3). There is limited information about 10-acetoxyligustroside regarding the biological activity, including the allelopathic activity, in the literature. This paper is the first report showing the allelopathic activity of 10-acetoxyligustroside. In addition, 10-acetoxyligustroside may be the main contributor to the allelopathic activity of *O. fragrans* leaves because the compound was isolated from the most active fraction in every separation step.

The extract of *O. heterophyllus* leaves was also partition with ethyl acetate (Table 2), and the ethyl acetate phase was separated using the chromatography of the silica gel, Sephadex LH-20, and C_{18} cartridge for the isolation of the main allelopathic substances, similar to other leaf extracts of *O. fragrans* and *O.* × *fortunei*. However, many active fractions were identified in every separation step, and we could not isolate the main allelopathic substances from the extracts. Several lignans, neolignan glycosides, and secoiridoid glycosides were isolated from *O. heterophyllus* leaves, but their biological activities have not yet been reported [47–49]. There has also been no information available on the allelopathic substances in *O. heterophyllus* leaves. Considering these investigations, many allelopathic substances including (+)-pinoresinol may be involved in the allelopathy of *O. heterophyllus*. Therefore, the extracts of the leaves of these *Osmanthus* species have allelopathic potential and contain allelopathic substances, which may be species specific. However, it is necessary to determine the concentrations of these compounds in the rhizosphere soil, and to clarify the contribution of these compounds to their allelopathy.

4. Materials and Methods

4.1. Plant Material

Mature leaves of *O. fragrans* and *O. heterophyllus*, and mature and fallen leaves of *O.* × *fortunei* were obtained from the campus of Faculty of Agriculture, Kagawa University, and the nearest botanical garden. Collected leaves were weighed and kept at −25 °C. The seeds of cress (*Lepidum sativum* L.) and alfalfa (*Medicago sativa* L.) were used to determine the allelopathic activity because of their stable germination. Weed species of *Vulpia myuros* (L.) C.C. Gmel. and *Lolium multiflorum* Lam. were also used to test the allelopathic activity.

4.2. Extraction

Mature leaves (50 g fresh weight) of *O. fragrans*, *O. heterophyllus* and *O.* × *fortunei* were cut into small pieces and extracted separately with 80% (*v*/*v*) aqueous methanol (300 mL) for 2 days. The extracts were filtered with a filter paper (No. 2; Toyo Ltd., Tokyo, Japan), and the residue of each extract was extracted again with methanol (300 mL) for 2 days and then filtered. Two filtrates of each species were mixed and concentrated under reduced pressure at 40 °C [50]. The yield of the extracts of *O. fragrans*, *O. heterophyllus*, and *O.* × *fortunei* was 0.62, 0.57 g, and 0.69 g, respectively.

4.3. Determination of Allelopathic Activity

The concentrated extracts of *O. fragrans*, *O. heterophyllus*, and *O.* × *fortunei* were dissolved with methanol, and a given quantity of the methanol solution was applied to a filter paper (No. 2) in a Petri dish (2.8 cm i.d.). After the methanol solution was completely evaporated in a draft chamber, 0.6 mL of Tween 20 solution (0.05%, Nacalai, Kyoto, Japan) was added onto the filter paper. Ten seeds of alfalfa and cress and 10 germinated seeds of *L. multiflorum* and *V. myuros*, which had been germinated for 48 h in the dark at 25 °C, were separately arranged on the filter paper in the Petri dishes. After 48 h of incubation in the dark at 25 °C, the length of the roots and hypocotyls/coleoptiles of these four test plants was measured using a ruler. The percentage length of the hypocotyls/coleoptiles and roots of the test plant species treated with the leaf extracts was determined against that of each of the control seedlings. The control seedlings were treated exactly same process without the extract application [50]. The concentrations of the applied extracts in the Petri dishes were 1, 3, 10, 30, 100, 300, and 1000 mg leaf equivalent extracts per mL. The concentrations of the extract caused for the 50% growth inhibition (defined as IC_{50}) of these test plant species were determined with GraphPad Prism 6.0 (GraphPad Software, Inc., La Jolla, CA, USA). The determination of the allelopathic activity was repeated four times using a completely randomized design with 10 plants for each determination.

4.4. Separation of *O.* × *fortunei* Extract

The leaves (100 g fresh weight) of *O.* × *fortunei* were extracted with 80% (*v*/*v*) aqueous methanol (600 mL) and methanol (600 mL), and were filtrated. Two filtrates were mixed and evaporated to obtain an aqueous solution, as described above. Then, the aqueous solution was made to pH 7.0 with the buffer of 0.5 M phosphate, and partitioned four times with ethyl acetate. Obtained ethyl acetate phase was concentrated to dryness, and applied to a silica gel (100 g, silica gel 60, 70–230 mesh; Nacarai, Kyoto, Japan) chromatography, which was eluted with 100 mL each of 20, 30, 40, 50, 60, 70, and 80% (*v*/*v*) ethyl acetate in *n*-hexane and ethyl acetate, and 200 mL of methanol. The allelopathic activity of those nine fractions was determined using cress seeds, as described in the Section 4.3. The activity was found in a fraction obtained by an elution of 80% ethyl acetate in *n*-hexane.

After the active fraction was evaporated, the residue was separated by a chromatography of Sephadex LH-20 (50 g, GE Healthcare, Uppsala, Sweden) and eluted with 100 mL each of 20, 30, 40, 50, 60, 70, 80, and 90% (*v*/*v*) aqueous methanol and 200 mL of methanol. An active fraction was obtained with the elution of 60% aqueous methanol, and the active residue was further separated using a reverse-phase ODS cartridge (YMC-Dispo Pack AT ODS-25; YMC Ltd., Kyoto, Japan), which was eluted with 150 mL each of 30, 40, 50, 60, 70, 80, and 90% (*v*/*v*) aqueous methanol. The active fraction was obtained by the elution with 50% aqueous methanol, and the residue was finally subjected to a reverse-phase HPLC (4.6 i.d. × 150 mm, μBondasphere 5μ C_{18}-100Å, Waters Co., Ltd. (Milford, MA, USA), detection 220 nm) and eluted with 45% aqueous methanol (flow rate; 1.5 mL/min). The allelopathic activity was found in a peak fraction eluted between 125–130 min, yielding active compound **1**. The chemical structure of compound **1** was characterized by the analyses of the HRESIMS (LCT Premier XE time-of-flight TOF mass spectrometer, Waters, Tokyo) and ^{1}H-NMR spectrum (400 MHz, CD_3OD; JNM-ECX400, JEOL, Tokyo, Japan), and the optical rotation.

4.5. Separation of *O. fragrans* Extract

The leaves (500 g fresh weight) of *O. fragrans* were extracted with 80% (*v*/*v*) aqueous methanol (3 L) and methanol (3 L), and separated by silica gel chromatography, as described above. An active fraction was obtained with the elution of methanol, and the active residue was applied to the second silica gel chromatography, which was eluted with 100 mL each of ethyl acetate, and 25, 50, and 75% (*v*/*v*) acetone in ethyl acetate, acetone, and methanol. The activity was obtained by the elution of 25% acetone in ethyl acetate, and the residue was applied to a chromatography of Sephadex LH-20, and eluted with 100 mL each of 20, 30, 40, 50, 60, 70, 80, and 90% (*v*/*v*) aqueous methanol. An active fraction obtained with 30% aqueous methanol was further purified using C_{18} cartridges (YMC Ltd.), and eluted with 15 mL each of 30, 40, 50, 60, 70, 80, and 90% (*v*/*v*) aqueous methanol. The active fraction was obtained by the elution of 30% aqueous methanol. The active residue was subjected to a reverse-phase HPLC (10 i.d. × 500 mm, ODS-AQ 5μ YMC-Pack, YMC, detection 220 nm) eluted with 45% aqueous methanol (1.5 mL/min). The allelopathic activity was found in a peak fraction eluted between 104–105 min, yielding active compound **2**. The chemical structure of compound **2** was characterized by the analyses of HRESIMS and ^{1}H-NMR (400 MHz, CD_3OD), ^{13}C-NMR (100 MHz, CD_3OD) spectra and the optical rotation.

4.6. Allelopathic Activity of Isolated Compounds 1 and 2

Each of isolated compounds **1** and **2** was dissolved with methanol, and a given quantity of the methanol solution was added onto a filter paper in the Petri dish. After the evaporation of methanol in the Petri dishes, the filter paper was moistened with 0.6 mL Tween 20. Then, 10 seeds of cress and 10 germinated seeds of *L. multiflorum*, were separately arranged on the filter paper and incubated in the dark at 25 °C for 48 h. The allelopathic activity of the compounds and its IC_{50} values were determined as described in Section 4.3.

4.7. Concentration of Compound 1 in the Leaves

Mature and fallen leaves (5 g) of *O.* × *fortune*, and mature leaves (5 g) of *O. fragrans* and *O. heterophyllus* were separately extracted and partitioned with ethyl acetate, as described above. The obtained residue was applied to a silica gel chromatography, and eluted with 100 mL each of 60, 70, and 80% (*v*/*v*) ethyl acetate in *n*-hexane. After evaporation of the fraction eluted with 80% ethyl acetate in *n*-hexane, the residue was applied to a reverse-phase C_{18} cartridge, and eluted with 15 mL each of 30, 40, and 50% (*v*/*v*) aqueous methanol. The fraction obtained by the elution of 40% aqueous methanol was subjected to a reverse-phase HPLC (4.6 i.d. × 250 mm, Cosmosil, Nacalai, detection 220 nm) eluted with 40% aqueous methanol (0.8 mL/min). The quantification of compound **1** was performed by the peak area on the HPLC chromatograms of samples. The recovery of compound **1** added to the leave extracts was 95 ± 11% (mean ± SE) as calculated from three replications.

4.8. Statistical Analysis

The data were subjected to a one-way analysis of ANOVA and subsequent post hoc analysis with Tukey's HSD test ($p < 0.05$). IC_{50} values were obtained using GraphPad Prism 6.0.

5. Conclusions

The leaf extracts of *O. fragrans, O. heterophyllus*, and *O.* × *fortunei* showed an allelopathic activity on the four test plant species, including the weed species of *Vulpia myuros* and *Lolium multiflorum*. The allelopathic substances of *O. fragrans* and *O.* × *fortunei* were isolated and identified as 10-acetoxyligustroside and (+)-pinoresinol, respectively, according to the concept of the total activity. Those compounds may be the main contributors to the allelopathy of *O. fragrans* and *O.* × *fortunei*, and provide a competitive advantage to these species through the inhibition of the growth of competing plant species in their vicinity, while several allelopathic substances, including (+)-pinoresinol, may be involved in the allelopathy of *O. heterophyllus*. The present investigation suggests that the extracts of the leaves of these *Osmanthus* species have allelopathic characteristics. Therefore, these extracts and the leaves themselves could potentially be useful for weed management options in some agriculture settings, as mulch, soil additive, and foliar spray, in order to reduce commercial herbicide dependency for the developing sustainable agriculture systems. However, it is necessary to evaluate the weed control activity of these leaves in the field condition.

Author Contributions: Conceptualization, H.K.-N.; methodology, Y.H., M.K., S.K., A.I. and H.K.-N.; software Y.H., M.K. and S.K.; validation, K.S. and H.K.-N.; formal analysis, Y.H., M.K. and S.K., investigation, Y.H., M.K., S.K. and A.I.; data curation, H.K.-N.; writing, H.K.-N.; visualization, H.K.-N.; supervision, H.K.-N. All authors have read and agreed to the published version of the manuscript.

Funding: This research received no external funding.

Institutional Review Board Statement: Not applicable.

Informed Consent Statement: Not applicable.

Data Availability Statement: No supporting data in this study.

Conflicts of Interest: The authors declare no conflict of interest.

References

1. Ren, Q.J.; Zhu, H.W.; Yu, J.P. The utilization value of *O. fragrans* resources. *Chin. Wild Plant Res.* **1999**, *18*, 34–35.
2. Wu, L.C.; Chang, L.H.; Chen, S.H.; Fan, N.C.; Ho, J.A. Antioxidant activity and melanogenesis inhibitory effect of the acetonic extract of *Osmanthus fragrans*: A potential natural and functional food flavor additive. *Food Sci. Technol.* **2009**, *42*, 1513–1519. [CrossRef]
3. Wang, B.; Luan, F.; Bao, Y.; Peng, X.; Rao, Z.; Tang, Q.; Zeng, N. Traditional uses, phytochemical constituents and pharmacological properties of *Osmanthus fragrans*: A review. *J. Ethnopharmacol.* **2022**, *293*, 115273. [CrossRef]
4. Liao, X.Y.; Hu, F.L.; Chen, Z.L. Identification and quantification of the bioactive components in *Osmanthus fragrans* roots by HPLC-MS/MS. *J. Pharm. Anal.* **2021**, *11*, 299–307. [CrossRef] [PubMed]
5. Casirola, D.M.; Ferraris, R.P. α-Glucosidase inhibitors prevent diet-induced increases in intestinal sugar transport in diabetic mice. *Metabolism* **2006**, *55*, 832–841. [CrossRef] [PubMed]
6. Fang, X.Y.; Zhou, J.L.; Wang, J.Q.; Tang, F.; Zhao, L.G. Preparation of flavonoids from *Osmanthus* dregs and enzymatic modification. *J. For. Eng.* **2020**, *5*, 99–105.
7. Li, G.H.; Wang, M.M.; He, W.; Liu, J.; Zhang, D.; Liu, J.Q.; Tang, W.Z. Iridoid glycoside extraction technology and anti-thrombotic activity of *Osmanthus fragrans* seeds. *Chin. Tradit. Herb. Drugs* **2018**, *49*, 3033–3041.
8. Li, Y.L.; Cai, Q.Y.; Lv, F.; Zhou, L.H.; Tang, X.Y.; Ren, D.X.; Li, Y. Study on immune function of *Osmanthus fragrans* extract Osthin. *Mod. Prev. Med.* **2021**, *48*, 1867–1870.
9. Wang, X.L.; Heraud, S.; Thepot, A.; Santos, M.O.; Luo, Z. The whitening properties of the mixture composed of pomegranate, *Osmanthus* and olive and the protective effects against ultraviolet deleterious effects. *Clin. Cosmet. Invest. Dermatol.* **2021**, *14*, 561–573. [CrossRef]
10. Fu, C.C.; Xu, F.Y.; Qian, Y.C.; Koo, H.L.; Duan, Y.F.; Weng, G.M.; Fan, T.P.; Chen, M.X.; Zhu, F.Y. Secondary metabolites of *Osmanthus fragrans*: Metabolism and medicinal value. *Front. Pharmacol.* **2022**, *13*, 922204. [CrossRef]
11. Li, F.; He, G.; Zhou, Q.; Liang, S. Compare the allelopathic potential of Osmanthus *fragrans* in different habitats. *J. Guangxi Norm. Univ. Nat. Sci.* **2013**, *31*, 88–93.
12. Rice, E.L. *Allelopathy*, 2nd ed.; Academic Press: Orlando, FL, USA, 1984; pp. 1–422.
13. Bais, H.P.; Weir, T.L.; Perry, L.G.; Gilroy, S.; Vivanco, J.M. The role of root exudates in rhizosphere interactions with plants and other organisms. *Annu. Rev. Plant Biol.* **2006**, *57*, 233–266. [CrossRef] [PubMed]
14. Bonanomi, G.; Sicurezza, M.G.; Caporaso, S.; Esposito, A.; Mazzoleni, S. Phytotoxicity dynamics of decaying plant materials. *New Phytol.* **2006**, *169*, 571–578. [CrossRef] [PubMed]
15. Belz, R.G. Allelopathy in crop/weed interactions—An update. *Pest. Manag. Sci.* **2007**, *63*, 308–326. [CrossRef]
16. Kato-Noguchi, H. Involvement of allelopathy in the invasive potential of *Tithonia diversifolia*. *Plants* **2020**, *9*, 766. [CrossRef] [PubMed]
17. Kato-Noguchi, H.; Kurniadie, D. Allelopathy and allelochemicals of *Leucaena leucocephala* as an invasive plant species. *Plants* **2022**, *11*, 1672. [CrossRef]
18. Kato-Noguchi, H. Allelopathy and allelochemicals of *Imperata cylindrica* as an invasive plant species. *Plants* **2022**, *11*, 2551. [CrossRef] [PubMed]
19. Narwal, S.S. Allelopathy in weed management. In *Allelopathy Update Basic and Applied Aspects*; Narwal, S.S., Ed.; Science Publishers Inc.: Enfield, NH, USA, 1999; Volume 2, pp. 203–254.
20. Duke, S.O.; Dayan, F.E.; Romagni, J.G.; Rimando, A.M. Natural products as sources of herbicide, current status and future trends. *Weed Res.* **2000**, *40*, 99–111. [CrossRef]
21. Mallik, A.U. Allelopathy and competition in coniferous forests. *Environ. For. Sci.* **1998**, *54*, 309–315.
22. Kato-Noguchi, H.; Kurniadie, D. Allelopathy of *Lantana camara* as an invasive plant. *Plants* **2021**, *10*, 1028. [CrossRef]
23. Kato-Noguchi, H. Phytotoxic substances involved in teak allelopathy and agroforestry. *Appl. Sci.* **2021**, *11*, 3314. [CrossRef]
24. Xie, L.H.; Akao, T.; Hamasaki, K.; Deyama, T.; Hattori, M. Biotransformation of pinoresinol diglucoside to mammalian lignans by human intestinal microflora, and isolation of *Enterococcus faecalis* strain PDG-1 responsible for the transformation of (+)-pinoresinol to (+)-lariciresinol. *Chem. Pharm. Bull.* **2003**, *51*, 508–515. [CrossRef]
25. Kikuchi, M. Studies on the constituents of *Osmanthus* species. On the components of the leaves of *Osmanthus fragrans* Lour. var. *Aurantiacus makino*. *Yakugaku Zasshi* **1984**, *104*, 535–539. [CrossRef]
26. Global Invasive Species Database, Species Profile: Polygonum cuspidatum. Available online: http://www.iucngisd.org/gisd/speciesname/Polygonum+cuspidatum (accessed on 19 October 2022).
27. Kato-Noguchi, H. Allelopathy of knotweeds as invasive plants. *Plants* **2022**, *11*, 3. [CrossRef]
28. Bailey, J.P.; Bímová, K.; Mandák, B. The potential role of polyploidy and hybridisation in the further evolution of the highly invasive *Fallopia* taxa in Europe. *Ecol. Res.* **2007**, *22*, 920–928. [CrossRef]
29. Parepa, M.; Schaffner, U.; Bossdorf, O. Help from underground soil biota facilitate knotweed invasion. *Ecosphere* **2013**, *4*, 31. [CrossRef]

30. Gaskin, J.F.; Schwarzländer, M.; Grevstad, F.S.; Haverhals, M.A.; Bourchier, R.S.; Miller, T.W. Extreme differences in population structure and genetic diversity for three invasive congeners: Knotweeds in western North America. *Biol. Invasions* **2014**, *16*, 2127–2136. [CrossRef]
31. Mandák, B.; Pyšek, P.; Lysák, M.; Suda, J.; Krahulcová, A.; Bímová, K. Variation in DNA-ploidy levels of *Reynoutria* taxa in the Czech Republic. *Ann. Bot.* **2003**, *92*, 265–272. [CrossRef]
32. Mandák, B.; Pyšek, P.; Bímová, K. History of the invasion and distribution of *Reynoutria* taxa in the Czech Republic: A hybrid spreading faster than its parents. *Preslia* **2004**, *76*, 15–64.
33. Hiradate, S.; Ohse, K.; Furubayashi, A.; Fujii, Y. Quantitative evaluation of allelopathic potentials in soils: Total activity approach. *Weed Sci.* **2010**, *58*, 258–264. [CrossRef]
34. Ward, R.S. Recent advances in the chemistry of lignans. In *Studies in Natural Products Chemistry*; Attaur, R., Ed.; Elsevier: Amsterdam, The Netherlands, 2000; pp. 739–798.
35. Ferrer, J.L.; Austin, M.B.; Stewart, C., Jr.; Noel, J.P. Structure and function of enzymes involved in the biosynthesis of phenylpropanoids. *Plant Physiol. Biochem.* **2008**, *46*, 356–370. [CrossRef]
36. Patyra, A.; Kołtun-Jasion, M.; Jakubiak, O.; Kiss, A.K. Extraction techniques and analytical methods for isolation and characterization of lignans. *Plants* **2022**, *11*, 2323. [CrossRef]
37. Fuentes-Gandara, F.; Torres, A.; Fernández-Ponce, M.T.; Casas, L.; Mantell, C.; Varela, R.; Martínez de la Ossa-Fernández, E.J.; Macías, F.A. Selective fractionation and isolation of allelopathic compounds from *Helianthus annuus* L. leaves by means of high-pressure techniques. *J. Supercrit. Fluids* **2019**, *143*, 32–41. [CrossRef]
38. Jong, T.T.; Chau, S.W. Antioxidative activities of constituents isolated from *Pandanus odoratissimus*. *Phytochemistry* **1998**, *49*, 2145–2148. [CrossRef]
39. Hwang, B.; Lee, J.; Liu, Q.H.; Woo, E.R.; Lee, D.G. Antifungal effect of (+)-pinoresinol isolated from *Sambucus williamsii*. *Molecules* **2010**, *15*, 3507–3516. [CrossRef]
40. Kim, H.Y.; Kim, J.K.; Choi, J.H.; Jung, J.Y.; Oh, W.Y.; Kim, D.C.; Lee, H.S.; Kim, Y.S.; Kang, S.S.; Lee, S.H.; et al. Hepatoprotective effect of pinoresinol on carbon tetrachloride-induced hepatic damage in mice. *J. Pharmacol. Sci.* **2010**, *112*, 105–112. [CrossRef]
41. Kato-Noguchi, H.; Takeshita, S.; Kimura, F.; Ohno, O.; Suenaga, K. A novel allelopathic active substance in *Ginkgo biloba*. *J. Plant Physiol.* **2013**, *170*, 1595–1599. [CrossRef] [PubMed]
42. Kato-Noguchi, H.; Kimura, F.; Ohno, O.; Suenaga, K. Involvement of allelopathy in inhibition of understory growth in red pine forests. *J. Plant Physiol.* **2017**, *218*, 66–73. [CrossRef]
43. Kato-Noguchi, H. Bioactive compounds involved in the formation of the sparse understory vegetation in pine forests. *Curr. Org. Chem.* **2021**, *25*, 1731–1738. [CrossRef]
44. Sakamoto, S.; Machida, K.; Kikuchi, M. Lignan glycosides from the leaves of *Osmanthus heterophyllus*. *J. Nat. Med.* **2008**, *62*, 362–363. [CrossRef]
45. Inouye, H.; Uesato, S. Biosynthesis of iridoids and secoiridoids. In *Progress in the Chemistry of Organic Natural Products*; Herz, W., Grisebach, H., Kirby, G.W., Tamm, C., Eds.; Springer: Wean, Australia, 1986; Volume 50, pp. 169–226.
46. Geu-Flores, F.; Sherden, N.; Courdavault, V.; Glenn, W.S.; Wu, C.; Nims, E.; Cui, Y.; O'Connor, S.E. An alternative route to cyclic terpenes by reductive cyclization in iridoid biosynthesis. *Nature* **2012**, *492*, 138–142. [CrossRef]
47. Kikuchi, M.; Yamauchi, Y. Studies on the constituents of *Osmanthus* species III. On the components of the leaves of *Osmanthus ilicifolius* (Hassk.) Mouillefert. *Yakugaku Zasshi* **1986**, *105*, 442–448. [CrossRef] [PubMed]
48. Machida, K.; Sakamoto, S.; Kikuchi, M. Structure elucidation and NMR spectral assignments of four neolignan glycosides with enantiometric aglycones from *Osmanthus ilicifolius*. *Magn. Reson. Chem.* **2008**, *46*, 990–994. [CrossRef] [PubMed]
49. Sakamoto, S.; Machida, K.; Kikuchi, M. Secoiridoid di-glycosides from *Osmanthus ilicifolius*. *Heterocycles* **2009**, *77*, 557–563.
50. Matuda, Y.; Iwasaki, A.; Suenaga, K.; Kato-Noguchi, H. Allelopathy and allelopathic substances of fossil tree species *Metasequoia glyptostroboides*. *Agronomy* **2022**, *12*, 83. [CrossRef]

plants

MDPI

Article

Modulating Expression Levels of TCP Transcription Factors by *Mentha x piperita* Volatiles—An Allelopathic Tool to Influence Leaf Growth?

Matthias Preusche [1,2], Marvin Vahl [1], Johanna Riediger [1,2], Andreas Ulbrich [1] and Margot Schulz [2,*]

[1] Department of Horticultural Production, University of Applied Science, 49090 Osnabrück, Germany
[2] Institute of Molecular Physiology and Biotechnology of Plants (IMBIO), University of Bonn, 53127 Bonn, Germany
* Correspondence: ulp509@uni-bonn.de

Abstract: Peppermint (*Mentha x piperita*) is a species with inhibitory allelopathic properties due to its high amounts of terpenes. Recent studies have disclosed dosage dependent growth promotion or defense reactions in plants when facing appropriate amounts of *Mentha* bouquet terpenes. These positive effects could be of interest for agricultural applications. To obtain more insights into leaf growth modulations, the expression of *Arabidopsis and Brassica rapa* TCP transcription factors were studied after fumigation with *M. x piperita* bouquets (*Arabidopsis*), with *M. x piperita* essential oil or with limonene (*Arabidopsis* and Chinese cabbage). According to qPCR studies, expression of *TCP3*, *TCP24*, and *TCP20* were downregulated by all treatments in *Arabidopsis*, leading to altered leaf growth. Expressions of *B. rapa TCPs* after fumigation with the essential oil or limonene were less affected. Extensive greenhouse and polytunnel trials with white cabbage and *Mentha* plants showed that the developmental stage of the leaves, the dosage, and the fumigation time are of crucial importance for changed fresh and dry weights. Although further research is needed, the study may contribute to a more intensive utilization of ecologically friendly and species diversity conservation and positive allelopathic interactions in future agricultural systems.

Keywords: *Mentha x piperita* volatiles; *Brassica oleracea*; *Arabidopsis thaliana*; TCP transcription factors; leaf growth; fumigation

Citation: Preusche, M.; Vahl, M.; Riediger, J.; Ulbrich, A.; Schulz, M. Modulating Expression Levels of TCP Transcription Factors by *Mentha x piperita* Volatiles—An Allelopathic Tool to Influence Leaf Growth? *Plants* **2022**, *11*, 3078. https://doi.org/10.3390/plants11223078

Academic Editor: Lígia R. Salgueiro

Received: 12 October 2022
Accepted: 9 November 2022
Published: 14 November 2022

1. Introduction

Terpenoids, one of the largest groups of plant and microbe secondary metabolites, can positively or negatively impact seed germination and plant growth. Depending on the dosage, many terpenoids have phytotoxic properties [1]. For instance, farnesene, a sesquiterpene, influences cell division, abolishes gravitropism in *Arabidopsis*, and downregulates auxin polar transport proteins [2]. Monoterpenes are described as interacting with transcriptional activities, cell division, and cytoskeletal elements in fumigated receiver plants, which results in growth inhibition or even plant death [3–7]. Volatile bouquets emitted from aromatic plants, such as *Mentha x piperita*, affect the transpiration of neighbored *Arabidopsis* plants by inducing stomatal opening [7]. Therefore, terpenes are considered as lead structures for the design of new herbicides. However, the effects of terpenoids are dosage dependent. Consistently, low dosages can have plant growth promoting properties.

Mentha x piperita L. belongs to the Lamiaceae, a family with numerous species containing high amounts of essential oil with antifungal, bactericidal, insecticidal, and antifeedant properties. More than 50 constituents have been identified in the essential oil fraction [8], which can quantitatively and qualitatively vary from cultivar to cultivar and from leaf to leaf. The composition is determined by genetic factors, depending on culture methods, plant age and organ, microbial interactions, further interkingdom communication, and other biotic and abiotic parameters such as climate, geographical region, altitude, and irradiation. The most abundant compounds in the essential oil fraction and in the bouquets emitted by *Mentha x piperita* are limonene, linalool, menthone, menthol, and pulegone, while the concentration of single compounds in the emitted bouquets differ from the essential oil fraction of the leaves [9]. Limonene is a precursor of many other monoterpenoids, such as pulegone, menthol, and menthone, and its contents vary over the course of the season [10,11].

In a field study, Sukegawa et al. [9] reported the strong upregulation of *Glycine max* (soybean) defense genes in the leaves when plants are exposed to *Mentha x piperita* volatiles. They found a lower damage of *Brassica rapa* by herbivores when co-cultured with peppermint. We found in a previous study that the volatile bouquet of *Mentha x piperita* (English Mint cv. Mitcham) positively influenced leaf development in young white cabbage (*Brassica oleracea* convar. *capitata* var. *alba*) plants during a 27-day-fumigation, while the influence on leaf growth seemed to depend on responsiveness during early stages of leaf development. A changed leaf shape but no enhanced leaf growth or dry weight was seen when young plants were exposed to menthol or to a combination of menthol and menthone [12]. The observation that the youngest leaves were sensitive to the treatment led to the assumption that fumigation with the *Mentha* volatile bouquet entails modulations of fundamental processes in early leaf development, which results in an enhanced growth of the young leaf. Thus, the responsiveness of transcription factors, in particular of TCPs, to *Mentha x piperita* monoterpenes could be causative for the modulation of leaf growth and development.

The differentiation process in the developing leaf is regulated by a network of hormones and transcription factors, while TCPs (TEOSINTEBRANCHED1/CYCLOIDEA/PROLIFERATING CELL FACTOR) are of particular importance. The roles and interactions of TCPs have been studied intensively during the last few decades [13,14], but many aspects of the complicated, hierarchically organized network with counteracting and redundant functions in controlling plant developmental processes are still elusive [15,16].

The TCP family is composed of two different classes, class I and class II. In *Arabidopsis thaliana*, the class II subfamily is further divided into the subclasses CYC/TB1 and CIN. While being under the control of miRNA 319 (miR319), TCP2, TCP3, TCP4, TCP5, TCP10, TCP13, TCP17, and TCP24 regulate leaf development. The miR319 interaction site is located near the 3′ part of the coding region and not within the *TCP* domain, a characteristic region of the *TCPs* [17]. This TCP domain is necessary for protein–protein interactions [18]. The miR319 interaction site is of importance for class II TCP mRNA abundance, as downregulation of class II TCPs seems to be regulated by miRNA-mediated degradation of their mRNAs [19]. Thus, high levels of miR319 downregulate these TCPs with effects on leaf morphogenesis in *Arabidopsis* [20].

Another group of TCPs have genes lacking the miRNA319 target sequence. The non-miR319-regulated class I TCPs, for instance TCP20 and TCP9, inhibit lipoxygenase 2 gene (*LOX2*) expression, which is crucial for jasmonate biosynthesis. *LOX2* expression is thought to be regulated by the ratios of class I and class II TCP (TCP4) levels [21]. While members of class II TCPs inhibit growth cell division and promote aging, members of class I TCPs have the opposite functions by controlling the expression of specific target genes [22,23]. Non-miR319-regulated TCP20 and TCP9 are also reported to control senescence. TCPs of the same class form homo- or hetero-dimers, which increases the flexibility in interacting with other proteins [19]. The TCP/miRNA interactions, the complex, not yet completely understood regulation processes by TCPs, their known target genes, as well as interaction with plant hormones and jasmonate synthesis, influences of light, temperature, and further factors on TCPs, have recently been reviewed [13].

In this study, the influence of *Mentha* volatiles on the gene expressions of CIN branch TCPs (TCP2, TCP3, TCP4, TCP10, and TCP24), the non-miR319-regulated class 1 TCP 20 and TCP9, and *LOX2* were investigated using the model plant *Arabidopsis thaliana*. These studies were performed to obtain clues concerning the essential interferences of the volatiles with fundamental processes in leaf development after 1, 3, and 7 days of fumigation. The expression of well-known TCP genes of *Brassica rapa* ssp. *Pekinensis*, [24,25] was studied after 7 days of fumigation. The major part of our study deals with the impacts of *Mentha piperita* bouquet volatiles on white cabbage plants in greenhouse and polytunnel trials to reveal conditions that may allow an ecologically acceptable utilization of a positive allelopathic interaction of *Mentha* bouquet terpenoids for the modulation of leaf development in horticulture.

2. Results

2.1. Effects of Fumigations on the Arabidopsis Phenotype

Fumigation with one *Mentha* plant of 10 shoots and 3 nodes or with larger plants possessing 5 nodes in the growth chamber severely altered the phenotype of *Arabidopsis* plants in comparison to the control plants. In the presence of the smaller *Mentha* plants, *Arabidopsis* developed larger rosettes with longer narrow leaves and longer petioles. A similar result was obtained with *Mentha* essential oil. Limonene led to longer petioles, whereas the leaf shape was not changed. With larger *Mentha* plants, the growth of *Arabidopsis* was heavily reduced (Figure 1), indicating a strong, negative influence of higher volatile concentrations. The negative growth effects are in agreement with the findings that many terpenes have phytotoxic properties when applied in high concentrations [1].

Figure 1. Effects of volatile bouquets emitted from different sized *Mentha* plants on the phenotype of *Arabidopsis thaliana* Col-0. (**A**): *Mentha* plant of 10 shoots and 3 nodes; (**B**): control; (**C**): *Mentha* plants of 10 shoots and 5 nodes. Fumigation time was one week.

2.2. Determinations of White and Chinese Cabbage Fresh and Dry Weights after Fumigation with Limonene and Mentha Essential Oil

Fresh and dry weights of the above-ground parts of white and Chinese cabbage plants were determined after fumigation with limonene or *Mentha* essential oil for one week to estimate threshold ranges of terpenoid concentrations causing inhibitory effects on cabbage growth. In contrast to *Arabidopsis*, no striking alterations of the phenotypes were observed. Nevertheless, white cabbage developed a significantly higher fresh and dry weight with limonene and a higher dry weight with *Mentha* essential oil, but a slightly reduced fresh weight (Figure 2). Regarding the fresh weight, Chinese cabbage was less responsive than white cabbage to limonene, although dry weight was also significantly enhanced. Here, *Mentha* essential oil increased fresh weight with high significance only with the 16 mL dosage.

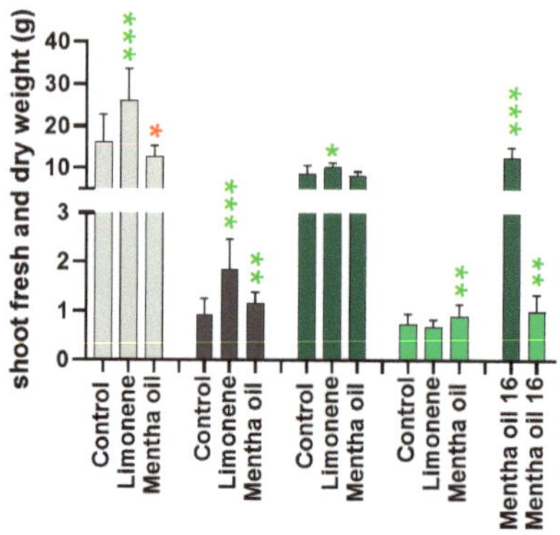

Figure 2. Fresh (FW) and dry weight (DW) of above-ground plant material after 1-week-fumigation in the greenhouse with high concentrations of limonene (8 mL/day) and *Mentha* essential oil (14 or 16 mL/day) and controls. White cabbage: light grey (FW), dark grey (DW); Chinese cabbage: dark green (FW), green (DW). *Mentha* oil 16 indicates 16 mL/day, for all other treatments 14 mL/day were used. Significance: * $p < 0.05$; ** $p < 0.005$; *** $p < 0.0005$), green asterisks: increase, red asterisk: reduction.

2.3. TCP Transcription Factors—Targets of Mentha x piperita Emitted Volatiles, Essential Oil Fraction, and Limonene

Because it was intended to address the growth promotion and phenotype modeling of *Arabidopsis* by *Mentha* volatile bouquets, TCP gene expressions were investigated with plants fumigated with the smaller *Mentha* plants. qPCR analysis with cDNA from growth-promoted *Arabidopsis* plants revealed a strong downregulation of *TCP2* and *TCP3* transcripts, most exhibited after 3 days of fumigation, and a downregulation of *TCP4*, strongest at day 7, while *TCP 9*, *TCP10*, *TCP 20*, and *TCP24* transcript levels showed a much weaker response and were only slightly, if at all, downregulated (Figure 3). The *LOX2* gene expression was ambivalent but tended towards an upregulation at day 7 without significance. Fumigation with the essential oil fraction led to similar results, i.e., a downregulation of *TCP2* transcripts after 1 and 3 day(s) and a slight upregulation at day 7. *TCP3* mRNA was downregulated at all time points. A weaker response of *TCP4* transcripts was stated, with the highest downregulation at day 3. Again, *TCP 9* and *TCP10* transcripts were only slightly affected, with a low upregulation of *TCP9* at day 1 and of *TCP10* at day 7, while *TCP20* and in particular *TCP24* were more strongly downregulated than was found after *Mentha* bouquet fumigation. The *LOX2* transcripts were downregulated. Using limonene for fumigation resulted in similar transcript profiles for *TCP3*, *TCP9*, and *TCP20*; *TCP10* was downregulated, but *TCP2* and *TCP4* were almost not affected. The downregulation of *TCP24* was between the one found for the bouquet and essential oil fumigation. Thus, fumigation with limonene had almost no effect on transcripts of *TCP2* and *TCP4*, which were responsive to the complex mixture of the volatiles present in the bouquet and the essential oil. The *LOX2* transcript level was similarly affected, as found after the essential oil fumigation. Most of the expression profiles showed a dependence on the fumigation time, in particular *TCP2*, *TCP3*, *TCP4*, and *LOX2* transcript abundances.

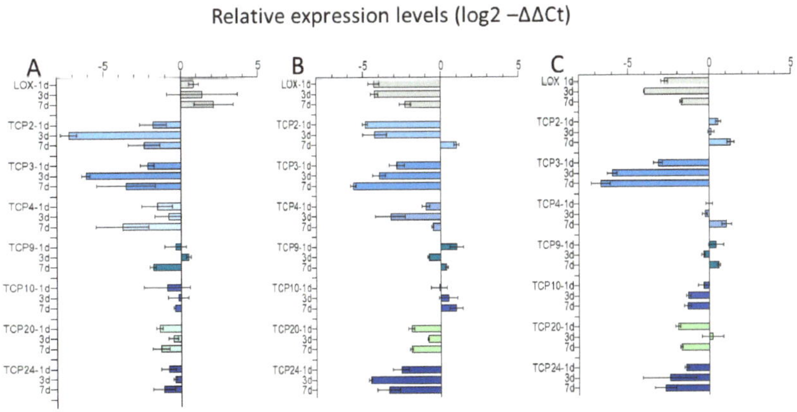

Figure 3. TCP gene expression in *Arabidopsis* is modulated by *Mentha* volatiles. Expression of class II *TCP2, TCP3, TCP4, TCP5, TCP10, TCP13, TCP17, TCP24*, and class I *TCP20* and *TCP9* genes, *LOX2* (LOX) after the different types of fumigation. **A**: *Mentha* plant; **B**: *Mentha* essential oil; **C**: limonene. The figure shows mean values and standard error of means (n = 2 samples, each sample prepared from 43–51 plants per time point).

2.3.1. Expression of TCP Transcription Factors in Chinese Cabbage after Exposure to Mentha Essential Oil Fraction and Limonene

In contrast to *Arabidopsis thaliana*, Chinese cabbage has two *BraTCP2*, one *BraTCP3*, three *BraTCP4*, one *BraTCP10*, and two *BraTCP24* genes [24,25]. When fumigated with the *Mentha* essential oil fraction or limonene, expressions of the tested *Brassica rapa* ssp. *Pekinensis* TCP transcription factors were considerably less affected than the *TCPs* of *Arabidopsis*. With limonene, all *TCPs* were downregulated in the rage of −0.5 to −2fold, except for *TCP2a*, which was slightly upregulated, and *TCP24a*, which was not responsive. With the essential oil fraction, the only downregulated TCP was *TCP4a* (−3 fold), while *TCP4b* was 4-fold, but *TCP4c* was only 0.5-fold upregulated (Figure 4). The latter range of upregulation was also found for *TCP2a*, *TCP3*, *TCP9a*, *TCP10*, and *TCP20b*. *TCP24b* was not responsive, and *TCP2b* showed a 2-fold, *TCP9b* and *TCP24a* each showed a 1-fold, and *TCP20a* showed an approximate 2-fold upregulation.

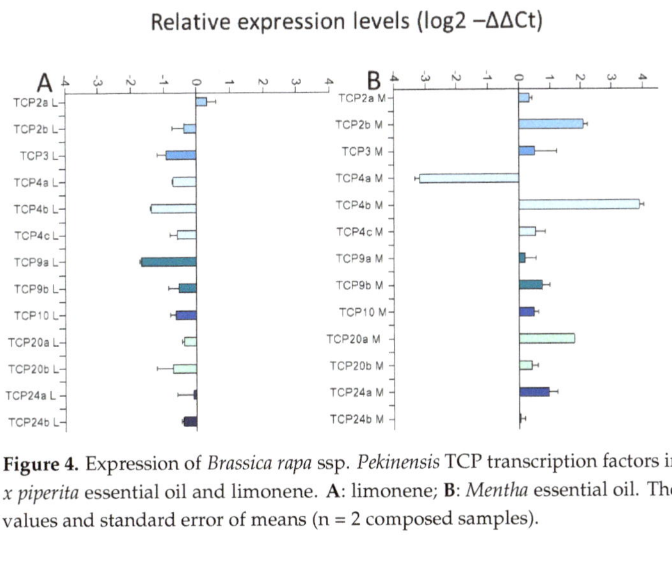

Figure 4. Expression of *Brassica rapa* ssp. *Pekinensis* TCP transcription factors in response to *Mentha x piperita* essential oil and limonene. **A**: limonene; **B**: *Mentha* essential oil. The figure shows mean values and standard error of means (n = 2 composed samples).

2.3.2. Choice of White Cabbage for Greenhouse and Polytunnel Trials

In turnips (*Brassica rapa* ssp. *rapa*), *Brr/BraTCP2* is involved in leaf development by regulating cell proliferation [25]. According to Efroni et al. [26], CIN-class TCPs have crucial functions, such as being negative regulators of cell proliferation, therefore causing leaf differentiation [24]. Figures 2 and 4 illustrate that the fumigation with limonene and *Mentha x piperita* essential oil have different impacts on (1) leaf fresh and dry weight and (2) TCP gene expressions. With Chinese cabbage, the study with *Mentha* essential oil disclosed that head shape modulating TCP4a was the only downregulated TCP gene; all others were upregulated or not responsive. Because downregulation of the CIN-class TCP abundance is a prerequisite for cell proliferation, enhanced leaf growth of Chinese cabbage may not occur when the plants were fumigated with *Mentha* plants as the fresh and dry weight determinations after limonene and *Mentha* essential oil fumigation indicated that Chinese cabbage needs higher volatile concentrations than white cabbage to achieve growth responses. Thus, the concentrations of *Mentha* bouquet compounds may not be high enough for modulating Chinese cabbage leaf growth, but could be high enough for white cabbage. Therefore, and by considering the results of the former pilot study [12], fumigation with *Mentha* plants in greenhouse and polytunnel experiments were performed with white cabbage and not with Chinese cabbage. The responsiveness of TCP transcription factors towards *Mentha* volatiles found for *Arabidopsis* was assumed to be similar in white cabbage.

2.4. Greenhouse and Polytunnel Trials with Brassica oleracea convar. capitata var. alba and Mentha plants

Greenhouse Trial—Leaf Length and Width, Growth Increments

Leaf length and width were measured after 1 (T1), 4 (T2), 8 (T3), 11 (T4), and 15 days (T5) after exposure to *Mentha* volatiles emitted by one or four plants (Figure S1). At T1 and T2, where only the first and the emerging (T1) or the further developed (T2) second leaves were present, growth showed no significant differences to the controls. At T3 (8 days), the difference in leaf length of control plants and those exposed to four *Mentha* plants (4M) was significant larger for leaf 2 when the distance to the *Mentha* plants was more than 35 cm. Leaf 1 was not responsive and leaf 3 tended towards an increased length growth but without significance. Leaf width was not affected at T3. The largest differences were seen with leaf 2 and leaf 4, but particularly with leaf 3 after 15 days (T5) when plants were exposed to four *Mentha* plants when the distance to the *Mentha* plants was more than 20 cm. Thus, the number of *Mentha* plants, dependence on the distance, and the dosage are crucial for leaf parameter modulations. The results indicate again that defined developmental stages of a leaf in combination with the dosage of the volatiles are important for the response to the treatments. To summarize, only the second and third leaves at T3, T4, and T5, and also the fourth leaf, clearly responded to the treatment with bouquet volatiles emitted from four *Mentha* plants. The highest increases in leaf length and width were obtained at distances exceeding 35 cm.

Measurements of the leaf increments support the observations (Figure 5). Compared to the controls, leaf growth increments show that the responses of leaves 2, 3, and 4 to the *Mentha* volatiles are significantly higher between 11 and 15 days (T4–T5) of fumigation when the distance to four *Mentha* plants was more than 25 cm. Increments of fumigated leaves at T1–T2, except for leaf 2, and T2–T3 were similar to the control values. The increments of leaf 2 were continuously larger after 8 days of fumigation but not during the early phase of the treatment.

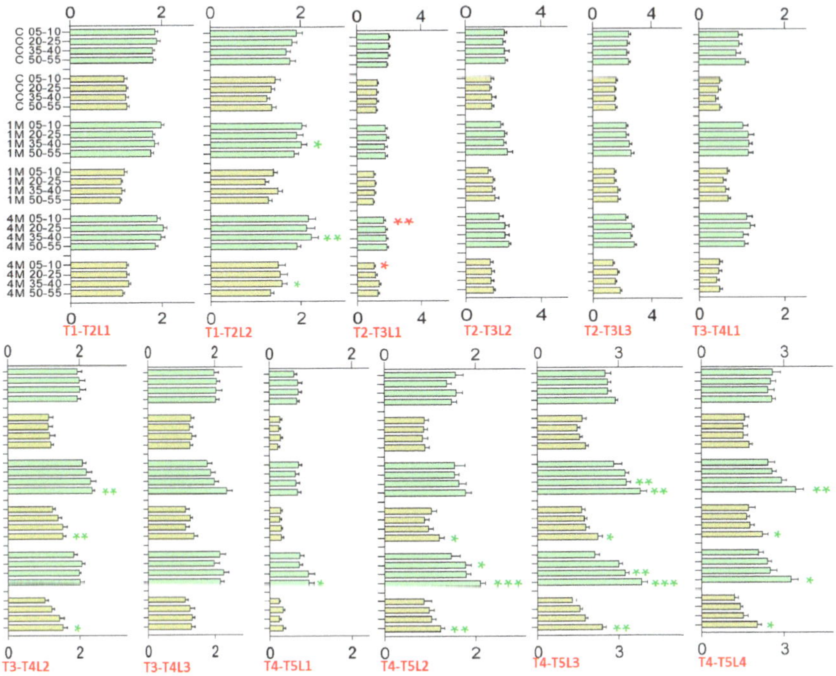

Figure 5. Greenhouse Trial: Increments (cm y-axis, consider graph right 90° rotation) in length (green) and width (olive) of the first two to four leaves (T1–T2, T2–T3, T3–T4, and T4–T5) at four difference distances (0.5–10, 20–25, 35–40 and 50–55 cm, x-axis, consider graph right 90° rotation) in control plants (no *Mentha*), plants exposed to one *Mentha* (1 M), and plants exposed to four *Mentha* (4 M) plants. Green asterisks indicate significantly increased growth, red asterisk significant lower growth compared to the controls, bars without asterisk: no significance. Enhanced increments occurred predominantly during the late period of fumigation with leaf 2 (T3–T4) and with leaves 2, 3, and 4 at T4–T5. The figure shows mean values and standard error of means (n = 16). Significance with reference to the appropriate cm control (no fumigation), (*t*-test): * $p < 0.05$; ** $p < 0.005$; *** $p < 0.0005$). Red labels give the increment (T1–T2, T2–T3, T3–T4, and T4–T5) and the leaf number (L1, L2, L3, L4).

2.5. Polytunnel Trial—Leaf Length and Width, Growth Increments

Leaf length and width were measured after 1 (T1), 4 (T2), 8 (T3), 12 (T4), and 15 days (T5) after exposure of the plants to volatiles emitted by four or nine *Mentha* plants. At all time-points, leaf 1 was comparable to the control, thus without response to the treatment. At T2, T3, and T4, the length growth of leaf 2 and, to a lesser degree, leaf 3 were enhanced by the bouquets of four and nine mint plants at both distances. At T5, the growth promoting effect of the volatiles was almost halted in leaf 2 and leaf 3. Leaf width was enhanced only at T3 with high significance and again only in leaf 2 and leaf 3. Thus, for leaf 2 and leaf 3, length growth was more affected than width growth and overall responsiveness, mainly restricted to leaf 2 and leaf 3, occurred between day 4 and day 12 of the treatment (Figure S2).

The increments of leaf growth, depictured in Figure 6, illustrate the dynamics of leaf growth more distinctly. Increments of leaves 2 and 3 occurred mainly between T1–T2 (leaf 2) and T2–T3 (both leaves), while at T3–T4, only four mints had a positive effect on leaf length at both distances. Leaf width did not increase and was even reduced at the short distance with nine *Mentha* plants. The reduction in increments, compared to control values or even lower values, continued at T4–T5 for all leaves (except for leaf 4) with four mint plants at the larger distance (60–75 cm). The increments demonstrate that the growth promoting effect of *Mentha* bouquet affected leaves in defined developmental stages only temporarily. No growth promotion was found with leaves that were developed during the late period of the treatment (T5, leaves 4, 5, and 6).

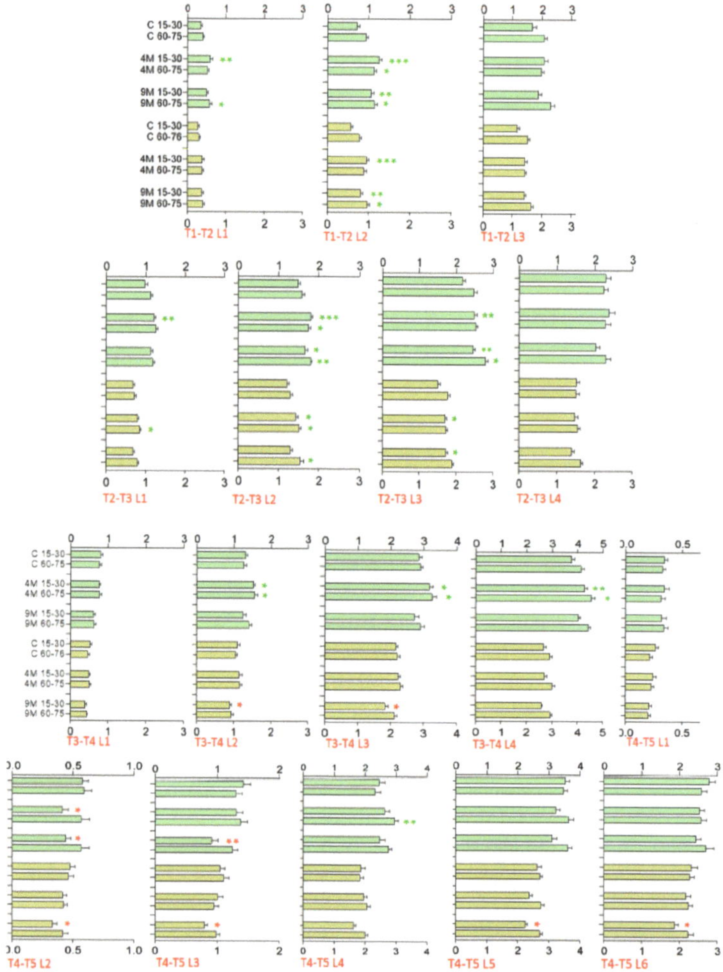

Figure 6. Polytunnel Trial: Consider graph right 90° rotation. Increments in length (green) and width (olive) of the first three to four leaves (T1–T2, T2–T3, T3–T4, T4–T5) at two difference distances (15–30 cm and 60–75 cm) in control plants (no *Mentha*), plants exposed to four *Mentha* (4 M), and plants exposed to nine *Mentha* (9 M) plants. Green asterisks indicate significantly increased growth, red asterisks indicates significantly lower growth compared to the controls, bars without asterisk: no significance; n= 24. Enhanced growth is primarily found during the early period of fumigation with leaves 2 and 3, while during the late period, a significantly inhibited or at least a reduced growth of these leaves was predominately seen. Significance with reference to the appropriate cm control (no fumigation), (*t*-test): * $p < 0.05$; ** $p < 0.005$; *** $p < 0.0005$). Red labels give the increment (T1–T2, T2–T3, T3–T4, and T4–T5) and the leaf number (L1, L2, L3, L4, L5, L6).

2.6. Greenhous and Polytunnel Trials—Fresh and Dry Weight

The fresh and dry weight of the above-ground plant material was determined after termination of the fumigation at day 21. Fresh and dry weights of the cabbage plants cultured in the greenhouse were enhanced when the distance to one mint was more than 35 cm; with four mint plants, more than 50 cm were necessary to achieve a similar increase

(Figure 7). Four *Mentha* plants led to lower dry masses at the smaller distances. Fresh and dry weights of the entire above-ground plant material mirror the overall growth, while increases must be attributed mainly to the second and third leaves. Nevertheless, the total increase of above-ground biomass fresh weight was in the range of 20% (calculated from mean values) of the above controls for plants growing with a 35–55 cm distance to one *Mentha* plant. The total dry weight was lower than control values, with the exception of the plants growing with a distance of 55 cm to one *Mentha* plant. Here, the increase in dry weight was approximately 10%.

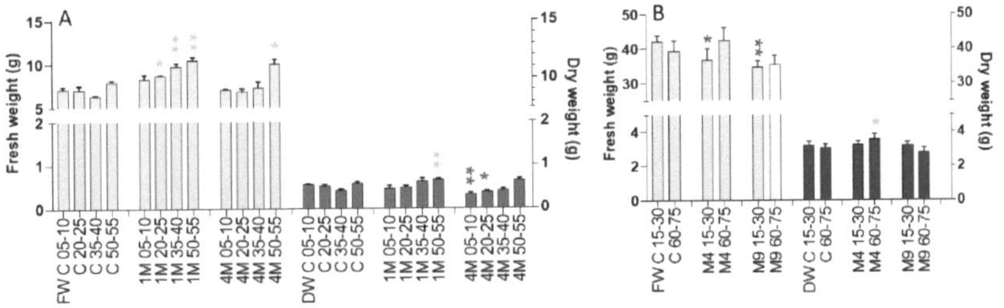

Figure 7. (**A**): Greenhouse trial: Fresh and dry weights after termination of the treatment (means + standard error of means). (**B**): Polytunnel Trial: Fresh and dry weights after termination of treatment (means + standard error of means). Light grey bars: fresh weight; dark grey bark dry weight. Green asterisks indicate increase, red asterisk inhibition, significance levels: (*t*-test): * $p < 0.05$; ** $p < 0.005$); bars without asterisk: no significance. Notice that the number of plants used for fresh and dry weight determinations was different compared to the greenhouse trial, due to the different arrangements.

As a consequence of the more or less stagnating leaf growth during the late period of the treatment (T5, 15 days) in the polytunnel trial, no enhanced fresh and dry weights of the above-ground plant material was found after treatment termination (21 days) (Figure 7). The cabbage plants cultured in the polytunnel developed more leaves compared to the plants in the greenhouse, due to the different culture conditions (pre-culture in the greenhouse in pots prior the planting in the polytunnel soil). Leaves 7 and 8 were very small and did not considerably contribute to fresh and dry weights. Because of their small size, these leaves were not evaluated further. Taken together, a higher biomass due to the volatile promoted growth of leaves 2 and 3 at T2, T3, and T4 was not a decisive factor when the fresh and dry weights of the above-ground biomass (Figure 7) was determined after 21 days. In comparison to the greenhouse trial with the strongest effects between 11 and 15 days (T4–T5), growth promoting effects occurred much earlier, namely between 4 and 12 days, when the plants were cultured in the polytunnel with 4 and 9 *Mentha* plants and was characterized by a subsequent decline (Figures 5 and 6).

3. Discussion

Presently, approaches to integrate positive allelopathic interactions in agricultural practices are rarely addressed, although the positive effects of many allelochemicals when applied in low concentrations are well known. This study disclosed some of the difficulties associated with the suitable application and maintenance of plant growth promoting concentrations of *Mentha x piperita* emitted volatiles and, on the other side, the dynamics in responses of the receiver plants.

The involvement of TCPs indicates that dosage dependent responsiveness to *Mentha* volatiles starts on superordinated levels that regulate the leaf growth process and is embedded in the entire life cycle of the plant. Causative molecular interactions that lead to modulations of transcriptional activities, resulting perhaps in shifts of the ratios of class I and class II TCP levels, could not be elucidated in this study, nor could the responsiveness of TCP target genes and hormone interactions or levels of miR319. Nevertheless, for a possible interpretation of the results, we would like to address some of the far-reaching downstream events of gene activity cascades that are presumably linked to *Mentha* volatile induced modulations of TCP transcript abundances.

Treatments with limonene, *Mentha* essential oil, and *Mentha* bouquet all downregulated *TCP3*, *TCP24* (both class II CIN branch TCPs), and *TCP20*, a class I non-miR319-regulated TCP in *Arabidopsis*. TCP3 negatively affects auxin responses [27]. Auxin is known to be crucial in leaf development for its functions in leaf initiation and influences on leaf size, shape, and form [28,29]. The effects of TCP3 on the auxin response and on flavonoid biosynthesis have been investigated by Li and Zachgo [30]. Thus, downregulation of TCP3 expression might have reverse effects on auxin triggered processes that are under the control of TCP3. Downregulation of TCP3 by *Mentha* volatiles should therefore influence effects linked with auxin responses. *TCP20* is another partner in the auxin network as it regulates the expression of *GH3.3*. GH3s are known to control the formation of IAA-amino acid conjugates, which are important in maintaining auxin homeostasis [31]. In our study, downregulation of TCP20 gene expression was, however, found to be less severe than that of *TCP3*. *TCP24* suppresses secondary wall thickening [32]; thus, a downregulation of *TCP24* transcription can lead to enhanced secondary cell wall formation, which would be in line with the increased dry weights of white cabbage after fumigation with high amounts of *Mentha* volatiles. Another group of transcription factors (growth-regulating factors (GRFs), which is under the control of miR396 [33], might also be affected, but the study of their expression profiles after *Mentha* fumigation remains to be carried out.

The results indicate that the positive effects of *Mentha* volatiles on leaf development probably proceed first from the interactions of volatile components with TCPs. Growth stimulation may later be due to interactions with the auxin network. Other downstream interactions with hormones and proteins are presumable, but further investigations must prove the assumption. The many antagonistic and synergistic interactions and functional redundancies raise difficulties in unambiguous assignments of the observed effects to defined developmental pathways. In addition, alterations in downstream events, performed as counteracting steps by the plant or its associated microorganisms, for instance monoterpene detoxification, may intercept responses to the volatiles.

Down regulation of the gene expression of other transcription factors, such as MYC-2, ANAC, and SCR-SHR, has been found after the application of the monoterpene citral. A binding of citral to these and further transcription factors (SSBPs WHY1, WHY2, and WHY3) was concluded from in silico molecular docking analyses [3]. Godard et al. [34] reported myrcene or ocimene induced expression of transcription factors, stress, and defense genes in *Arabidopsis*. The gene expression, at least of several plant transcription factors but also of some genes involved in defense, is apparently sensitive to terpenoid volatiles. Thus, it is possible that interactions of terpenoids with transcription factors and perhaps with many other proteins are common features. With MYC-2, ANAC, SCR-SHR, SSBPs WHY1, WHY2, and WHY3, the molecular backgrounds of the modulation are also not yet understood. Sukegawa et al. [11] concluded that there was a fumigation-time dependent epigenetic regulation of the JA- and SA-signaling-dependent defense genes TI and PR1 by *Mentha x piperita* volatiles in soybean. Here, the transient chromatin acetylation levels of the regulatory promoter region of these genes were immediately changed after starting the fumigation [11]. It cannot be discounted that epigenetic regulation of volatile binding is important at certain stages of the plant's life cycle, allowing defined gene responses or hindering them, leading to different gene expression profiles and finally to other protein interactions [35]. Some TCPs also seem to be under circadian clock control [36]; thus, transcriptional responses upon fumigation may succumb to diurnal variations. Transient changes in the gene expressions of the TCPs, in combination with the leaf life cycle and the abundance of miR319, and perhaps also epigenetic regulations, could be one reason for the decrease of the growth promoting effect of *Mentha* volatiles found with the plants grown in the polytunnel during the late phase of fumigation and may explain the failed growth stimulation in oldest leaves.

Further explanations are given by the dosage effects. The greenhouse and the polytunnel trials indicate the possibility of transient TCP-involved interactions. Concentrations of volatiles able to elicit growth effects were reached earlier with a higher number of *Mentha* plants, and thus the threshold concentration necessary for leaf growth induction in the greenhouse is later in place. Leaf surface can be influential, particularly wax layers. Monoterpenoids have been found to precipitate on wax layers and permeate them subsequently [7]; thus, local concentrations on the leaf may be spotted and increase during the course of fumigation. Because wax deposition on cabbage leaves was found to increase with aging [37], inhibitory amounts of terpenoids can finally accumulate in the wax layer, serving as a local source for the release of the molecules that infiltrate leaf cells underneath. We assume that plants with a thinner wax layer than cabbage, such as *Arabidopsis*, are more quickly affected by low monoterpene concentrations, which would explain the rapidly altered expression profiles of sensitive TCPs. More for cabbage than for *Arabidopsis*, the shape of the shoot and the positioning of the leaves are properties that influence the receiving of volatiles by leaves. Nevertheless, with the exception of *TCP24*, the expression profiles of affected TCPs were surprisingly similar in *Arabidopsis* when fumigated with small *Mentha* plants or with *Mentha* essential oil. Are some of the volatiles better trapped in the wax layers than others? In this case, subsequently different TCP gene expression profiles in cabbage would occur, if their expression is compound specific. For volatile trapping, the structure of the wax crystals, the wax composition, and immersed secondary metabolites, for instance flavonoids, may be of importance for species-dependent differences in the responsiveness to *Mentha* volatiles. In addition, many plant species have low quantities of di- and triterpenes, sometimes rare ones, deposited in their wax layers, including *Arabidopsis* and *Brassica oleracea* [38–40], indicating that wax layers are suitable sites for terpenoid deposition. At present, it is speculative to assume an involvement of TCPs, perhaps TCP4, in wax layer modulations and functions.

On the other hand, the amounts of volatiles emitted by the *Mentha* plants can vary as the terpenoid biosynthesis is modulated by many factors such as season, temperature, wind, leaf and whole plant age, growth of the plant, and, as important modulators, by soil and plant microbiomes. The temperature dependency of *Mentha x piperita* essential oil composition is known [41]. Menthol and menthone decrease at high temperature, menthylacetate and pulegone increase, and others, such as limonene, are not affected. Consistently, the maintenance of identical volatile compositions over time is unlikely. Altered volatile compositions can be an additional, important reason for shaping the leaf responsiveness. Alterations should be more presumable in the polytunnel trial, with consequences for the leaf growth, because, in this trial, temperature was not controlled.

In summary, the study disclosed the leaf growth promoting and dosage dependent effect of *Mentha* volatiles in cabbage and *Arabidopsis* at defined developmental stages. Greenhouse cultures may be more suitable for utilizing the positive effects of *Mentha* volatiles on the leaf growth of young cabbage plants, and also on other important vegetables. The maintenance of optimal volatile concentrations over time is, however, an essential problem as the concentration depends on air circulation and temperature as well as alterations in the bouquet composition, and is also due to the different chemical properties, such as the volatility, of the constituents.

Although optimization of culture and fumigation conditions are expected to be species dependent and need further extensive research, the ecologically and environmentally friendly utilization of allelopathic volatiles in suitable concentrations for leaf growth promotion seems to be feasible, when harvest times are adapted. The results presented here are in line with our former pilot study [12]. Moreover, the growth promoting effects of *Mentha* species, when applied in low amounts, have been described by Islam and Kato-Noguchi [42]. Finally, the use of plant volatile bouquets for plant growth promoting effects does not have the potential risk to animal health and the environment as may occur with high amounts of terpenoids [43].

4. Material and methods

4.1. Plant Material

4.1.1. *Mentha x piperita* Variety English Mint cv. "Mitcham"

For the fumigation of receiver plants, *Mentha x piperita* L. variety English Mint cv. "Mitcham" was used. The young plants were purchased from a local gardener and grown in a greenhouse (20/16 °C day/night) for approximately one and a half month until the start of the experiments. Two weeks after delivery, *Mentha* plants were propagated form cuttings. The uniformly-sized *Mentha* plants were grown for one month under the same conditions as described. The *Mentha* plants were cut again one week before being grown (4 main shoots with 5 nodes each, if not otherwise noted).

4.1.2. *Arabidopsis thaliana* Col-0

Seeds of *A. thaliana* Col-0 were surface sterilized and germinated in petri dishes with standard MS media and 1% sucrose (Murashige & Skoog, without any additives) under sterile conditions. After two days of stratification at 4 °C, the plates were transferred into a growth chamber for germination and growth at 23 °C, 16/8 h day/night. After 10 days, the young seedlings were transferred into small pots (Ø 5 cm) filled with water-saturated soil ("Seedlingsubstrat" from Klaasmann-Deilmann, Germany). Treatments were started one day after pricking.

4.1.3. *Brassica oleracea* cv. *capitata* var. *alba* convar. LENNOX F1 (white cabbage) and *Brassica rapa* (Chinese cabbage)

White cabbages (*Brassica oleracea* cv. *capitata* var. *alba* convar. LENNOX F1, Bejo Germany) were grown as described in [12]. In short, cabbages were sown in pressed peat pots (4.5 × 4 × 5 cm, cf. Figure 1) and germinated under 16/10 °C day/night conditions in a greenhouse and grown for 14–21 days until the second leaf was approximately 1.5–2 cm long. During this period, the plants were watered once a day. The substrate used for germination and growth was "Potground P" (Klaasmann-Deilmann, Germany). Chinese cabbages (*Brassica rapa* ssp. *Pekinensis*, Bilko F1, Bejo, Germany) were similarly cultured.

4.2. Fumigation Experiments

4.2.1. Fumigation of *Arabidopsis thaliana* with *Mentha x piperita* Mitcham

For determination of *TCP* gene expression profiles, standardized *Mentha* plants were used. *A. thaliana* seedlings were co-cultured with one *Mentha* plant of 10 shoots and 3 nodes or with larger plants possessing 5 nodes. *A. thaliana* control groups were cultured without *Mentha* plants. These pre-experiments revealed the growth promoting and phenotype modeling properties of smaller *Mentha* plants, while the larger plants led to growth inhibition of the co-cultured *Arabidopsis* plants (Figure 1). The following experiments were therefore performed with the smaller *Mentha* plants. Co-culturing with *A. thaliana* after one day of acclimation was performed for 1, 3, and 7 days in the growth chamber.

4.2.2. Fumigation of *A. thaliana* and *Brassica* plants with Mentha essential oil and limonene for Gene Expression Experiments

A. thaliana plants were treated with steam distillates of English mint essential oil, free of organic solvents (Norfolk Essential Oils, Wisbech, UK) or limonene (Merck Chemicals GmbH, Darmstadt, Germany; companies' information regarding the products are given in the Supplementary File). Limonene is one of the known components of the *Mentha* essential oil fractions (in the used essential oil charge: 2.4%) and also of the plant-emitted bouquet. Limonene was also separately applied, because it is the precursor of further monoterpenes. The treatment was performed for one week by injecting the essential oil fraction or limonene into cotton balls swinging approximately 5 cm above the plants (Figures 8 and 9).

Figure 8. One-week-fumigation of cabbage and *Arabidopsis* for gene expression experiments. (**A**): Cabbage plants were fumigated with volatile-loaded cotton balls. Controls were with volatile-free cotton balls. (**B**): Fumigation arrangement of *Arabidopsis* with one *Mentha* plant in the growth chamber. (**C**): Control, (**D**): Limonene-fumigated *Arabidopsis* after 7 days. (**E**,**F**): Harvested shoot material for gene expression studies. The innermost parts of the *Arabidopsis* rosettes and shoot apex containing material of cabbage plants (white rings) were harvested and used for RNA extraction. Pictures (**E**) and (**F**) show control plants. (**E**): *Arabidopsis*; (**F**): Chinese cabbage.

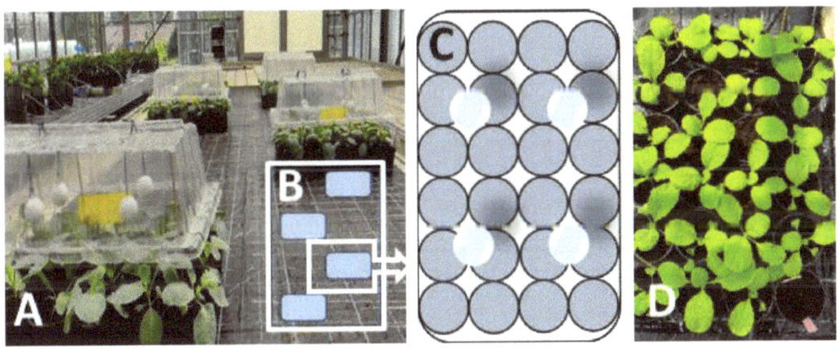

Figure 9. Experimental setup of cabbage plants in 4 culture boxes for limonene and *Mentha* essential oil fumigations in the greenhouse. Distances between culture boxes: 60 cm. (**A**,**C**): plants and cotton balls in one box ((**B**): inserted scheme in the picture of culture box arrangements in the greenhouse). (**D**): cabbage plants used for fresh and dry weight determination.

These arrangements allow efficient fumigation of the plants with the terpenoids emitted from the cotton balls. Injection was carried out at the start of the experiment and after 3½ days. In total, 4 mL of the essential oil fraction or limonene were applied to *Arabidopsis* plants per week.

4.3. Real Time PCR

Expression studies were performed to reveal the alterations of *TCP* transcript abundances during the first week of fumigation. The material for qPCR were obtained by harvesting the innermost part of the *Arabidopsis* rosettes with the stem apex and emerging leaves. Each sample comprised 43–51 individual plants per time point, i.e., 1, 3, and 7 days after co-cultivation with *Mentha* plants, application of *Mentha* essential oil and limonene, respectively, and the corresponding controls. The plant material was collected in liquid nitrogen and stored at −80 °C, if not immediately used for RNA extraction. Material from *Brassica rapa* (Chinese cabbage) was likewise harvested after 7 days of fumigation with *Mentha* essential oil or with limonene (Figure 8).

Two replicates were obtained at all three dates, resulting in 6 *Arabidopsis* samples for the treated and control groups. For mRNA preparation, 100 mg of plant material was used. RNA was extracted with the NucleoSpin® RNA Plant" extraction kit (Macherey and Nagel, Düren, Germany) according to the manufacturer's instructions. RNA quantity was determined by NanoDrop measurements. For cDNA synthesis, the RevertAid First Strand cDNA Synthesis Kit (Thermo Fisher Scientific) was utilized according to the manufacturer's protocol.

Real time PCR was performed as described in [44], with the qPCR-Mix "my-Budget 5× EvaGreen® QPCR-Mix II (ROX)" (Bio-Budget Technologies GmbH, Krefeld, Germany) and the thermocycler Applied Biosystems 7300 Real Time PCR System (Thermo Fisher Scientific, München, Germany). For all repetition setups with *Arabidopsis* and for all experiments with Chinese cabbage, the thermocycler CFX Connect (Bio-Rad, Feldkirchen, Germany) was used. Quantitative analysis was carried out for *TCP2, TCP3, TCP4, TCP9, TCP10, TCP20, TCP24,* and *LOX2* expression. Actin (*ACT2*) was used as a reference gene. Primers were synthesized by Eurofins Genomics (Ebersberg, Germany) and are listed in Table S1 [24,45]. The qPCR reaction mixtures (20 µL) contained µL 5× EvaGreen® QPCR-Mix II (ROX, Bio-Budget Technologies GmbH, Krefeld, Germany), 5 µL cDNA sample (15.52 ng/µL), 1 µL forward and reverse Primer Mix (1:1, v/v; 10 µM each), and 10 µL DNase free H_2O. For evaluation of RT-qPCR results, the ΔΔCT method was applied.

4.4. Determination of Cabbage Dry and Fresh Weights after Fumigation with High Mentha Essential Oil and Limonene Concentrations

For determinations of fresh and dry weights after applications of high amounts of the volatiles, 18-day-old plants were arranged into groups of 24 plants and set up in rows of 4 × 6 under a transparent hood equipped with four cotton balls, as shown in Figure 9. To ensure that the concentration of chemicals was constantly high throughout the experimental period, the cotton balls were soaked every day with either 8 mL limonene or 14 and 16 mL *Mentha* essential oil. Cotton balls of the control groups had no volatile injections. This experiment was performed to elucidate threshold ranges of inhibitory volatile concentrations.

Watering of the plants was performed automatically by an ebb and flow system, which was adjusted to flood the table once a day for 8 min. The heating temperature in the greenhouse was set to 18/15 °C day/night, and the aeration temperature to 21 °C. Trials were started in May and lasted until the end of July. The duration of the treatments was 7 days; afterwards, the plants were cut above ground and fresh weight (FW) and dry weight (DW) after drying at 60 °C for two days were determined.

4.5. Greenhouse Trial with Mentha piperita Plants

In early May 2018, treatments were started by placing pots with *Mentha* plants next to boxes with 120 ten-day-old white cabbage plants for 21 days (Figure 10). The *Mentha* treatment was run in three plot arrangements with one, four, or no mint plants, the latter as the control. Each plot was placed at a different table with two replicates for each treatment group. The distance between the boxes was 80 cm. The distances of the cabbage plants placed in rows in the boxes to the *Mentha* plants were: 1st row: 5–10 cm; 2nd row: 20–25 cm; 3rd row: 35–40 cm; 4th row: 50–55 cm. Control plants were placed in a neighbored greenhouse and grown under the same conditions (18/15 °C day/night, 16 °C venting temperature, shading when luminous flux was > 60 klx).

Watering was ensured by an automated ebb and flow system (9 min flow, triggered via light integral of 0.6 kWh). Each table (i.e., plot) had its individual immersion pump powering the ebb and flow system. The watering regime was adjusted to 0.4 kWh after four days. The watering conditions for the control plants were correspondingly adjusted. After 11 days, the light integral was changed to 0.9 kWh until the end of the experiment.

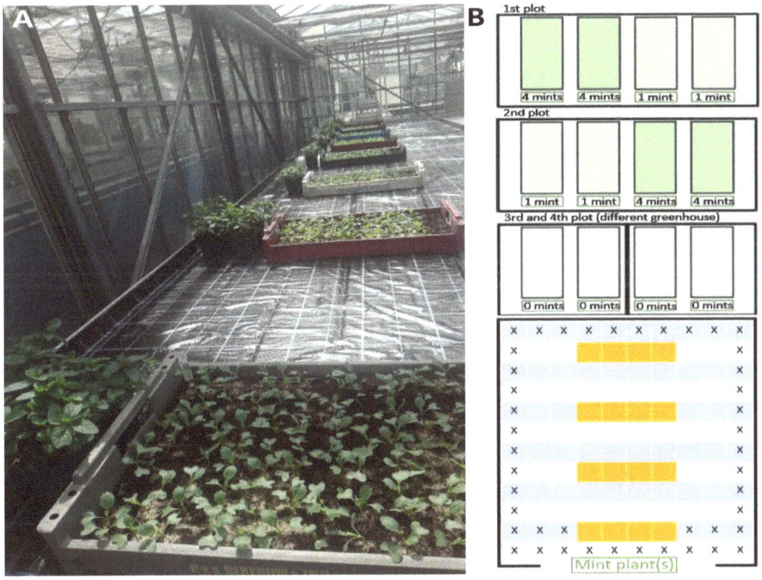

Figure 10. Greenhouse trial. Arrangement of plots (**A**,**B** above) and experimental setup of a single box (8 replications) containing 120 plants (**B** below). The setups in two different greenhouses were carried out in a completely randomized design. Orange box = scored plant (repeated measurements); X = unused plant.

4.6. Polytunnel Trial

Three independent sets were conducted in May and July 2018, following the same design as described for the greenhouse experiments. Four blocks were designed, each consisting of three plots: one untreated control and two co-cultivated plots, here with 4 and 9 mint plants, respectively, taking into account that the air exchange was faster than in the greenhouse. The blocks were arranged as shown in Figure 11, and each block was orthogonally aligned towards the prevailing wind direction.

Mentha plants were placed in front of their respective plots, allowing the air flow to carry the volatiles into the bulk of the cabbage plants. The squarely designed plots had an edge length of 1.05 m with 8 × 8 cabbage plants (Figure 11). Each set was conducted in the same polytunnel. Between the sets, weeds were removed and the roots of the *Mentha* plants were cut to prevent growing towards the experimental plots. In addition, *Mentha* plants were cut to maintain equal size and uniformity. The smallest *Mentha* plant served as reference for the mode of cutting.

After sowing, the young cabbage seedlings of each set were grown for 21 days in a greenhouse, as described above, before planting into the soil covered by the polytunnel. Each set of experiments was run for 22 days before cutting the plants for fresh and dry matter determination. Set 2 was started 8 days after set 1, and set 3 was started 6 days after set 2.

Watering in the tunnel was carried out by a sprinkler system; once a day for at least 30 min in the first three days after bedding out, then when necessary but at least two to three times a week.

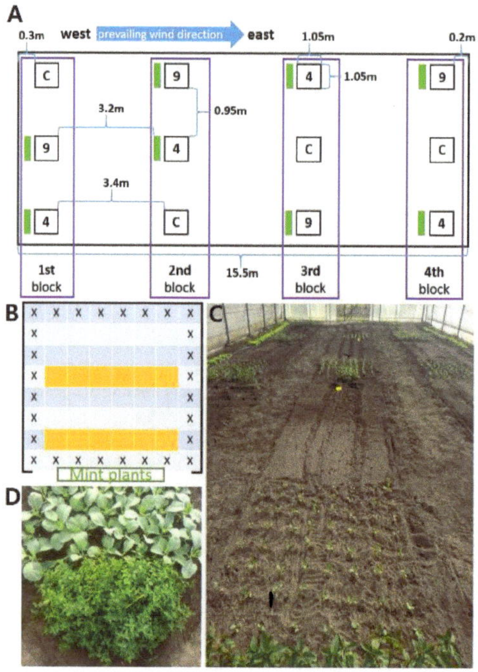

Figure 11. Polytunnel trial. Schema of the experimental setup: total experimental design (**A**) and of a single plot (**B**). Dimensions, the prevailing wind direction, and the positions of individual plots and mints are indicated; (**C**): control plants, 9 and 4: plants exposed to 9 and 4 mints (green bars). orange box = scored plant (repeated measurements); X = unused plant. (**C**): Arrangements of cabbage and *Mentha* plants in the Polytunnel soil; (**D**): picture of a single plot.

4.7. Measurements of Leaf Parameters

Leaf parameters were determined with leaves longer than 15 mm. Measurement of leaf length and width was performed after 1, 4, 8, 11, and 15 days after exposure to mint volatiles (T1–T5) with plants at 4 different distances (1st row: 5–10 cm; 2nd: 20–25 cm; 3rd: 35–40 cm; 4th: 50–55 cm), and scored as shown in Figure 10. Regarding the polytunnel trial, the leaf parameters length and width were determined for 12 individual plants/plot and at two different distances (1st row: 15–30 cm; 2nd row: 60–75 cm) once a week, 1 (T1), 8 (T3), and 15 (T5) days after the start of co-cultivation (set 1). In the second set, an additional measurement was taken on day 12 (T4) after treatment, and in the third set additional measurements were conducted after 5 (T2) and 12 (T4) days. The methods were the same as described for the greenhouse trial. Evaluations were performed with data obtained from at least 2 plots of identical conditions. Rates of leaf growth increments were determined for T1–T2, T2–T3, T3–T4, and T4–T5. For calculations, zero growth rates were not considered; the quantity of plants/plots showing no increment are mentioned in the text (or Table S1).

4.8. Fresh and Dry Weight

To determine dry matter percentage, plants used for leaf parameter determinations were cut above ground 21 days after treatment. After fresh weight measurements, the plant material was placed in an oven at 105 °C for five days and then used for dry weight determinations.

4.9. Statistics

All measured variables were subjected to one-way analysis of variance (ANOVA) or to the Kruskal–Wallis test in the case of no Gaussian distribution to verify the differences between the means (GraphPad Software, Inc., La Jolla, CA, USA). Normality was tested according to Anderson–Darling, DÁgostino–Pearson, Shapiro–Wilk, and Kolmogorow–Smirnov. Results are presented in the figures as means + standard error of means. Student's t-tests were performed in addition and p-values provided in the results.

Supplementary Materials: The following supporting information can be downloaded at: https://www.mdpi.com/article/10.3390/plants11223078/s1, Figure S1: Greenhouse trial leaf growth; Figure S2: Polytunnel trial leaf growth, Table S1: Primer Used for Real Time PCR. Suppliers Volatile Oil Analyses and Certificates.

Author Contributions: Conceptualization, M.S. and A.U.; methodology, M.S. and A.U.; investigation, M.P., M.V. and J.R.; writing—original draft preparation, M.S.; writing—review and editing, M.S., M.P. and A.U.; supervision, M.S. and A.U. All authors have read and agreed to the published version of the manuscript.

Funding: This research received no external funding.

Data Availability Statement: Not applicable.

Acknowledgments: The authors thank the technical staff in the greenhouse and polytunnel trials for assistance.

Conflicts of Interest: The authors declare no conflict of interest.

References

1. Verdeguer, M.; Sánchez-Moreiras, A.M.; Araniti, F. Phytotoxic Effects and Mechanism of Action of Essential Oils and Terpenoids. *Plants* **2020**, *9*, 1571. [CrossRef] [PubMed]
2. Araniti, F.; Bruno, L.; Sunseri, F.; Pacenza, M.; Forgione, I.; Bitonti, M.B.; Abenavoli, M.R. The allelochemical farnesene affects *Arabidopsis thaliana* root meristem altering auxin distribution. *Plant Physiol. Biochem.* **2017**, *121*, 14–20. [CrossRef] [PubMed]
3. Graña, E.; Díaz-Tielas, C.; Sánchez-Moreiras, A.M.; Reigosa, M.J.; Celeiro, M.; Abagyan, R.; Teijeira, M.; Duke, M.V.; Clerk, T.; Pan, Z.; et al. Transcriptome and binding data indicate that citral inhibits single strand DNA-binding proteins. *Physiol. Plant.* **2020**, *169*, 99–109. [CrossRef]
4. Sarheed, M.M.; Rajabi, F.; Kunert, M.; Boland, W.; Wetters, S.; Miadowitz, K.; Kazmierczak, A.; Sahi, V.P.; Nick, P. Cellular Base of Mint Allelopathy: Menthone Affects Plant Microtubules. *Front. Plant Sci.* **2020**, *11*, 546345. [CrossRef] [PubMed]
5. Chaimovitsh, D.; Shachter, A.; Abu-Abied, M.; Rubin, B.; Sadot, E.; Dudai, N. Herbicidal activity of monoterpenes is associated with disruption of microtubule functionality and membrane integrity. *Weed Sci.* **2017**, *65*, 19–30. [CrossRef]
6. Kriegs, B.; Jansen, M.; Hahn, K.; Peisker, H.; Samajova, O.; Beck, M.; Braun, M.; Ulbrich, A.; Baluska, F.; Schulz, M. Cyclic monoterpene mediated modulations of *Arabidopsis thaliana* phenotype—Effects on the cytoskeleton and on the expression of selected genes. *Plant Signal. Behav.* **2010**, *5*, 832–838. [CrossRef]
7. Schulz, M.; Kussmann, P.; Knop, M.; Kriegs, B.; Gresens, F.; Eichert, T.; Ulbrich, A.; Marx, F.; Fabricius, H.; Goldbach, H.; et al. Allelopathic monoterpenes interfere with *Arabidopsis thaliana* cuticular waxes and enhance transpiration. *Plant Signal. Behav.* **2007**, *2*, 231–239. [CrossRef]
8. Cappellari, L.D.R.; Santoro, M.V.; Schmidt, A.; Gershenzon, J.; Banchio, E. Improving Phenolic Total Content and Monoterpene in *Mentha x piperita* by Using Salicylic Acid or Methyl Jasmonate Combined with *Rhizobacteria* Inoculation. *Int. J. Mol. Sci.* **2020**, *21*, 50. [CrossRef]
9. Sukegawa, S.; Shiojiri, K.; Higami, T.; Suzuki, S.; Arimura, G.-I. Pest management using mint volatiles to elicit resistance in soy: Mechanism and application potential. *Plant J.* **2018**, *96*, 910–920. [CrossRef]
10. Grulova, D.; De Martino, L.; Mancini, E.; Salamon, I.; De Feo, V. Seasonal variability of the main components in essential oil of *Mentha × piperita* L. *J. Sci. Food Agric.* **2015**, *95*, 621–627. [CrossRef]
11. Kjonaas, R.; Croteau, R. Demonstration that Limonene is the First Cyclic Intermediate in the Biosynthesis of Oxygenated p-Menthane Monoterpenes in *Mentha piperita* and Other *Mentha* Species. *Arch. Biochem. Biophys.* **1983**, *220*, 79–89. [CrossRef]
12. Ulbrich, A.; Kahle, H.; Kramer, P.; Schulz, M. *Mentha x piperita* volatiles promote *Brassica oleracea*—A pilot study for sustainable vegetable production. *Allelopath. J.* **2018**, *43*, 93–104. [CrossRef]
13. Lan, J.; Qin, G. The Regulation of CIN-like TCP Transcription Factors. *Int. J. Mol. Sci.* **2020**, *21*, 4498. [CrossRef]
14. Schommer, C.; Debernardi, J.M.; Bresso, E.G.; Rodriguez, R.E.; Palatnik, J.F. Repression of cell proliferation by miR319-regulated TCP4. *Mol. Plant.* **2014**, *7*, 1533–1544. [CrossRef] [PubMed]
15. Urano, K.; Maruyama, K.; Koyama, T.; Gonzalez, N.; Inzé, D.; Yamaguchi-Shinozaki, K.; Shinozaki, K. CIN-like TCP13 is essential for plant growth regulation under dehydration stress. *Plant Mol. Biol.* **2022**, *108*, 257–275. [CrossRef] [PubMed]
16. Hao, J.; Lou, P.; Han, Y.; Chen, Z.; Chen, J.; Ni, J.; Yang, Y.; Jiang, Z.; Xu, M. GrTCP11, a Cotton TCP Transcription Factor, Inhibits Root Hair Elongation by Down-Regulating Jasmonic Acid Pathway in *Arabidopsis thaliana*. *Front. Plant Sci.* **2021**, *12*, 769675. [CrossRef]

17. Baulies, J.L.; Bresso, E.G.; Goldy, C.; Palatnik, J.F.; Schommer, C. Potent inhibition of TCP transcription factors by miR319 ensures proper root growth in *Arabidopsis*. *Plant Mol. Biol.* **2022**, *108*, 93–103. [CrossRef]
18. Danisman, S. TCP transcription factors at the interface between environmental challenges and the plant's growth responses. *Front. Plant Sci.* **2016**, *7*, 1930. [CrossRef]
19. Palatnik, J.F.; Allen, E.; Wu, X.; Schommer, C.; Schwab, R.; Carrington, J.C.; Weigel, D. Control of leaf morphogenesis by microRNAs. *Nature* **2003**, *425*, 257–263. [CrossRef]
20. Bresso, E.G.; Chorostecki, U.; Rodriguez, R.E.; Palatnik, J.F.; Schommer, C. Spatial Control of Gene Expression by miR319-Regulated TCP Transcription Factors in Leaf Development. *Plant Physiol.* **2018**, *176*, 1694–1708. [CrossRef]
21. Danisman, S.; van der Wal, F.; Dhondt, S.; Waites, R.; de Folter, S.; Bimbo, A.; van Dijk, A.D.; Muino, J.M.; Cutri, L.; Dornelas, M.C.; et al. *Arabidopsis* class I and class II TCP transcription factors regulate jasmonic acid metabolism and leaf development antagonistically. *Plant Physiol.* **2012**, *159*, 1511–1523. [CrossRef] [PubMed]
22. Li, Y.; An, S.; Cheng, Q.; Zong, Y.; Chen, W.; Guo, W.; Zhang, L. Analysis of Evolution, Expression and Genetic Transformation of TCP Transcription Factors in Blueberry Reveal That VcTCP18 Negatively Regulates the Release of Flower Bud Dormancy. *Front. Plant Sci.* **2021**, *12*, 697609. [CrossRef] [PubMed]
23. Li, D.; Zhang, H.; Mou, M.; Chen, Y.; Xiang, S.; Chen, L.; Yu, D. *Arabidopsis* Class II TCP Transcription Factors Integrate with the FT–FD Module to Control Flowering. *Plant Physiol.* **2019**, *181*, 97–111. [CrossRef] [PubMed]
24. Mao, Y.; Wu, F.; Yu, X.; Bai, J.; Zhong, W.; He, Y. microRNA319a-Targeted *Brassica rapa* ssp. *pekinensis* TCP Genes Modulate Head Shape in Chinese Cabbage by Differential Cell Division Arrest in Leaf Regions. *Plant Physiol.* **2014**, *164*, 710–720. [CrossRef]
25. Du, J.; Hu, S.; Yu, Q.; Wang, C.; Yang, Y.; Sun, H.; Yang, Y.; Sun, X. Genome-Wide Identification and Characterization of BrrTCP Transcription Factors in *Brassica rapa* ssp. *rapa*. *Front. Plant Sci.* **2017**, *8*, 1588. [CrossRef]
26. Efroni, I.; Blum, E.; Goldshmidt, A.; Eshed, Y. A Protracted and Dynamic Maturation Schedule Underlies *Arabidopsis* Leaf Development. *Plant Cell* **2008**, *20*, 2293–2306. [CrossRef] [PubMed]
27. Koyama, T.; Mitsuda, N.; Seki, M.; Shinozaki, K.; Ohme-Takagi, M. TCP transcription factors regulate the activities of ASYMMETRIC LEAVES1 and miR164, as well as the auxin response, during differentiation of leaves in *Arabidopsis*. *Plant Cell* **2010**, *22*, 3574–3588. [CrossRef]
28. Xiong, Y.; Jiao, Y. The Diverse Roles of Auxin in Regulating Leaf Development. *Plants* **2019**, *23*, 243. [CrossRef]
29. Saini, K.; Markakis, M.N.; Zdanio, M.; Balcerowicz, D.M.; Beeckman, T.; De Veylder, L.; Prinsen, E.; Beemster, G.T.S.; Vissenberg, K. Alteration in Auxin Homeostasis and Signaling by Overexpression of PINOID Kinase Causes Leaf Growth Defects in *Arabidopsis thaliana*. *Front. Plant Sci.* **2017**, *8*, 1009. [CrossRef]
30. Li, S.; Zachgo, S. TCP3 interacts with R2R3-MYB proteins, promotes flavonoid biosynthesis and negatively regulates the auxin response in *Arabidopsis thaliana*. *Plant J.* **2013**, *76*, 901–913. [CrossRef]
31. Kong, Q.; Low, P.M.; Lim, A.R.Q.; Yang, Y.; Yuan, L.; Ma, W. Functional Antagonism of WRI1 and TCP20 Modulates GH3.3 Expression to Maintain Auxin Homeostasis in Roots. *Plants* **2022**, *11*, 454. [CrossRef] [PubMed]
32. Wang, H.; Mao, Y.; Yang, J.; He, Y. TCP24 modulates secondary cell wall thickening and anther endothecium development. *Front. Plant Sci.* **2015**, *6*, 436. [CrossRef] [PubMed]
33. Omidbakhshfard, M.A.; Proost, S.; Fujikura, U.; Mueller-Roeber, B. Growth-Regulating Factors (GRFs): A Small Transcription Factor Family with Important Functions in Plant Biology. *Mol. Plant* **2015**, *8*, 998–1010. [CrossRef]
34. Godard, K.-A.; White, R.; Bohlmann, J. Monoterpene-induced molecular responses in *Arabidopsis thaliana*. *Phytochemistry* **2008**, *69*, 1838–1849. [CrossRef]
35. Onosato, H.; Fujimoto, G.; Higami, T.; Sakamoto, T.; Yamada, A.; Suzuki, T.; Ozawa, R.; Matsunaga, S.; Seki, M.; Ueda, M.; et al. Sustained defense response via volatile signaling and its epigenetic transcriptional regulation. *Plant Physiol.* **2022**, *189*, 922–933. [CrossRef]
36. Nie, Y.; Han, F.; Ma, J.; Chen, X.; Song, Y.-T.; Niu, S.-H.; Wu, H.X. Genome-wide TCP transcription factors analysis provides insight into their new functions in seasonal and diurnal growth rhythm in *Pinus tabuliformis*. *BMC Plant Biol.* **2022**, *22*, 167. [CrossRef] [PubMed]
37. Laila, R.; Robin, A.H.K.; Yang, K.; Park, J.-I.; Suh, M.C.; Kim, J.; Nou, I.-S. Developmental and Genotypic Variation in Leaf Wax Content and Composition, and in Expression of Wax Biosynthetic Genes in *Brassica oleracea* var. *capitata*. *Front. Plant Sci.* **2017**, *7*, 1972. [CrossRef]
38. Martelanc, M.; Vovk, I.; Simonovska, B. Determination of three major triterpenoids in epicuticular wax of cabbage (*Brassica oleracea* L.) by high-performance liquid chromatography with UV and mass spectrometric detection. *J. Chromatogr. A* **2007**, *1164*, 145–152. [CrossRef]
39. Buschhaus, C.; Jetter, R. Composition and Physiological Function of the Wax Layers Coating *Arabidopsis* Leaves: β-Amyrin Negatively Affects the Intracuticular Water Barrier. *Plant Physiol.* **2012**, *160*, 1120–1129. [CrossRef]
40. Hadacek, F. Cuticular Wax Terpenoids in Plants: Functional Diversity of. In *Encyclopedia of Lipidomics*; Wenk, M., Ed.; Springer: Dordrecht, The Netherlands, 2017. [CrossRef]
41. Heydari, M.; Zanfardino, A.; Taleei, A.; Shahnejat Bushehri, A.A.; Hadian, J.; Maresca, V.; Sorbo, S.; Di Napoli, M.; Varcamonti, M.; Basile, A.; et al. Effect of Heat Stress on Yield, Monoterpene Content and Antibacterial Activity of Essential Oils of *Mentha x piperita* var. Mitcham and *Mentha arvensis* var. piperascens. *Molecules* **2018**, *23*, 1903. [CrossRef]

42. Islam, A.K.M.M.; Kato-Noguchi, H. *Mentha sylvestris*: A Potential Allelopathic Medicinal Plant. *Int. J. Agric. Biol.* **2013**, *15*, 1313–1318.
43. Maes, C.; Meersmans, J.; Lins, L.; Bouquillon, S.; Fauconnier, M.-L. Essential Oil-Based Bioherbicides: Human Health Risks Analysis. *Int. J. Mol. Sci.* **2021**, *22*, 9396. [CrossRef]
44. Girel, S.; Schütz, V.; Bigler, L.; Dörmann, P.; Schulz, M. Bioactive Nitrosylated and Nitrated N-(2-hydroxyphenyl)acetamides and Derived Oligomers: An Alternative Pathway to 2-Amidophenol-Derived Phytotoxic Metabolites. *Molecules* **2022**, *27*, 4786. [CrossRef] [PubMed]
45. Danisman, S.; van Dijk, A.D.J.; Bimbo, A.; van der Wal, F.; Hennig, L.; de Folter, S.; Angenent, G.C.; Immink, R.G.H. Analysis of functional redundancies within the *Arabidopsis* TCP transcription factor family. *J. Exp. Bot.* **2013**, *64*, 5673–5685. [CrossRef] [PubMed]

plants
MDPI

Article

Is the Phytotoxic Effect of Digestive Fluid of *Nepenthes x ventrata* on Tomato Related to Reactive Oxygen Species?

Pawel Staszek *, Maciej Piekarniak, Agnieszka Wal, Urszula Krasuska and Agnieszka Gniazdowska *

Department of Plant Physiology, Institute of Biology, Warsaw University of Life Sciences-SGGW, 02-776 Warsaw, Poland
* Correspondence: pawel_staszek1@sggw.edu.pl (P.S.); agnieszka_gniazdowska@sggw.edu.pl (A.G.)

Abstract: The digestive fluid of pitcher plants is a rich source of enzymes and secondary metabolites, but its impact on higher plant growth and development remains unknown. The aim of the study was to determine the phytotoxicity of the digestive fluid of the pitcher plant (*Nepenthes x ventrata*) on the germination of tomato (*Solanum lycopersicum* L.) seeds, elongation growth and cell viability of roots of tomato seedlings. The digestive fluid was collected from pitchers before feeding and four days after feeding; the pH and electrical conductivity of the fluid were determined. Undiluted and 50% fluids were used in the study. An inhibition of germination of tomato seeds, by around 30% and 55%, was respectively observed in 50% and 100% digestive fluids collected before and after feeding. Digestive fluid did not affect the root growth of tomato seedlings; a slight (6%) inhibition was only observed after the application of 100% digestive fluid from an unfed trap. The roots of the tomato seedlings treated with undiluted fluid were characterized by reduced cell viability. Reactive oxygen species (H_2O_2 and $O_2^{\bullet -}$) were mainly localized in the root apex regardless of the used phytotoxic cocktail, and did not differ in comparison to control plants.

Keywords: carnivorous plant; ROS; root growth; seed germination; allelopathy

Citation: Staszek, P.; Piekarniak, M.; Wal, A.; Krasuska, U.; Gniazdowska, A. Is the Phytotoxic Effect of Digestive Fluid of *Nepenthes x ventrata* on Tomato Related to Reactive Oxygen Species? *Plants* **2023**, *12*, 755. https://doi.org/10.3390/plants12040755

Academic Editors: Margot Schulz and Vincenzo Tabaglio

Received: 9 January 2023
Revised: 2 February 2023
Accepted: 6 February 2023
Published: 8 February 2023

1. Introduction

Nepenthes x ventrata belongs to the Nepenthaceae family and is a natural hybrid of *Nepenthes alata* Blanco and *Nepenthes ventricosa* Blanco. This species is endemic to forests in the Philippines [1]. *N. ventrata* forms jug-shaped containers that function as passive pitcher traps. Pitchers are filled with digestive fluid, being the mixture that enables the release of nutrients from the bodies of the captured prey [2].

Many enzymes have been found in the digestive fluid of pitcher plants: proteases, peptidases, phosphatases, esterases, ribonucleases and chitinases [3–6]. The aspartyl proteinases Nepenthesin-1 and Nepenthesin-2 have also been isolated from the digestive fluid [7]. The digestive fluid also contains a thaumatin-like protein with antibacterial and antifungal properties [8]. In addition, pitcher fluid contains compounds of low molecular weight, e.g., two naphthoquinones: droseron (3,5-dihydroxy-2-methyl-1,4-naphthoquinone) and its derivative 5-O-methyldroserone (2-methyl-3-hydroxy-5-methoxy-1,4- naphthoquinone) [9], of antifungal activity against a wide range of human and plant fungal pathogens [10]. Another low-molecular substance is plumbagin, isolated from *N. alata* [11] and characterized by antimalarial, antibacterial, antifungal and anticancer properties [12–14] and 7-methyl-juglone [11]. The digestive fluid of the Nepenthes species is rich in mineral nutrient composition (boron, calcium, copper, iron and nickel) and anions Cl^-, Br^- and SO_4^{2-} [11]. Prey digestion (feeding process) is accompanied by alteration in fluid composition, including reactive oxygen species (ROS) and reactive nitrogen species (RNS) levels [15].

The physicochemical properties of the digestive fluid, such as lower surface tension against water and high viscoelasticity, play an important role in preventing the escape of captured insects [16].

Application of allelopathic cocktails in cultivation can effectively influence weed populations and insect pests. Allelopathy thus offers an attractive environmentally friendly alternative to pesticides in agriculture [17]. Volatile organic compounds belonging to the group of allelochemicals could repel insects or mask the host's odor to confuse them [18,19].

Digestive fluid as a mixture of many compounds with a possible phytotoxic potential may prove to be a key element in research on natural herbicides that would replace artificial and less specific compounds currently used in agriculture. The exact composition of the digestive fluid is unknown, but the digestive fluid of some species is known to contain compounds that have phytotoxic potential.

The aim of our study was to characterize some properties of phytotoxic cocktails and to determine the effect of the digestive fluid from pitchers of *N. ventrata*, collected before and after plant feeding, on tomato (*Solanum lycopersicum* L.) seed germination and root growth.

Based on the mode of action of many allelochemicals and phytotoxins [20], we hypothesized that pitcher plants' digestive fluid impacts the seed germination and root growth of tomato seedlings by the induction of oxidative stress.

2. Results

2.1. Electrical Conductivity

The electrical conductivity and pH of the phytotoxic cocktail depended on the origin of the fluid (from fed or unfed traps) and the concentration. The highest pH value was found in distilled water used as a control, and the lowest in heat-treated, 100% fluid from fed traps (Table 1). The pH of the fluid from fed pitchers was lower than that from unfed ones, regardless of concentration or heat-treatment (Table 1).

Table 1. Electrical conductivity and pH of digestive fluid solutions used in the experiments as phytotoxic cocktails.

		pH	Electrical Conductivity [mS cm^{-1}]
Control		6.29 ± 0.07 c	0.021 ± 0.017 A
		Crude fluid	
Unfed trap	50% fluid	5.96 ± 0.17 bc	0.786 ± 0.078 B
	100% fluid	5.82 ± 0.03 b	1.333 ± 0.012 C
Fed trap	50% fluid	3.78 ± 0.19 a	0.704 ± 0.042 B
	100% fluid	3.38 ± 0.08 a	1.247 ± 0.017 C
		Heat-treated fluid	
Unfed trap	50% fluid	6.09 ± 0.29 bc	0.771 ± 0.026 B
	100% fluid	5.54 ± 0.10 b	1.364 ± 0.052 C
Fed trap	50% fluid	3.83 ± 0.31 a	0.630 ± 0.040 B
	100% fluid	3.36 ± 0.16 a	1.127 ± 0.039 C

Table includes average values ± SD of three replicated experiments. Homogenous groups (a–c or A–C) were evaluated using Fisher's LSD post hoc test with $p < 0.5$.

The highest electrical conductivity (1.364 mS cm^{-1}) was recorded for heat-treated undiluted fluid from unfed pitchers, and the lowest (0.0209 mS cm^{-1}) for the distilled water (control) (Table 1). Digestive fluids (100%) were characterized by higher electrical conductivity compared to diluted ones, irrespective of the origin of the fluid and heat-treatment.

2.2. Germination Tests

Two-way Anova showed that there were no statistically significant differences between the germination of tomato seeds in crude fluid and heat-treated fluid ($p = 0.73$, F = 0.13). Due to the lack of associations, the dataset for crude and heat-treated fluids was not separated in the remaining analyses.

Tomato seeds germinated in 100% fluid after three days of imbibition in water (Figure 1). Undiluted digestion fluids from unfed or fed traps inhibited the germination of tomato seeds by 59% or 53%, respectively. In diluted phytotoxic cocktails of fed and unfed pitchers, tomato seeds germinated in 67% and 75%, respectively (Figure 1).

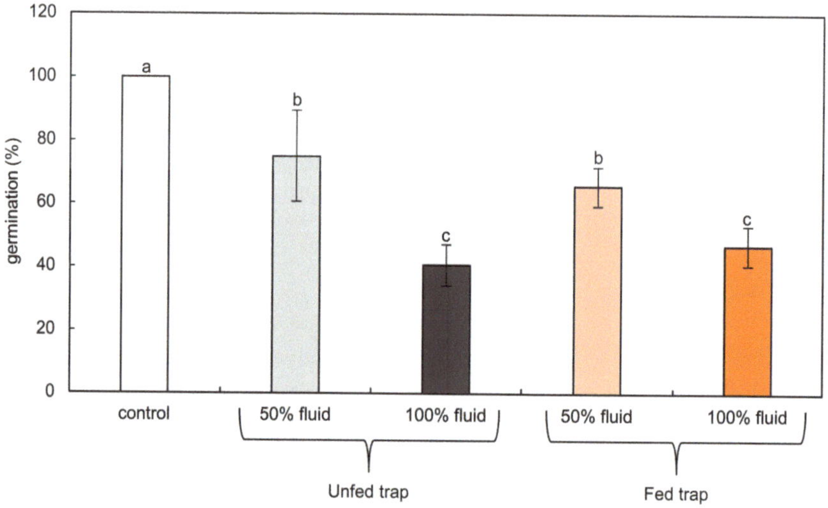

Figure 1. Germination of tomato seeds in water (control) or in phytotoxic cocktails of 100% or 50% digestive fluids from fed and unfed traps of pitcher plants. Due to the lack of significant differences, we did not separate data obtained after treatment with crude and heat-treaded fluids. Values are average ± SD of three replications. Homogenous groups (a–c) were evaluated using Fisher's LSD post hoc test with $p < 0.5$.

2.3. *Root Growth Biotest*

Two-way Anova showed that there were no statistically significant differences between the growth of tomato seedlings in crude fluid and heat-treated fluid ($p = 0.33$, $F = 0.95$). Due to the lack of associations, the dataset for crude and heat-treated fluids was not separated in the remaining analyses.

Roots of the control tomato seedlings growing in water for 24 h were about 16 mm long. Fluids from traps of pitcher plants repressed the growth of tomato roots in only 6% (Figure 2). The longest roots were observed in seedlings treated with the diluted fluid from unfed traps. The strongest inhibition of root growth was noted for seedlings treated with 100% fluids from unfed traps. Root lengths measured in seedlings treated with diluted and undiluted fluid from fed traps differed by 0.25 mm and 0.5 mm from the control, respectively, but the differences were not statistically significant.

2.4. *Test of Root Cell Viability*

The viability of the root cells of seedlings treated for 24 h with crude and heat-treated phytotoxic cocktails slightly differed (Table 2). Excluding the control plants, the highest viability was estimated for roots of seedlings treated with crude, diluted fluid from fed traps, and the lowest viability was noticed for the roots of seedlings grown in 100% crude digestion fluid from unfed pitchers, which was very similar to 100% heat-treated fluid from fed traps (31,8–31,3% of damages, respectively). The roots of seedlings treated with 50% fluid were characterized by a higher viability than the roots of seedlings treated with undiluted fluid, irrespective of heat treatment.

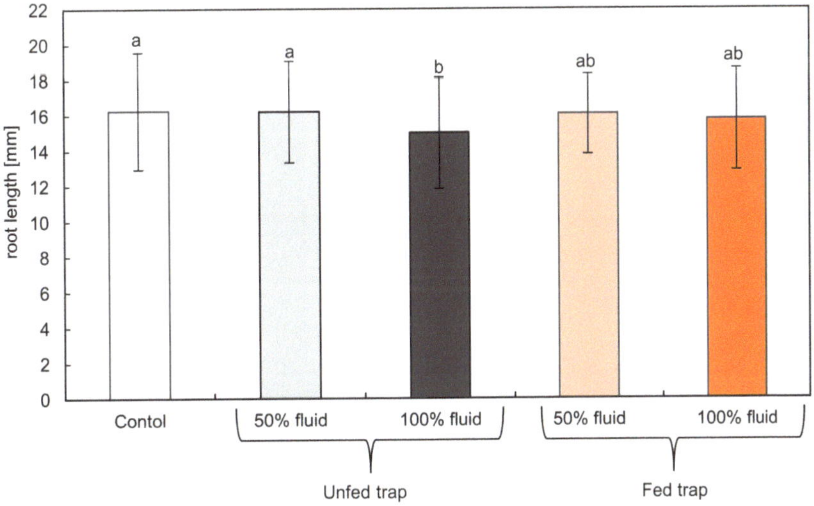

Figure 2. Length of roots of tomato seedlings growing for 24 h in water (control) or in phytotoxic cocktails of 100% or 50% digestive fluids from fed and unfed traps of pitcher plants. Due to the lack of significant differences, we did not separate data obtained after treatment, the division into crude and heat-treaded fluids was not included. Values are average ± SD of three replications. Homogenous groups (a, b) were evaluated using Fisher's LSD post hoc test with $p < 0.5$.

Table 2. Cell viability (Evans blue test) and cell damage of roots of tomato seedlings cultured in water or treated with digestive fluids from Nepenthes traps (crude, heat-treated, concentrated (100%) or diluted (50%)) after 24 h of the culture.

		Evans Blue Uptake to the Roots [mg dye g^{-1} FW]	Cell Damage [%]
Positive control (heat-treated roots)		0.476 ± 0.011 c	100 E
Control (seedlings grown in water)		0.086 ± 0.021 a	18.1 A
		Crude fluid	
Unfed trap	50% fluid	0.122 ± 0.064 b	25.7 BC
	100% fluid	0.151 ± 0.053 b	31.8 D
Fed trap	50% fluid	0.100 ± 0.048 b	20.9 B
	100% fluid	0.140 ± 0.079 b	29.3 BC
		Heat-treated fluid	
Unfed trap	50% fluid	0.109 ± 0.056 b	22.9 BC
	100% fluid	0.113 ± 0.055 b	23.7 D
Fed trap	50% fluid	0.135 ± 0.008 b	28.4 B
	100% fluid	0.149 ± 0.084 b	31.3 BC

Table includes average values ± SD of three replicated experiments. Homogenous groups (a–c or A–E) were evaluated using Fisher's LSD post hoc test with $p < 0.5$.

2.5. *In Situ H_2O_2 Localization in Tomato Seedlings Cultured*

H_2O_2 was localized using DAB staining as brown precipitates of polymerized DAB (Figure 3). For all tested seedlings, the most visible staining was detected in the root tips. Stronger coloration of the upper part of the roots, on the border with the hypocotyl, was observed in seedlings treated with heat-treated fluid, regardless of the concentration of the fluids or the feeding of the traps (from fed or unfed ones).

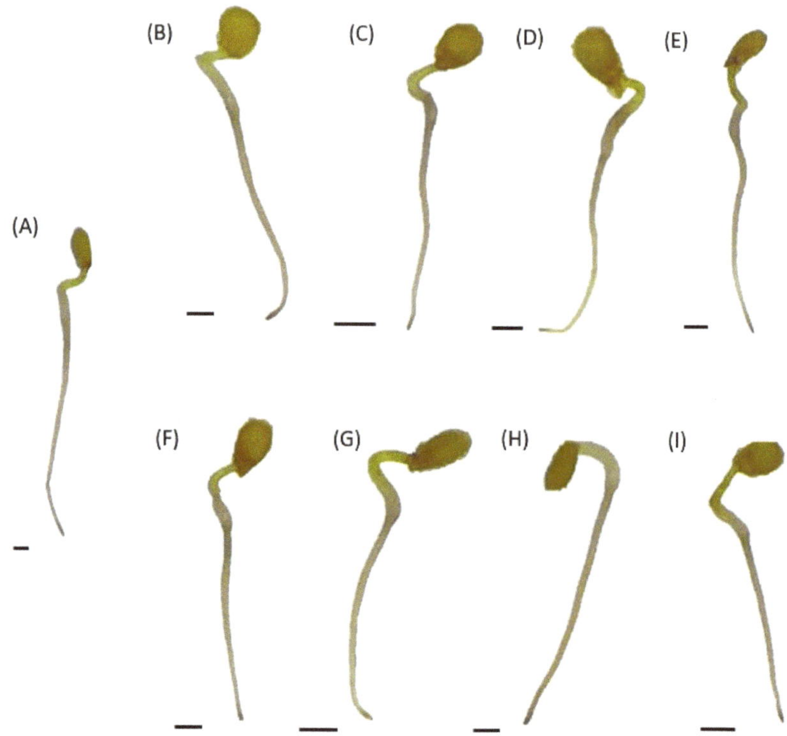

Figure 3. In situ H_2O_2 localization in tomato seedlings cultured for 24 h in water (the control) or in phytotoxic cocktails of 100% or 50% fed or unfed traps of pitcher plants. Control seedlings (**A**), seedlings growing in crude concentrated fluids of unfed traps (**B**), treated with crude diluted cocktail of unfed traps (**C**), cultured in crude concentrated fluids of fed traps (**D**), exposed to crude diluted cocktails of unfed traps (**E**), growing in heat-treated concentrated fluids of unfed traps (**F**), heat-treated diluted cocktails of unfed traps (**G**), heat-treated concentrated fluids of fed traps (**H**) and heat-treated diluted cocktails of fed traps (**I**). Bar = 2 mm.

2.6. *In Situ* $O_2^{\bullet -}$ *Localization*

The histochemical localization of $O_2^{\bullet -}$ was carried out using NBT staining. A dark blue coloration of formazan was observed in the root tips of all the tested seedlings (Figure 4). Only the outer layers of the roots were colored above the tips. The roots of seedlings treated with 50% digestive fluid were characterized by a slightly higher color intensity, regardless of the heating of the phytotoxic cocktails (Figure 4). However, in general, no differences in NBT staining were noticed between the roots of tomato seedlings treated with digestive fluids and those growing in water (Figure 4).

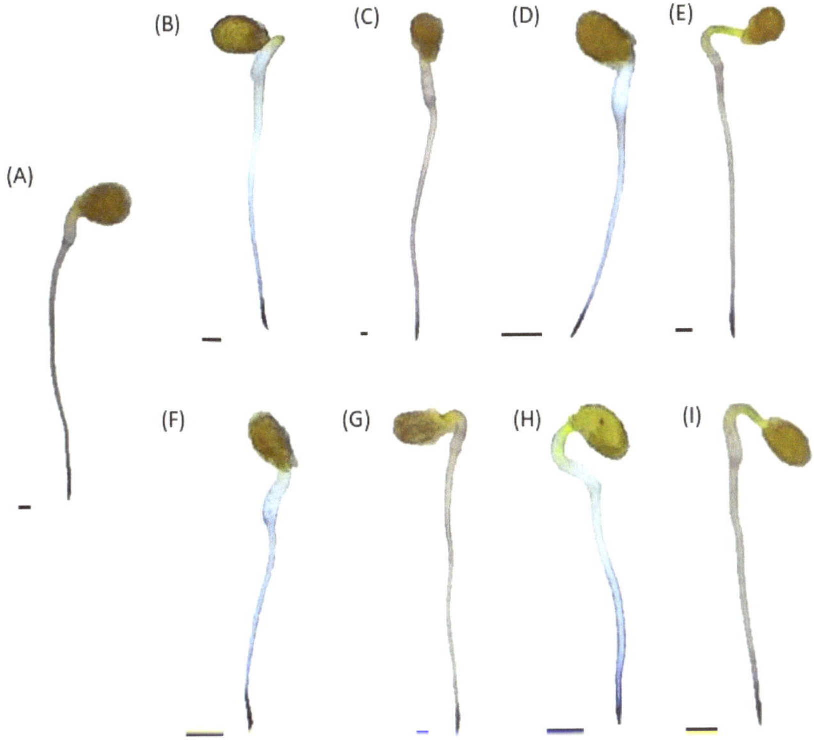

Figure 4. In situ $O_2^{\bullet -}$ localization in tomato seedlings cultured for 24 h in water (control) or in phytotoxic cocktails of 100% or 50% fed or unfed traps of pitcher plants. Control seedlings (**A**), seedlings growing in crude concentrated fluids of unfed traps (**B**), treated with crude diluted cocktail of unfed traps (**C**), cultured in crude concentrated fluids of fed traps (**D**), exposed to crude diluted cocktails of unfed traps (**E**), growing in heat-treated concentrated fluids of unfed traps (**F**), heat-treated diluted cocktails of unfed traps (**G**), heat-treated concentrated fluids of fed traps (**H**) and heat-treated diluted cocktails of fed traps (**I**). Bar = 2 mm.

3. Discussion

Nepenthes is a plant which, by adapting to specific environmental conditions, in the process of evolution, developed a carnivorous syndrome. Although it is known where the components of the fluid are produced and how the substances resulting from the digestion of the prey are absorbed by the plant, knowledge about the composition of the digestive fluid is limited. The phenomenon of carnivorous syndrome in plants has been described for over 150 years; the isolation and identification of enzymes involved in digestion is still only fragmentary. It refers to both the composition of enzymes and other molecules in the digestive fluid. Their identification, structure and biosynthesis may be an interesting target for further research [9]. Until now, the digestive fluid of pitcher plants has been mostly examined in the context of plant nutrition, but its putative toxicity on neighboring plants should not be omitted. Our study is a unique observation of the impact of the digestive fluid on the germination and growth of other plants; acceptors, according to allelopathic terminology.

At the start of this study, we determined the impact of pitcher plant feeding and the inactivation of the fluid by heating on its phytotoxicity. No influence of the temperature inactivation on the toxicity of the digestive fluid on the germination of tomato seeds and the elongation growth of roots suggests temperature-resistant compounds of the fluid, rather than enzymes of the fluid, are responsible for its negative action. The feeding of the pitcher plant resulted in decreased pH of the fluid, which is a typical reaction in the digestion process [21], although it did not affect tomato seed germination. The germination of seeds was inhibited by phytotoxic cocktails with no respect to the pH of the fluid. Tomato seeds similarly germinated in the digestive fluid with a pH range of 3.3–3.8, taken from fed traps and in the digestive fluid of unfed traps (pH 5.5 to 6.1). The low pH of the digestive fluid also did not influence the elongation growth of the tomato roots. The pH value also seems to be correlated with bacterial community composition [22] and feeding. Taking to account no differences in cocktail phytotoxicity in the tomato seedlings before and after heating, it can be concluded that microorganisms from the fluid are not toxic to acceptor tissues; however, perhaps the experiments should be performed longer to observe such effects. The electrical conductivity of the digestive fluid showed no difference in dependence on feeding, nor heating; it only decreased as the dilution of the liquid was performed. Electrical conductivity in a range of 0.8–1.6 mS cm^{-1}, observed in the research, indicates moderately saline soils, where only plants resistant to salinity can grow [23,24]. This data confirms the information that the pitcher fluid of *Nepenthes* plants (*N. alata, N. fusca, N.gracilis, N. mirabilis, N. superba, N. thorelii* and *N. ventricosa*) is a 25 mM KCl solution with few additional ions [11].

During the germination tests, solutions with a similar pH, but different digestive fluid concentration, were applied, on which the results of the experiment mainly depended. Undiluted digestive fluid, taken from a pitcher of *N. ventrata* fed with the proteins, inhibited tomato seed germination as compared to the 50% cocktail. Some compounds that may adversely affect the growth and development of other plants have been isolated from carnivorous plants. Studies conducted with the use of Portuguese sundew ((*Drosophyllum lusitanicum* L. (Link)) extracts have shown that the extracts inhibited the germination of lettuce (*Lactuca sativa* L.) and wheat (*Triticum aestivum* L.) seeds. Plumbagin has been identified as the main compound responsible for the inhibition of lettuce sprouting [25]. Other compounds, isolated from carnivorous plants, may show a toxic effect; these include two naphthoquinones, droserone and its derivative 5-O-methyldroserone, which have beem found in the digestive fluid of *N. khasiana* [10] or 7-methyl-juglone isolated from *N. alata* [11]. Previous reports from our laboratory indicated the presence of phenolic compounds in the digestive fluid, the total content of which increased in fed traps [15]. Phenolics are regarded as strong allelochemicals of various modes of action in acceptor plants [26].

Investigating the mode of action of allelopathic compounds is complicated due to the multitude of potential molecular targets. Biological tests using plants or plant tissues successfully detect the biological activity of many synthetic and natural compounds [27]. One of the first visible effects of phytotoxin's action is the reduction of seed germination and/or seedling growth [28]. Diverse phytotoxins (like cyanamide, citral, coumarin) act as inhibitors of root tip cell division, influencing root growth [29–31]. Treatment of tomato seedlings with digestive fluids showed that components of these cocktails had no effect on the growth and development of the young seedlings. Morphology and length of the roots of tomato plants exposed to 100% and 50% digestive fluids of *N. ventrata* traps did not differ and were similar to those observed in control seedlings. It suggests that the impact of the phytotoxic cocktails of the digestive fluids on growth of the roots is weaker than on seed germination. This observation is atypical in allelopathy research, as most allelochemicals applied as the mixture or extract originated from the tissue of allelopathic plant are more toxic in growth tests than germination tests [32]. Many allelochemicals affect seed germination by inhibition of the activity of the enzymes crucial for the process. Extracts from the leaves and flowers of catnip (*Nepeta meyeri* Benth.) decreased α-amylase, ß-1,3 glucanase and protease activities in seeds of field dodder (*Cuscuta campestris* Yunck) [33]. Allelochemicals could disrupt the action of phytohormones in germinating seeds. The inhibition of germination of lettuce (*Lactuca sativa* L.) by myrigalone A, C-methylated dihydrochalcone from sweet gale (*Myrica gale* L.) was accompanied by alterations in gibberellins metabolism by inhibiting GA3 oxidase, as well as by interfering with the gibberellin signaling pathway [34]. Perturbations in ROS metabolism and the induction of oxidative stress were identified as the cause of inhibition of seed germination. Sunflower (*Helianthus annus* L.) extracts increased H_2O_2 content in mustard (*Sinapsis alba* L.) seeds, resulting in the restriction of their germination [35].

Although evidence for allelopathic interactions and/or the potential of allelochemicals is based on the evaluation of seed germination and seedling growth of the target species [28], these tests only provide general information regarding the biological effects of phytotoxins. They do not indicate a mechanism of their action [36,37]. In many studies, it has been demonstrated that, for allelopathy stress, as for other stresses, the induction of oxidative stress is the common response of acceptor plants [20,28,29,38]. Such observations are typical for both crude extracts of allelopathic plants, as well as for single allelopathic compounds. An accumulation of ROS, a modification of their localization and a disruption of cellular antioxidant systems were shown for various plants exposed to different allelochemicals (for review, see [20,29,38]. In our study, phytotoxic cocktails of digestive fluids did not lead to increased levels of H_2O_2 or $O_2^{\bullet-}$. Moreover, no alteration in the localization of ROS in roots in response to digestive fluids were shown. It may be proposed that compounds in the fluids from *N. ventrata* traps do not impact ROS production. Although investigations of the activity of enzymes of the antioxidant system are still required, the presence of ROS and reactive nitrogen species in the digestion liquid of these plants has been recently revealed [15].

4. Materials and Methods

4.1. Experimental Model

In our experiments, we used tomato (*Solanum lycopersicum* L. cv Malinowy Ożarowski) seedlings as the test plant. A digestive fluid from pitchers of Nepenthes was applied in the tests as a phytotoxic cocktail.

Pitcher plants *Nepenthes x ventrata* Hort. ex Fleming [–(*N. ventricosa* Blanco × *N. alata* Blanco)] were grown in the greenhouse under conditions of constant high humidity (60%) and temperature (28 °C); light was supplemented with sodium lamps to obtain a 16-/8-h day/night photoperiod. A mixture of acid peat, perlite and sphagnum moss was used as a growing medium and plants were watered with distilled water every other day. The traps of pitcher plants, immediately after opening, were covered with gauze to avoid contamination by accidental insects. Digestive fluid was collected from fed or unfed mature pitchers (aerial type, length approximately 9 cm). Traps were fed with 40 μg hen egg white solution (1 μg μL^{-1}) introduced with a pipette directly into a pitcher. Digestive fluids from fed traps were collected four days after feeding. The fluids (from fed and unfed pitchers) were shortly centrifuged (5522× *g*, 5 min at 4 °C), and the supernatants were transferred to the sterile tubes, frozen and used for biotests.

Before biotesting, digestive fluids from several traps were combined (fluid separately from fed and unfed traps) to maximize the volume of the phytotoxic cocktail. One portion of the fluid was boiled at 100 °C for 10 min (heat-treated fluid) and the rest was left untreated (crude fluid). Then, part of each fluid (heat-treated or crude) was diluted twice to obtain a 50% solution. The electrical conductivity and pH of phytotoxic cocktails were determined using a pH/conductivity meter (CPC-505 Elmetron, Zabrze, Poland, with suitable electrodes ECF-60 and EPS-1, respectively).

4.2. Germination Tests

Tomato seeds (*Solanum lycopersicum* L. cv. Malinowy Ożarowski) (commercially obtained from PNOS Sp. z o.o., Poland) were surface sterilized with 0.5% sodium hypochloride for 10 min at room temperature. Then, seeds were rinsed three times with distilled water, and 8 seeds were placed on Petri dishes (ø 3.5 cm) filled with filter paper wetted with 2.5 mL phytotoxic cocktail or 2.5 mL distilled water (control). The culture was carried out in a growth chamber (Fitotron Versatile Environmental Test Chamber-PAR-10Mn, by Sanyo,

Osaka, Japan, model MLR-350H), in darkness at 20 °C for 3 days. Seeds were considered germinated when the radicle had emerged through the seed coat.

To investigate the impact of pitcher fluids on the growth of tomato seedlings, biotests were performed. Tomato seeds were germinated in water in darkness at 20 °C for 3–4 days. Then, 8 seedlings with roots of equal length (0.5 cm) were selected and transferred to Petri dishes (ø 3.5 cm) filled with filter paper wetted with 2.5 mL phytotoxic cocktails or 2.5 mL distilled water (control). Control seedlings and seedlings treated with digestive fluid were cultured in a growth chamber at 23/20 °C, 12-/12-h day/night regime, and a light intensity of 90 µmol PAR m^{-2} s^{-1} for 24 h as described by Krasuska et al. [39]. The root length of seedlings was measured after 24 h of the culture.

4.3. Test of Root Cell Viability

The viability of tomato root cells was determined using Evans blue staining [40]. The whole control seedlings or seedlings treated with pitcher fluids for 24 h were incubated in a 0.25% solution of Evans blue for 30 min at room temperature. Then, seedlings were washed three times in distilled water and roots were isolated, weighed and homogenized in 1 mL of a 1% solution of sodium dodecylsulphate (SDS). After 10 min centrifugation at 21,000× *g* at 4 °C, the supernatant was collected and absorbance was measured at 600 nm (Sunrise, Tecan, Männedorf, Switzerland). For the positive control (100% cell damage), the roots were boiled for 10 min, cooled and then treated as described above. Three roots were used for one repetition. The concentration of the extracted dye was estimated from the standard curve prepared with Evans blue in 1% SDS and expressed as mg g^{-1} FW or % of cell damage. The data were means of three measurements from each of three sets of experiments.

4.4. In Situ H_2O_2 Localization

The histochemical localization of H_2O_2 was carried out with 3,3′-diaminobenzidine (DAB) staining [39]. The roots of the control seedlings and seedlings treated with pitcher fluids for 24 h were washed twice in distilled water and incubated with DAB solution (1 mg mL^{-1}) with 2 mM DMSO. Staining was carried out for 4 h at room temperature, in darkness.

4.5. In Situ $O_2^{\bullet-}$ Localization

The histochemical localization of $O_2^{\bullet-}$ was performed according to Beyer and Fridovich [41] using nitroblue tetrazolium (NBT) staining. Roots isolated from the control seedlings and seedlings treated with pitcher fluids for 24 h were washed twice in distilled water and incubated for 20 min in darkness at room temperature in 2 mM NBT (Sigma–Aldrich, Tokyo, Japan), freshly prepared in 10 mM Tris-HCl buffer pH 7.4 with 2 mM DMSO. After staining, the roots were cleaned for 24 h in chloral hydrate to avoid artifacts by removing oxidized phenolics [42]. $O_2^{\bullet-}$ was visualized as a deposit of a dark blue insoluble formazan compound.

Images of seedlings stained with DAB and NBT were taken with the TAGARNO FHD TREND digital microscope (magnification x 5.2, F4 diaphragm, enhancement 10.2 dB, exposure time 1/15 s).

Data were obtained in at least three independent experiments with at least three repetitions each. The plants for experiments were randomly selected from 25 individuals. Data were analyzed using Statistica Software. Mean values were calculated and SD was provided. After Anova, homogenous groups were evaluated using Fisher's LSD post hoc test.

5. Conclusions

We demonstrated the negative, dose-dependent effect of the digestive fluid of the pitcher plant on tomato seed germination; whereas its impact on root elongation was less evident, but still significant. Application of the digestive fluid decreased cell viability but did not alter ROS (H_2O_2 and $O_2^{\bullet-}$) distribution in the roots of tomato seedlings. The explanation of the phytotoxic potential of the digestive fluid requires further study, in which the compounds of the trap fluid, being mainly secondary metabolites, are determined prior to their use in phytotoxicity tests. The context of perturbation in ROS levels in acceptor plants as the target of digestive fluid action needs examination of the activity of the enzymatic antioxidants in the tissue of the acceptor plants.

Author Contributions: Conceptualization, P.S. and U.K.; methodology, P.S. and U.K.; validation, P.S. and A.G.; formal analysis, P.S. and A.G.; investigation, A.W., M.P. and P.S.; data curation, P.S.; writing—original draft preparation, P.S. and A.G.; writing—review and editing, P.S., A.W. and A.G.; visualization, P.S. and A.W.; supervision, P.S. All authors have read and agreed to the published version of the manuscript.

Funding: This research received no external funding.

Institutional Review Board Statement: Not applicable.

Informed Consent Statement: Not applicable.

Data Availability Statement: Upon request, the data will be provided by the corresponding author.

Conflicts of Interest: The authors declare no conflict of interest.

References

1. Clarke, C. *Nepenthes of Sumatra and Peninsular Malaysia*; Natural History Publications (Borneo): Kota Kinabalu, Malaysia, 2001.
2. Adlassnig, W.; Peroutka, M.; Lendl, T. Traps of Carnivorous Pitcher Plants as a Habitat: Composition of the Fluid, Biodiversity and Mutualistic Activities. *Ann. Bot.* **2011**, *107*, 181–194. [CrossRef] [PubMed]
3. An, C.; Fukusaki, E.I.; Kobayashi, A. Aspartic Proteinases Are Expressed in Pitchers of the Carnivorous Plant *Nepenthes Alata* Blanco. *Planta* **2002**, *214*, 661–667. [CrossRef] [PubMed]
4. Eilenberg, H.; Pnini-Cohen, S.; Schuster, S.; Movtchan, A.; Zilberstein, A. Isolation and Characterization of Chitinase Genes from Pitchers of the Carnivorous Plant Nepenthes Khasiana. *J. Exp. Bot.* **2006**, *57*, 2775–2784. [CrossRef] [PubMed]
5. Płachno, B.J.; Adamec, L.; Lichtscheidl, I.K.; Peroutka, M.; Adlassnig, W.; Vrba, J. Fluorescence Labelling of Phosphatase Activity in Digestive Glands of Carnivorous Plants. *Plant Biol.* **2006**, *8*, 813–820. [CrossRef] [PubMed]
6. Stephenson, P.; Hogan, J. Cloning and Characterization of a Ribonuclease, a Cysteine Proteinase, and an Aspartic Proteinase from Pitchers of the Carnivorous Plant *Nepenthes Ventricosa* Blanco. *Int. J. Plant Sci.* **2006**, *167*, 239–248. [CrossRef]
7. Athauda, S.B.P.; Matsumoto, K.; Rajapakshe, S.; Kuribayashi, M.; Kojima, M.; Kubomura-Yoshida, N.; Iwamatsu, A.; Shibata, C.; Inoue, H.; Takahashi, K. Enzymic and Structural Characterization of Nepenthesin, a Unique Member of a Novel Subfamily of Aspartic Proteinases. *Biochem. J.* **2004**, *381*, 295–306. [CrossRef]
8. Hatano, N.; Hamada, T. Proteome Analysis of Pitcher Fluid of the Carnivorous Plant *Nepenthes alata*. *J. Proteome Res.* **2008**, *7*, 809–816. [CrossRef]
9. Mithöfer, A. Carnivorous Pitcher Plants: Insights in an Old Topic. *Phytochemistry* **2011**, *72*, 1678–1682. [CrossRef]
10. Eilenberg, H.; Pnini-Cohen, S.; Rahamim, Y.; Sionov, E.; Segal, E.; Carmeli, S.; Zilberstein, A. Induced Production of Antifungal Naphthoquinones in the Pitchers of the Carnivorous Plant *Nepenthes khasiana*. *J. Exp. Bot.* **2010**, *61*, 911–922. [CrossRef]
11. Buch, F.; Rott, M.; Rottloff, S.; Paetz, C.; Hilke, I.; Raessler, M.; Mithöfer, A. Secreted Pitfall-Trap Fluid of Carnivorous Nepenthes Plants Is Unsuitable for Microbial Growth. *Ann. Bot.* **2013**, *111*, 375–383. [CrossRef]
12. Likhitwitayawuid, K.; Kaewamatawong, R.; Ruangrungsi, N.; Krungkrai, J. Antimalarial Naphthoquinones from *Nepenthes thorelii*. *Planta Med.* **1998**, *64*, 237–241. [CrossRef] [PubMed]
13. Sandur, S.K.; Ichikawa, H.; Sethi, G.; Ahn, K.S.; Aggarwal, B.B. Plumbagin (5-Hydroxy-2-Methyl-1,4-Naphthoquinone) Suppresses NF-KB Activation and NF-KB-Regulated Gene Products Through Modulation of P65 and IκBα Kinase Activation, Leading to Potentiation of Apoptosis Induced by Cytokine and Chemotherapeutic Agents. *J. Biol. Chem.* **2006**, *281*, 17023–17033. [CrossRef] [PubMed]
14. Shin, K.-S.; Lee, S.-K.; Cha, B.-J. Antifungal Activity of Plumbagin Purified from Leaves of Nepenthes Ventricosa x Maxima against Phytopathogenic Fungi. *Plant Pathol. J.* **2007**, *23*, 113–115. [CrossRef]
15. Wal, A.; Staszek, P.; Pakula, B.; Paradowska, M.; Krasuska, U. ROS and RNS Alterations in the Digestive Fluid of Nepenthes × Ventrata Trap at Different Developmental Stages. *Plants* **2022**, *11*, 3304. [CrossRef] [PubMed]
16. Gaume, L.; Forterre, Y. A Viscoelastic Deadly Fluid in Carnivorous Pitcher Plants. *PLoS ONE* **2007**, *2*, 1185. [CrossRef]
17. Macías, F.A.; Molinillo, J.M.; Varela, R.M.; Galindo, J.C. Allelopathy—A Natural Alternative for Weed Control. *Pest Manag. Sci.* **2007**, *63*, 327–348. [CrossRef]
18. Piesik, D.; Wenda-Piesik, A. Sitophilus Granarius Responses to Blends of Five Groups of Cereal Kernels and One Group of Plant Volatiles. *J. Stored Prod. Res.* **2015**, *63*, 63–66. [CrossRef]

19. Piesik, D.; Wenda-Piesik, A.; Krasińska, A.; Wrzesińska, D.; Delaney, K.J. Volatile Organic Compounds Released by *Rumex Confertus* Following *Hypera Rumicis* Herbivory and Weevil Responses to Volatiles. *J. Appl. Entomol.* **2016**, *140*, 308–316. [CrossRef]
20. Staszek, P.; Krasuska, U.; Wal, A.; Zak, J.; Gniazdowska, A. NO and Metabolic Reprogramming under Phytotoxicity Stress. In *Nitric Oxide in Plant Biology*; Elsevier: Amsterdam, The Netherlands, 2022; pp. 297–318.
21. Chan, X.-Y.; Hong, K.-W.; Yin, W.-F.; Chan, K.-G. Microbiome and Biocatalytic Bacteria in Monkey Cup (*Nepenthes* Pitcher) Digestive Fluid. *Sci. Rep.* **2016**, *6*, 20016. [CrossRef]
22. Bittleston, L.S.; Wolock, C.J.; Yahya, B.E.; Chan, X.Y.; Chan, K.G.; Pierce, N.E.; Pringle, A. Convergence between the Microcosms of Southeast Asian and North American Pitcher Plants. *Elife* **2018**, *7*, e36741. [CrossRef]
23. Hossain, M.S.; Rahman, G.K.M.M.; Solaiman, A.R.M.; Alam, M.S.; Rahman, M.M.; Mia, M.A.B. Estimating Electrical Conductivity for Soil Salinity Monitoring Using Various Soil-Water Ratios Depending on Soil Texture. *Commun. Soil Sci. Plant Anal.* **2020**, *51*, 635–644. [CrossRef]
24. Huliusz, P. Quantitative and Qualitative Diferentiation of Soil Salinity in Poland. *Ber. Dtsch. Bodenkd. Ges.* **2008**, *1*, 1–4.
25. Gonçalves, S.; Ferraz, M.; Romano, A. Phytotoxic Properties of Drosophyllum Lusitanicum Leaf Extracts and Its Main Compound Plumbagin. *Sci. Hortic.* **2009**, *122*, 96–101. [CrossRef]
26. Li, Z.-H.; Wang, Q.; Ruan, X.; Pan, C.-D.; Jiang, D.-A. Phenolics and Plant Allelopathy. *Molecules* **2010**, *15*, 8933–8952. [CrossRef] [PubMed]
27. Macías, F.A.; Castellano, D.; Molinillo, J.M.G. Search for a Standard Phytotoxic Bioassay for Allelochemicals. Selection of Standard Target Species. *J. Agric. Food Chem.* **2000**, *48*, 2512–2521. [CrossRef]
28. Gniazdowska, A.; Bogatek, R. Allelopathic Interactions between Plants. Multi Site Action of Allelochemicals. *Acta Physiol. Plant* **2005**, *27*, 395–407. [CrossRef]
29. Staszek, P.; Krasuska, U.; Ciacka, K.; Gniazdowska, A. ROS Metabolism Perturbation as an Element of Mode of Action of Allelochemicals. *Antioxidants* **2021**, *10*, 1648. [CrossRef]
30. Soltys, D.; Rudzińska-Langwald, A.; Gniazdowska, A.; Wiśniewska, A.; Bogatek, R. Inhibition of Tomato (*Solanum Lycopersicum* L.) Root Growth by Cyanamide Is Due to Altered Cell Division, Phytohormone Balance and Expansin Gene Expression. *Planta* **2012**, *236*, 1629–1638. [CrossRef]
31. Graña, E.; Sotelo, T.; Díaz-Tielas, C.; Araniti, F.; Krasuska, U.; Bogatek, R.; Reigosa, M.J.; Sánchez-Moreiras, A.M. Citral Induces Auxin and Ethylene-Mediated Malformations and Arrests Cell Division in *Arabidopsis Thaliana* Roots. *J. Chem. Ecol.* **2013**, *39*, 271–282. [CrossRef]
32. Haugland, E.; Brandsaeter, L.O. Experiments on Bioassay Sensitivity in the Study of Allelopathy. *J. Chem. Ecol.* **1996**, *22*, 1845–1859. [CrossRef]
33. Shekari, F.; Shekari, F.; Najafi, J.; Abassi, A.; Radmanesh, Z.; Bones, A.M. Phytotoxic Effects of Catnip (Nepeta Meyeri Benth.) on Early Growth Stages Development and Infection Potential of Field Dodder (Cuscuta Campestris Yunck). *Plants* **2022**, *11*, 2629. [CrossRef]
34. Oracz, K.; Voegele, A.; Tarkowská, D.; Jacquemoud, D.; Turečková, V.; Urbanová, T.; Strnad, M.; Sliwinska, E.; Leubner-Metzger, G. Myrigalone A Inhibits Lepidium Sativum Seed Germination by Interference with Gibberellin Metabolism and Apoplastic Superoxide Production Required for Embryo Extension Growth and Endosperm Rupture. *Plant Cell Physiol.* **2012**, *53*, 81–95. [CrossRef] [PubMed]
35. Oracz, K.; Bouteau, H.E.-M.; Farrant, J.M.; Cooper, K.; Belghazi, M.; Job, C.; Job, D.; Corbineau, F.; Bailly, C. ROS Production and Protein Oxidation as a Novel Mechanism for Seed Dormancy Alleviation. *Plant J.* **2007**, *50*, 452–465. [CrossRef] [PubMed]
36. Inderjit; Weston, L.A.; Duke, S.O. Challenges, Achievements and Opportunities in Allelopathy Research. *J. Plant Interact.* **2005**, *1*, 69–81. [CrossRef]
37. Williams, R.; Hoagland, R. Bioassays- Useful Tools for the Study of Allelopathy. In *Allelopathy*; CRC Press: Boca Raton, FL, USA, 2003; pp. 315–351.
38. Gniazdowska, A.; Krasuska, U.; Andrzejczak, O.; Soltys, D. Allelopathic Compounds as Oxidative Stress Agents: Yes or NO. In *Reactive Oxygen and Nitrogen Species Signaling and Communication in Plants, Signaling and Communication in Plants*; Springer: Cham, Switzerland, 2015; Volume 23, pp. 155–176.
39. Krasuska, U.; Andrzejczak, O.; Staszek, P.; Borucki, W.; Gniazdowska, A. Toxicity of Canavanine in Tomato (*Solanum Lycopersicum* L.) Roots Is Due to Alterations in RNS, ROS and Auxin Levels. *Plant Physiol. Biochem.* **2016**, *103*, 84–95. [CrossRef]
40. Baker, C.; Mock, M. An Improved Method for Monitoring Cell Death in Cell Suspension and Leaf Disc Assays Using Evans Blue. *Plant Cell Tissue Organ Cult.* **1994**, *39*, 7–12. [CrossRef]
41. Beyer, W.F.; Fridovich, I. Assaying for Superoxide Dismutase Activity: Some Large Consequences of Minor Changes in Conditions. *Anal. Biochem.* **1987**, *161*, 559–566. [CrossRef]
42. Ciacka, K.; Tyminski, M.; Gniazdowska, A.; Krasuska, U. Cold Stratification-Induced Dormancy Removal in Apple (Malus Domestica Borkh.) Seeds Is Accompanied by an Increased Glutathione Pool in Embryonic Axes. *J. Plant Physiol.* **2022**, *274*, 153736. [CrossRef]

plants

MDPI

Article

Phytochemical and Biological Study of Trophic Interaction between *Pseudosphinx Tetrio* L. Larvae and *Allamanda Cathartica* L.

Linda Matignon [1], Mame Marietou Lo [2], Magneric Monpierre [3], Mauro Vicentini Correia [4], Drochss Pettry Valencia [5], Marie-Noëlle Sylvestre [3], Ludovic Pruneau [3], Muriel Sylvestre [1], Antonio Domenech [7], Zohra Benfodda [2], Patrick Meffre [2] and Gerardo Cebrián-Torrejón [1,*]

1 COVACHIM-M2E Laboratory EA 3592, Department of Chemistry, Fouillole Campus, University of the French West Indies, UFR SEN, CEDEX, 97157 Pointe-à-Pitre, France
2 CHROME Laboratory, EA7352, University of Nîmes, CEDEX 1, 30021 Nîmes, France
3 Institut de Systématique, Evolution, Biodiversité (ISYEB), Department of Chemistry, Fouillole Campus, University of the French West Indies, UFR SEN, CEDEX, 97157 Pointe-à-Pitre, France
4 Instituto de Química, Campus Universitário Darcy Ribeiro, Universidade de Brasília, Brasília 70910-900, Brazil
5 Departamento de Ciencias Naturales y Matemáticas, Pontificia Universidad Javeriana sede Cali, Calle 18 No. 118-250, Cali 760031, Colombia
6 Instituto de Química, Universidade Federal Fluminense, Outeiro S. João Batista S/N, Niterói 24020-141, Brazil
7 Departament de Química Analítica, Facultat de Química, Universitat de València, Dr. Moliner 50, 46100 Valencia, Spain
* Correspondence: gerardo.cebrian-torrejon@univ-antilles.fr

Citation: Matignon, L.; Lo, M.M.; Monpierre, M.; Correia, M.V.; Valencia, D.P.; Palmeira-Mello, M.V.; Sylvestre, M.-N.; Pruneau, L.; Sylvestre, M.; Domenech, A.; et al. Phytochemical and Biological Study of Trophic Interaction between *Pseudosphinx Tetrio* L. Larvae and *Allamanda Cathartica* L.. *Plants* **2023**, *12*, 520. https://doi.org/10.3390/plants12030520

Academic Editors: Margot Schulz and Vincenzo Tabaglio

Received: 6 December 2022
Revised: 13 January 2023
Accepted: 16 January 2023
Published: 23 January 2023

Abstract: In this article, we propose to explore the chemical interaction between *Pseudosphinx tetrio* L. and *Allamanda cathartica* L. using different analytical methods, including an innovative electrochemical approach (called electrochemical ecology) and multivariate analysis, and we investigate the potential antimicrobial effects (antibacterial and antifungal activities) of this interaction in order to gain a better understanding of their specific interaction. The analytical study presents a similar chemical profile between the leaves of healthy and herbivorous *A. cathartica* and the excretions of the caterpillars. The similar analytical profile of the leaves of *A. cathartica* and the excretions of *P. tetrio*, and the difference with the caterpillar bodies, suggests a selective excretion of compounds by the caterpillar. The measured antimicrobial activities support the physicochemical tests. The natural products found selectively in the excretions (rather than in the body) could explain the ability of *P. tetrio* to feed on this toxic Apocynaceae species.

Keywords: chemical ecology; electrochemical ecology; aposematism; allelopathy; plant–insect interactions; *Pseudosphinx tetrio*; *Allamanda cathartica*; phytochemical analysis

1. Introduction

The tetrio sphinx, *Pseudosphinx tetrio* L. (Lepidoptera: Sphingidae), is a moth widely distributed in tropical and subtropical regions of the Americas and the Caribbean basin, where suitable host plants for the development of its larvae (large, conspicuous caterpillars) are present [1–10].

Adult females of *P. tetrio* [11] lay 50–100 eggs on the underside of the host plant's leaves, from which they hatch in about three days [3]. After hatching, the gregarious larvae live in colonies for at least the first three developing stages, feeding initially on the upper surface of leaves and in later stages on all leaf tissues except leaf veins and midribs as they progress to the later instars. Adult larvae (fifth or sixth stages) feed singly or in small groups on entire leaves can reach a length of 63–69 mm [3,4]. The color of larvae is characteristic of aposematism (antipredation strategy: warning predators of bad taste or toxicity). According to an aposematic hypothesis, it becomes toxic to its predators when it ingests toxic phytocomponents from the milky sap of some plants [12]. The body is velvety black with yellow rings on each thoracic and abdominal segment, the head is dark orange-red, and the prothoracic shield, prominent legs and anal segment are orange-red with dark markings. The spine on the eighth abdominal segment is black and emerges from a raised, orange, button-like structure [4,13–15]. The aposematic larvae have few predators due to their venomous nature. However, the squirrel cuckoo, *Piaya cayana* Linn., of Belize, consumes *P. tetrio* by taking larvae, beating them against a branch until the poisonous gut contents are gone, and then swallowing the remains [16]. The larvae wrap themselves in a cocoon of silk or dead leaves and molt into new pupae about 7.0 cm long. The adult moths, colored in shades of gray, hatch from the pupae 53 days after egg laying and have a wingspan of 12.7–14.0 cm (eggs hatch in 3 days or more; average duration of the 5–6 larval stages is 23–4 days; the prepupal stage lasts about 4 days and the pupal stage about 22 days [3]). *P. tetrio* has the ability to eat up to twice its weight in food during a day, hence its name, "glutton caterpillar". The larva of *P. tetrio* feeds preferentially on the leaves of the Apocynaceae family's plants, such as *Allamanda cathartica* L. and *Plumeria alba* L., which are rich in toxic latex [17,18].

A. cathartica is commonly known as yellow allamanda, golden trumpet or "*Liane à lait*" [19]. This is a common name for the species. In the review paper by Ghosh and coworkers, it was described as a hermaphroditic, vine-like, woody shrub used primarily in landscaping, more spreading than tall, and with perennial leaves [20]. *A. cathartica* is a fast-growing species that is widely distributed worldwide, but mainly in tropical and subtropical regions, where the caterpillar *P. tetrio* is present [21,22]. *A. cathartica* is one of the fifteen species of the Apocynaceae family according to the "WFO Plant List" [23].

The plant is native to Brazil [24] and its various parts such as leaves, flowers, roots and stems are used in traditional medicine [25] for the treatment of jaundice [26], and have various pharmacological activities such as wound healing [26], antioxidant [27–30], antimicrobial [30–32], antifungal [33], antimalarial [34], anti-inflammatory [35]; anticancer [20,36,37] and gastrointestinal effects [38]. A dichloromethane (CH_2Cl_2) extract of the whole plant tested on two pathogenic dermatophytes, *Trichophyton rubrum* and *Microsporum gypseum*, showed moderate activity at a concentration of 50 µg/disk, but very strong activity at a concentration of 200 µg/disk, suggesting that *A. cathartica* may possess antidermatophyte constituents that could be useful in the treatment of ringworm [39,40].

In general, insects are not strictly restricted to the leaves and reproductive parts of living plants. Most studies conducted on strict herbivores [41] indicate that larval hosts of sphingids are concentrated on a few plant families whose foliage is particularly rich in alkaloids, milky sap, essential oils, irritant acids or other small toxic natural products (e.g., Rubiaceae, Euphorbiaceae, Apocynaceae, Vitaceae, Bignoniaceae, etc.).

However, all parts of *A. cathartica* are rich in secondary metabolites, which are exhaustively listed in the review by Petricevich and Abarca-Vargas [42], with plumieride, plumericin, and allamandin being the most characteristic. A total of 151 compounds were distributed as follows: 3 hydrocarbons in flowers; 7 alcohol compounds, 9 esters, 1 ether, 6 aldehydes, and 1 ketone in flowers, leaves, and stems extracts; 37 fatty acids and phospholipids; 43 volatile compounds mostly in flowers and leaves; 5 phenolic compounds and 6 flavonoids in the flowers and stems; 2 alkaloids in stems; 11 steroids and terpenes in leaves, stems and flowers; and 14 lactones in roots, stems, leaves, flowers and bark; 6 carbohydrates in leaves, stems, and nectar. Of the compounds identified, leaves contain: ursolic acid, β-amyrin; β-sitosterol, sesquiterpenes, plumericin, plumieride, long chain esters, flavonoids, polyphenols, allamandin, alkaloids, saponins and carbohydrates; stem and bark contain: ursolic acid, β-amyrin, β-sitosterol, triterpenoids, glucosides, alkaloids, flavonoids and polyphenols; flowers contain: quercetin, quercitrin, kaempferol, hesperetin, flavonoids, polyphenols and polysaccharides; and roots contain: lactones, allamandin, allamandicin, allamdin, plumieride iridoids, triterpenoids, alkaloids and several glucosides [36,38,43] (cited by [44]); [19,21,27,42]. Like *A. cathartica*, various parts of *P. rubra* are widely used in traditional medicine [45]. *P. rubra* contains many chemical constituents such as glycosides, phenolic compounds (phenolic acids and flavonoids), alkaloids, amino acids and terpenoids (iridoids) which give the flower antibacterial and antifungal activities. However, the main chemical constituents responsible for the pharmacological activities of *P. rubra* are: plumieride, fulvoplumierine, lupeol, rubrinol, stigmasterol, oleanolic acid, taraxasteryl acetate, rubranonoside, plumieride-p-E-coumarate, isoplumericin, rubrajalellol and plumericin [46–49]. These molecules form plant may have antimicrobial activities and effects and may be of interest for use against pathogenic microorganisms.

In nature, predatory insects are highly host-specific and exhibit a variety of traits. Direct associations between plants and insects are ubiquitous, and in the context of tropical chemical ecology, many different plant–herbivore interactions are highly specialized [50]. What attracts an egg-laying insect to one plant and prevents it from laying its eggs on another is often an aspect of plant chemistry that is recognized by the insect [51]. Nevertheless, plants can biosynthesize protective secondary metabolites (in the milky sap or via vascular tissue or in specialized tissues) in a self-defense process to reduce herbivory. In turn, herbivores may respond to these compounds and attempt to metabolize them, selectively excrete them, or use them (after ingestion into the body) for their own defense [52,53]. In 2022, McCoy and colleagues showed that for Eocene herbivorous insects that eat leaves with such defense mechanisms, a distinct burrowing or cutting behavior that disrupts the supply of protective compounds distal to the plant tissue allowed the insect to eat the leaf [54].

The coevolution of specialized plant–insect herbivore interactions has been the subject of decades of study [55–57]. Lepidoptera caterpillars, for example, often specialize on toxic plants and are able to either sequester [58–60], metabolize [61,62] or excrete chemical compounds from their hosts in an unmodified form [63]. Understanding how herbivores deal with toxic phytochemicals in their diet is necessary to understanding the evolution of the two most biodiverse groups of multicellular organisms: plants and insects

In this article, we propose to explore the chemical interaction between *P. tetrio* and *A. cathartica* using different analytical methods, including an innovative electrochemical approach (called electrochemical ecology) and multivariate analysis, and we investigate the potential antimicrobial effects (antibacterial and antifungal activities) of this interaction, in order to gain a better understanding of their specificity.

2. Results and Discussion

2.1. Phytochemical Tests

Thin layer chromatography (TLC). First, a TLC study was conducted. The different results of this approach are presented in Table 1. In our first hypothesis (Hyp. 1), we propose that the caterpillars take up and store certain metabolites (for example, the metabolite with an Rf (Retention factor) of 0.16 on TLC 3 (orange color)). The second hypothesis proposed (Hyp. 2), also listed in Table 1, refers to the selective excretion of an unidentified molecule by the caterpillars (for example, the metabolite with an Rf of 0.95 on TLCs 2 and 3 (blue color)). Finally, the third hypothesis (Hyp. 3) refers to the response of the healthy plant to herbivory. For example, the metabolite with an Rf of 0.28 on TLC 1 (pink color), since this metabolite is present in healthy leaves and not in predated leaves. The opposite case is observed with the metabolite with an Rf value of 0.18 on TLC 4 (green color), which is present only in predated leaves and could therefore be considered a metabolic marker for herbivores.

Table 1. Organic phases obtained with TLC results. (A) Healthy *A. cathartica* leaves, (B) herbivored *A. cathartica* leaves, (C) caterpillar bodies and (D) caterpillars' feces. Solvent percentages are presented as highest polarity in solution (MeOH and EtOAc) *.

Elution Solvent	% *	TLC	R_f	A (1.2)	B (1.3)	C (1.4)	D (1.5)
CH_2Cl_2	100	1	0.16				
			0.28	Hyp. 1 and 3		Hyp. 1	
			0.58				
CH_2Cl_2/MeOH	5	2	0.05				
			0.19				
			0.22				
			0.60				
			0.91				
			0.95	Hyp. 2	Hyp. 2		Hyp. 2
	10	3	0.16	Hyp. 1	Hyp. 1	Hyp. 1	
			0.20				
			0.57				
			0.90				
			0.95	Hyp. 2	Hyp. 2		Hyp. 2
	1	4	0.18		Hyp. 1 and 3	Hyp. 1	
			0.30	Hyp. 1 and 3		Hyp. 1 and 3	
			0.79				
			0.93				
			0.94		Hyp. 2 and 3		Hyp. 2
CH_2Cl_2	100	5	0.08				
			0.17				
			0.37				
			0.62				
			0.93				
C_6H_{14}	100	6	0.03				
			0.37	Hyp. 1	Hyp. 1	Hyp. 1	
$C_6H_{14}/C_4H_8O_2$	20	7	0.17				
			0.30				
			0.34	Hyp. 1	Hyp. 1	Hyp. 1	
			0.64				
			0.94		Hyp. 2 and 3		Hyp. 2
			0.97				
			0.98				
			1				
	50	8	0.48		Hyp. 1 and 3	Hyp. 1	
			0.95				
			0.97				
	80	9	0.50				
			0.91				
			0.94				

* Similarly, metabolites between two extracts or more are described in colors: orange: A, B and C; blue: A, B and D; pink: A and C; yellow: B and D. Hypotheses that can be illustrated by each compound are cited.

Nuclear magnetic resonance spectroscopy ^1H NMR. The ^1H NMR data were obtained only from the organic extracts of the plants and caterpillars using DMSO-d^6 as a solvent. Solvent signals from deuterated DMSO and residual water at 2.5 and 3.3 ppm, respectively, were removed (Figure 1). The analysis shows minor differences between the spectra of healthy and predated *Allamanda* leaves. This result is inconsistent with hypothesis 3 (in which a plant response to aggression was observed).

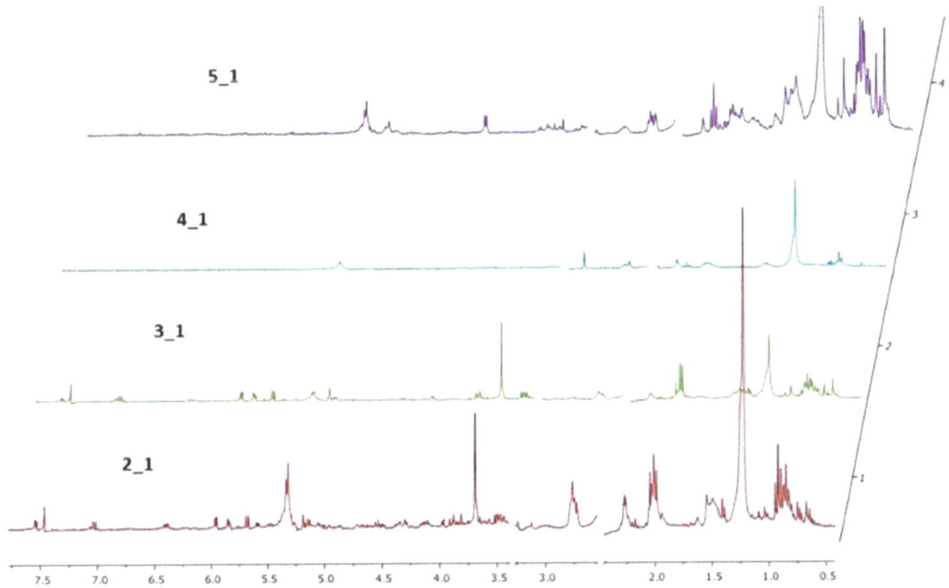

Figure 1. ^{1}H NMR of organic extracts in DMSO-d^6. (2_1) Healthy *Allamanda* leaves; (3_1) predated *Allamanda* leaves; (4_1) caterpillar bodies and (5_1) caterpillar feces.

The ^{1}H NMR spectra of caterpillars' bodies and feces show large differences between the samples, and the chemical profile of healthy *Allamanda* leaves and the caterpillar feces is very similar. This result confirms hypothesis 2 and shows the similarity between the chemical profile of *A. cathartica* leaves and caterpillar feces compared to the caterpillar body. These similarities and dissimilarities confirm our hypothesis (Hyp. 3) of the selective excretion of toxic compounds.

High performance liquid chromatography coupled with mass spectrometry (HPLC-MS). In this study, the samples containing the healthy and eaten leaves of *A. cathartica*, as well as the caterpillars and their feces, were analyzed using high-performance liquid chromatography techniques [19,64–66] (Figure 2). The similar profile observed in these spectra indicates the presence of the same main compounds in the healthy (A) and predated (B) *A. cathartica* leaves, with similar retention times (RT) of 20.00 min and 20.50 min, respectively. A second compound can be identified in these spectra with comparable retention times to hypothesis 3 (RTs: 25.52 min and 26.02 min). These results are in contrast to hypothesis 3 (as we can see in the TLCs and ^{1}H NMR profiles). The analysis shows the presence of this compound in the caterpillar's feces (D, RT: 25.73 min) but not in its body (C), which also confirms hypothesis 2 (in agreement with the ^{1}H NMR and TLC results). Finally, the results show that the compounds found in *P. tetrio*'s bodies (C, RT: 20.43 min and 25.66 min) are significantly different from other compounds found in the feces (also confirming hypothesis 2).

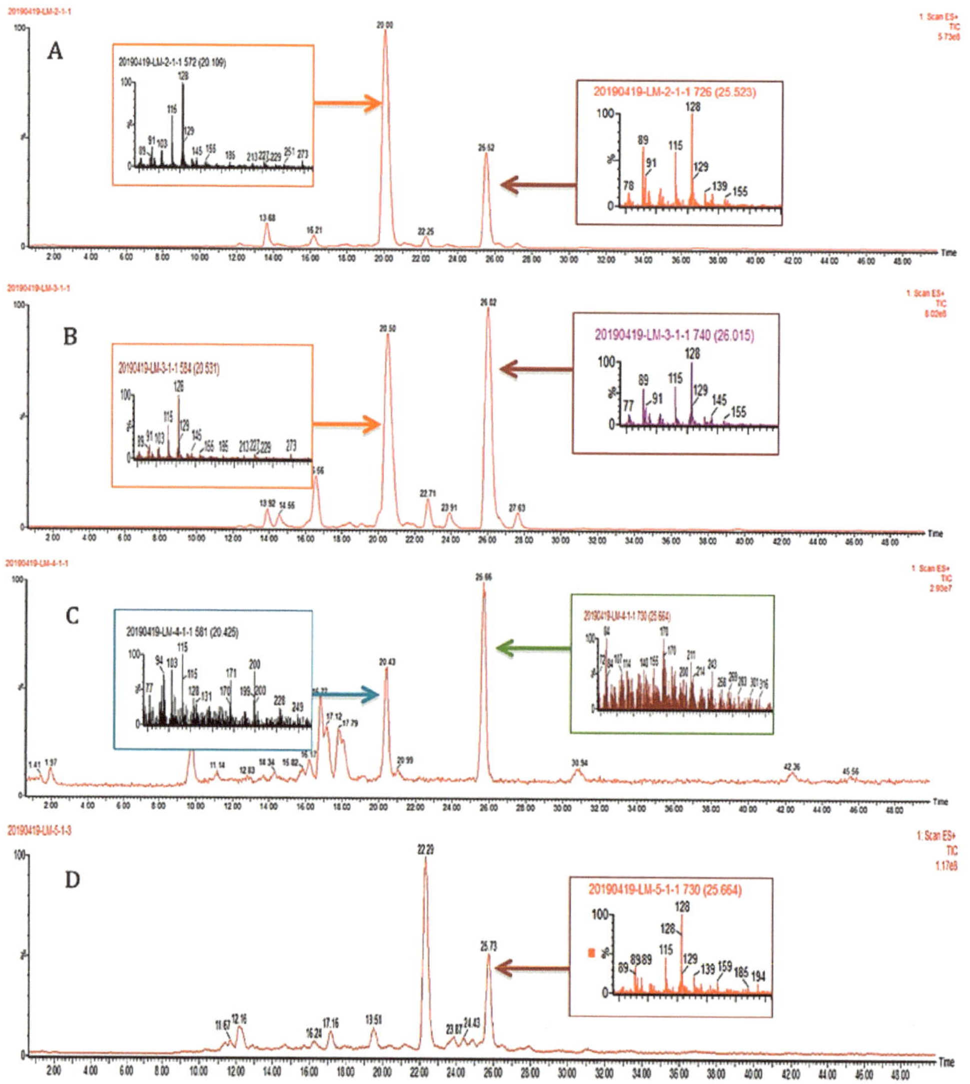

Figure 2. HPLC-MS profiles of (**A**) healthy and (**B**) predated leaves of *A. cathartica* and (**C**) caterpillar bodies and (**D**) caterpillar feces.

Electrochemical investigation. Cyclic voltammetry (CV) studies have been used to investigate possible interactions between different compounds and were previously used to monitor plant defenses against external stressors [67–70]. Here, the predation of *A. cathartica* by *P. tetrio* has been monitored electrochemically using CV. We studied the electrochemical response of films (on the surface of a glassy carbon electrode) of organic and aqueous extracts of healthy and predated leaves of the plant and samples of bodies and feces of *P. tetrio* in 0.10 mol L^{-1} phosphate solution buffer at pH 7.0. The reported methodology is ultimately based on the voltammetry of microparticles (VMP) technique [71].

The behavior of films on glassy carbon electrodes of healthy and predated leaves of *A. cathartica* is shown in Figure 3. A series of chemically irreversible oxidation processes: A_1, A_2 and A_3, were observed (0.28 V, 0.65 V and an undefined shoulder at 1.00 V, respectively). While the peaks A_1 and A_2 decrease with cycling until a fully passivated and stable state is reached at about the second cycle, an increase in the A_3 peak was observed. Thus, the variations in the ratio of peaks A_1 and A_3 suggest differences in the nature of the redox species of *A. cathartica* and in leaves that are strengthened by them. These oxidation processes represent a profile observed in several plant species and are due to the oxidation of polyphenolic organic compounds. Indeed, the voltammogram of the predated leaves exhibits the same signals as healthy leaves, but peak A_1 is depleted while the A_3 signal is enhanced. Finally, a reduction process can be observed at about -1.0 V when the potential sweep is reversed to less positive potentials. Although these processes are slightly more negative in healthy leaves than in predated leaves, both are consistent with the typical behavior of the oxygen reduction reaction (ORR) in water. This result basically contradicts hypothesis 3 about the chemical response to the herbivory, but using this technique, we can identify slight differences between healthy and predated leaves, possibly related to polyphenols (reaction against oxidative stress) [72].

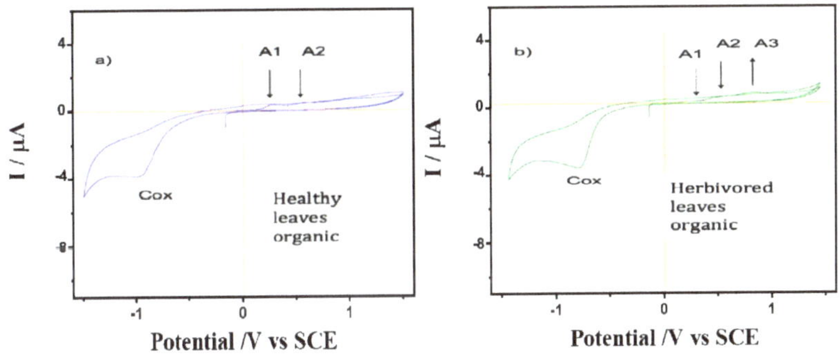

Figure 3. Cyclic voltammograms of films from (**a**) healthy and (**b**) predated leaves of *A. cathartica* on glassy carbon electrode in contact with air-saturated 0.10 M phosphate buffer at pH 7.0. Scan rate 10 mV.s^{-1}.

The CV of films on glassy carbon electrodes of *P. tetrio*'s bodies and its excrement after predation of leaves of *A. cathartica* are shown in Figure 4. CV shows no clear oxidation process in the initial anodic scan for the body of *P. tetrio*, but three oxidation processes for its excreta at 0.22, 0.45 and 0.78 V for A_1, A_2 and A_3, respectively, which are similar to those found for the *A. cathartica* analysis. Again, a typical oxygen reduction reaction (ORR) in water was observed in the cathodic region in both analyses. In addition, a characteristic signal C_1 appears in both samples, indicating the presence of oxidized metabolite of *P. tetrio*. This result confirms hypothesis 2, in agreement with the other analytical approaches.

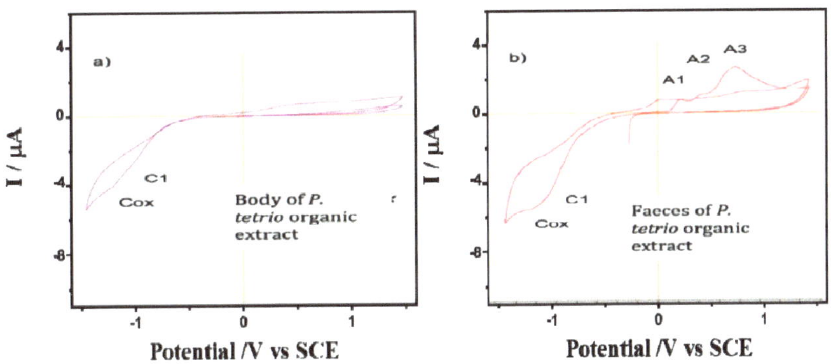

Figure 4. Cyclic voltammograms of films from (**a**) *P. tetrio*'s bodies and (**b**) *P. tetrio*'s excrement after predation on leaves of *A. cathartica* on glassy carbon electrode in contact with air-saturated 0.10 M phosphate buffer at pH 7.0. Potential scan rate 10 mV.s^{-1}.

A comparison was made between the CV of films from the water extracts of healthy and spoiled leaves of *A. cathartica* in contact with air-saturated 0.10 mol L^{-1} phosphate buffer at pH 7.0 (Figure 5). Here, signal A_1 appears clearly recorded in the healthy leaves and is accompanied by a well-defined cathodic peak at—1.25 V (C_2). Signal C_2 disappears in the CV of the damaged leaves, while signal A_1 decreases and signal A_3 is clearly visible at 0.78 V. This result confirms hypothesis 3, which shows a reaction of leaves against herbivory; the disappearance of signal C_2 and the appearance of signal A_3 are a result of herbivory. This CV is reproduced to some extent reproduced in the aqueous extracts of *P. tetrio* and its excrement, as shown in Figure 6, thus suggesting the existence of common electroactive compounds.

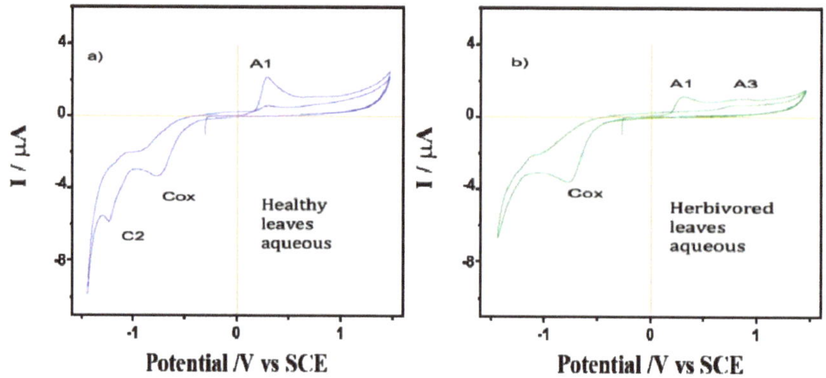

Figure 5. Cyclic voltammograms of films from the water extracts of (**a**) healthy and (**b**) predated leaves of *A. cathartica* on glassy carbon electrode in contact with air-saturated 0.10 M phosphate buffer at pH 7.0. Potential scan rate 10 mV.s^{-1}.

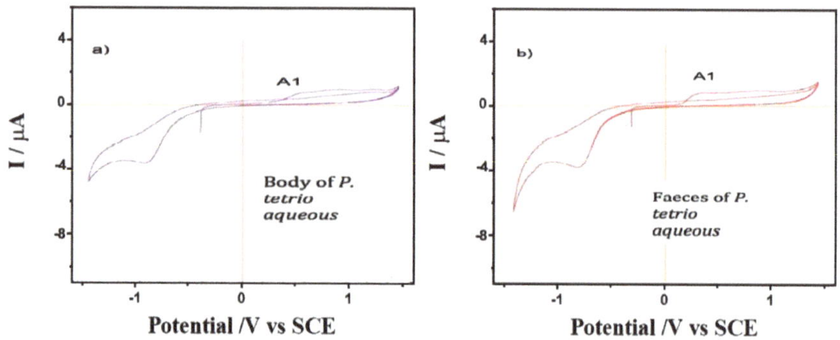

Figure 6. Cyclic voltammograms of films from water extracts of (**a**) *P. tetrio*'s bodies and (**b**) its excrement after predation on leaves of *A. cathartica* on glassy carbon electrode in contact with air-saturated 0.10 M phosphate buffer at pH 7.0. Potential scan rate 10 mV.s^{-1}.

Principal Component Analysis (PCA). A total of 24 chromatograms were acquired using HPLC-EI-MS, 12 from the aqueous fraction and 12 from the organic fraction (three biological replicates for each species). To identify differences in the metabolites or even in their concentration in the samples, the set of 24 samples was analyzed, obtained from XCMS and used for data analysis. Two spreadsheets were obtained after data processing of the HPLC-MS analysis, with 168 and 412 variables from the organic and aqueous fractions, respectively (retention time m/z). In the PCA analysis, two plots were generated, namely score and loading plots. The first one shows the sample groupings, whereas the second one indicates the contribution of each variable to these samples. The PCA analysis was focused on the detection of any inherent pattern within the data. As observed in Figure 7, samples of healthy and predated *Allamanda* leaves are closely related, with small differences from each other, when compared with the caterpillars' bodies and feces. Further, when all samples are compared, the predated and not predated leaves from *Allamanda* are very similar to each other, and different from the other samples. In addition, the caterpillar body extracts present a great difference. In Figure 8 (loadings plot), the variables that influence the grouping observed are presented.

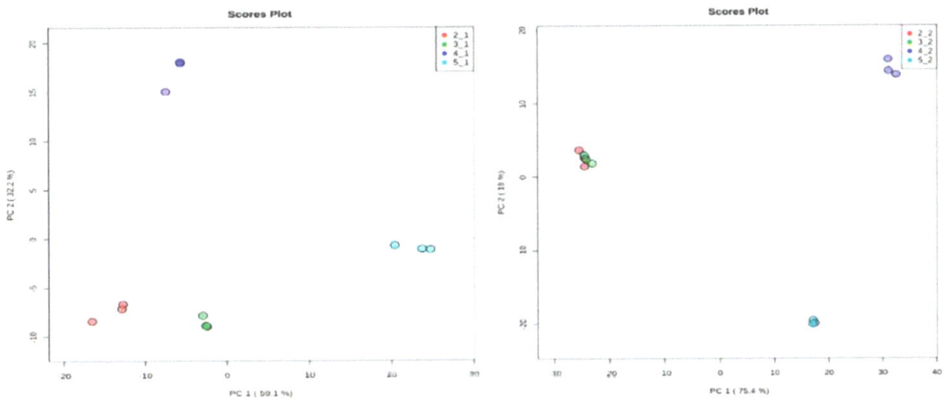

Figure 7. Score plot of 12 samples of organic and aqueous extracts of all manipulation, obtained from 168 (**left**) and 412 (**right**) variables, respectively. (2_1and 2_2) red circle, healthy *Allamanda* leaves; (3_1and 3_2) green circle, predated *Allamanda* leaves; (4_1and 4_2) blue circle, caterpillar bodies and (5_1and 5_2) light blue circle, caterpillar feces.

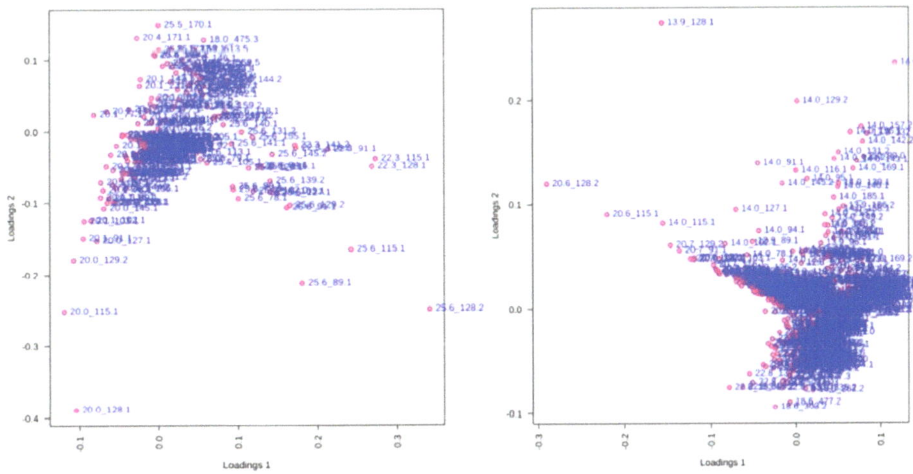

Figure 8. Score plots of 12 samples of organic and aqueous extracts of all manipulation, obtained from 168 (**left**) and 412 (**right**) variables, respectively. Each dot represents one variable (retention time m/z).

2.2. Microbial Activities

The potential antimicrobial activities of *P. tetrio*/*A. cathartica* interaction are shown in Table 2. The yellow boxes show the effects of the consumption of the *A. cathartica* leaves by the caterpillar. The comparison between the healthy leaves (A) and the leaves eaten by the caterpillar (B) shows an increase or occurrence of inhibitory activity in favor of the leaves eaten by the caterpillar (B) on the microorganisms, regardless of the type of extraction (Aqueous ($_A$); Organic ($_O$)), A_O versus B_O for *Escherichia coli* and *Candida albicans* (diameter of inhibition: 7 mm versus 0 mm; 9,5 mm versus 7 mm); and A_A versus B_A for *Escherichia coli* and *Candida albicans* (diameter of inhibition: 7 mm versus 0 mm; 9 mm versus 8 mm)). The occurrence or enhancement of this bioactivity in the predated leaf could be explained by the induction of a chemical or biological defense reaction of the host plant against herbivory (proposed hypothesis 3).

Table 2. Microbial activity results of aqueous and organic extracts, obtained from *Allamanda cathartica* (*A. cathartica*) leaves, caterpillar bodies and caterpillar feces on three bacterial (*Escherichia coli; Staphylococcus aureus; Pseudomonas aeruginosa*) and two fungal (*Candida albicans; Aspergillus fumigatus*) strains.

Extracts (100 µg/disc) Microorganisms	Caterpillar Feces (D)		Caterpillar Bodies (C)		Healthy *A. cathartica* Leaves (A)		Predated *A. cathartica* Leaves (B)	
	D_A	D_o	C_A	C_o	A_A	A_o	B_A	B_o
Escherichia Coli	-	+++ (11)	+ (7)	-	-	-	+ (7)	+ (7)
Staphylococcus aureus	-	++ (9.5)	-	-	-	+ (8.5)	-	+ (8)
Pseudomonas aeruginosa	-	-	-	++ (9)	-	+ (7)	-	+ (8,5)
Candida albicans	++ (8,5)	++ (9)	+ (7)	+ (7)	+ (8)	+ (7)	++ (9)	++ (9,5)
Aspergillus fumigatus	-	+++ (10,5)	-	-	-	+++ (10)	-	+++ (11)

Legend: Aqueous extracts ($_A$); organic ($_O$) extracts; healthy *A. cathartica* leaves (A); predated *A. cathartica* leaves (B); caterpillar bodies (C); caterpillar feces (D), on three bacterial (*Escherichia coli; Staphylococcus aureus; Pseudomonas aeruginosa*) and two fungal (*Candida albicans; Aspergillus fumigatus*) strains. (mm): size of the inhibition halos in millimeters. (-) no inhibitory activity; (+ to +++) more or less intense inhibitory activity.

The green and blue boxes, respectively, represent the non-uptake and the uptake of bioactive compounds from the leaves into the bodies of the caterpillars. The samples obtained after extraction with organic solvents ($_O$) show the same evolution of the inhibitory profile in three of the five microorganisms tested (*Escherichia coli, Staphylococcus aureus* and *Aspergillus fumigatus*). Our results suggest a selective excretion of bioactive compounds from ingested leaves into the feces without being retained or ingested in the caterpillar's body (green boxes). Indeed, the feces of *P. tetrio*'s (D_O) show inhibitory activity that is as strong or stronger than that of the eaten *A. cathartica* leaves (B_O) [diameter of inhibition fecal matter (D_O) versus eaten leaves (Bo): 11 mm versus 7 mm for *Escherichia coli;* 9.5 mm versus 8 mm for *Staphylococcus aureus;* and 10,5 mm versus 11 mm for *Aspergillus fumigatus*) whereas caterpillar body extracts (C_O) showed no bio-inhibition (proposed hypothesis 2). The inhibitory effect of the organic and aqueous samples on *Pseudomonas aeruginosa* and *Escherichia coli*, respectively (blue boxes), shows an opposite profile. Indeed, the eaten leaves and the caterpillar body show a bio-inhibition that is not found in the caterpillar feces. This suggests that certain bioactive compounds in the eaten leaves, different from the previous ones, could be selectively taken up and stored specifically in the caterpillar body.

There are few data in the literature on the potential antimicrobial activity of predated leaves and the interaction between caterpillars and plants. Our samples showed higher antimicrobial activity than that reported by Islam et al. (2010) [31] for a concentrated methanol extract of healthy *A. cathartica* leaves fractionated into fractions soluble in petroleum ether, carbon tetrachloride, chloroform and water. The zones of inhibition observed by them have a diameter of less than 10 mm for *E. coli, S. aureus, C. albicans* and *P. aeruginosa*, at a tested concentration of 400 µg disk^{-1} against the 100 µg disk^{-1} in our case. At the same time, they showed higher antimicrobial activity with their methanolic and ethyl acetate extracts from healthy leaves [73]. Quercitrin was explicitly attributed antifungal and antibacterial activity [43].

Our qualitative results on biological activity may explain why *P. tetrio* can feed on this toxic Apocynaceae species. Indeed, the spectral (NMR) profiles of the eaten leaves and the caterpillar droppings are very similar, but very different from those of the caterpillar bodies. There is an association between the plant and caterpillar (specific herbivory). The plant produces organic chemical compounds for defense when the caterpillar attacks, and the caterpillar excretes or assimilates these molecules or derivatives by its own mechanism. The inhibition may be due to a synergistic effect of the different molecules contained in the extracts for all the microorganisms, hence the interest in using different analytical techniques. The genetic and cellular mechanisms by which metabolite diversity arises are becoming better understood, but the evolutionary explanations for the continued diversification of plant secondary metabolites have received less attention. Speed et al. (2019) [74] show a fundamental coevolutionary asymmetry between plants and their herbivores, which is that herbivores must resist all plant toxins, whereas plants must challenge and override only a single resistance trait.

Sequestration likely evolved as a protection from predators [75], [76] because caterpillars with chemical protection are less attractive to predators [77]. *P. tetrio* feeds on toxic species of the Apocynaceae and is aposematically colored, leading to the hypothesis that *P. tetrio* caterpillars may sequester compounds from their host. In many plant species, herbivory leads to the increased production of a secondary metabolite [78]. In particular, herbivores specialized in sequestration are thought to benefit from the induction of intermediate levels of chemical defenses, while plants eaten by such herbivores would benefit most from weak or strong induction. On the other hand, specialists that do not specialize in sequestration may not be deterred by induction unless it is high [55]. Caterpillars also have the ability to metabolize some organic chemical compounds and excrete others intact from the same host plant [63]. This is the case in Monarch butterflies, *Danaus plexippus* Linn. [58]. The study of Ramos et al., 2015 [79], compared the proteolytic system of the gut of *P. tetrio* and *D. plexippus* with the proteolytic system of the milky sap of their respective host plants (*P. rubra* and *Calotropis procera* Aiton). This revealed that the ability of the insect proteolytic systems (serine and cysteine peptidase inhibitors) to digest the milk sap proteins (in vivo) appears to be an important event favoring the caterpillars in overcoming plant chemical defenses.

3. Materials and Methods

3.1. Plant Material

Sample preparation. Healthy and eaten leaves of *A. cathartica*, as well as *P. tetrio* caterpillars were collected on the island of Guadeloupe (French West Indies), specifically in Le Gosier (16°13′00.5″ N 61°31′09.9″ W). The leaves were cleaned (with distilled water) and lyophilized. The caterpillars were kept in a cage for 24 h, to collect their feces, and starved for 48 h to use their corpse in the evaluation of biomolecule incorporation into the bodies of the caterpillars, excluding the digestive material. Finally, the dried leaves, the caterpillars' bodies and feces were lyophilized and powdered. Afterwards, 50 g of each sample was extracted by maceration extraction for 48 h. Maceration was performed in a ternary mixture of dichloromethane/methanol/distilled water [80,81] (1:1:1 *v/v*, 200 mL). After the complete extraction, the solutions were filtered and extracted by liquid–liquid extraction and two phases were obtained for each manipulation (Table 3). Finally, these phases were dried by rotary evaporation.

Table 3. Phases obtained for each manipulation after liquid–liquid extractions.

	Organic Phase	Aqueous Phase
Healthy *A. cathartica* leaves	1.2	2.2
Hervivored *A. cathartica* leaves	1.3	2.3
Caterpillars	1.4	2.4
Caterpillar feces	1.5	2.5

3.2. Phytochemical Tests

Analysis and quantification. Analyses were performed using several techniques such as Thin Layer Chromatography (TLC), Nuclear Magnetic Resonance Spectroscopy (1H-NMR), High Performance Liquid Chromatography coupled with Mass Spectrometry (HPLC-MS) and Cyclic Voltammetry (CV).

Thin layer chromatography (TLC). The thin layer chromatography of all the organic phases was carried out with a gradient of two different mixtures of elution solvents composed of CH_2Cl_2/MeOH or nhexane/AcOEt with a volume of 10 mL in different proportions (80/20, 50/50, 90/10, 95/5, 99/1, 50/50). Further, TLCs of all aqueous phases were made on cellulose TLC with a mixture of H2O/MeOH elution solvent in different proportions (50/50–100) as a mobile phase.

Nuclear magnetic resonance spectroscopy (1H-NMR). 15 mg of each extract was solubilized in 650 µL of DMSO and transferred to NMR tubes. 1D 1H nuclear magnetic resonance spectra were recorded with a BRUKER Avance 300 MHz spectrometer equipped with a BBO probe and automatic tube changer. Chemical shifts (δ) were expressed in ppm relative to tetramethylsilane (TMS) taken as the external reference, with internal calibration performed on the solvent signal. All spectra were processed using Topspin 2.1 software. The classical 1D proton with a 90° pulse width was performed. The spectra were acquired using 256 scans and 2 dummy scans of 32 K data points with a spectral width of 5411.255 Hz.

High performance liquid chromatography coupled with mass spectrometry (HPLC-MS). Triplicates of each extract were made and dissolved in 750 µL of methanol (LCMS grade) and 750 µL of ultra-pure water (ElgaPurelab Classic). The mixtures were sonicated in the ultrasonic bath then filtered on 13 mm and 0.45 µm PTFE filters and placed in amber vials. The analysis was performed on a Waters (Milford, MA, USA) Alliance e2695 liquid chromatographic system, equipped with a Waters 2996 photodiode array detector (PDA), coupled with a orthogonal quadrupole mass spectrometer (Micromass ZQ, Manchester, UK). The systems were controlled by MassLynx v.4.1 software (Micromass, Manchester, UK). The mass was equipped with an electrospray ionization ESI (Waters) source; the ionization was performed in positive mode. A double detection was carried out via mass spectrometry in ESI (range 71 to 1200 Da) and using a PDA diode array detector (UV detection between 210 and 400 nm). The analytical column was an XTERRA MS C18 column (2.1 x 100 mm, 3.5 µm) (Waters). The elution was performed with a mobile phase flow rate of 0.2 mL/min consisting of a mixture of ultra-pure water containing 0.1% formic acid (A) and methanol containing 0.1% formic acid (B). The program started at t = 0 min with a ratio (A: B) of 95:5, at t = 2 min (95:5), at t = 5 min (70:30), at t = 15 min (60:40), at t = 40 min (50:50), at t = 55 min (45:55), at t = 60 min (0:100), at t = 65 min (0:100), at t = 75 min (95:5) and at t = 80 min (95:5). The injection volume was 10 µl and detection was at 280 nm and at 320 nm. The ESI-MS parameters were as follows: desolvation gas (N_2) flow rate: 650 L/h; cone gas flow rate: 40 L/h; drying gas temperature: 450 °C; source temperature: 120 °C; capillary voltage: 3 kV; cone voltage: 85 V; and RF lens voltage: 0.1 V.

Cyclic Voltammetry (CV). Cyclic voltammograms were recorded at 298 ± 1 K in a conventional three electrode cell using a platinum wire auxiliary electrode and an Ag/AgCl (3M NaCl) reference electrode. Measurements were carried out with CH I660 equipment using 0.10 M potassium phosphate buffer at pH 7.0 as a supporting electrolyte. The working electrode was prepared by evaporating 50 µL of an ethanol (EtOH) suspension of the extract of interest, ground leaves or insect samples, under air on a glassy carbon electrode (GCE, BAS MF 2012, geometrical area 0.071 cm^2). In order to mimic the natural environment, no degasification of the electrolyte was performed.

Principal Component Analysis (PCA). The raw data from HPLC-MS were exported as netCDF files, using DataBridge software (Waters, USA), and pre-processed using XCMS Online [82,83] for feature detection, retention time correction and alignment of metabolites detected on HPLC-MS analysis. The dataset was created with 12 samples from each organic and aqueous extract fraction (3 samples from each manipulation (see Table 1)). Peak detection was performed using cent Wave peak detection ($\Delta m/z$ = 10 ppm; minimum peak width, 5 s; maximum peak width, 20 s) and mzwid = 0.015; minfrac = 0.5, and bw = 5 were used for the retention time alignment. The processed data (csv file) were further exported to MetaboAnalyst 4.0 [84]. All data variables were scaled by the pareto method prior to PCA. PCA is an unsupervised method commonly used to identify patterns between multivariate samples [85,86] and was recently employed on the chemical variability of Allamanda cathartica extracts [87].

3.3. Antimicrobial Activities

To investigate the potential antimicrobial effects of the interaction between the caterpillar *Pseudosphinx tetrio* L. and one of its host plants, *Allamanda cathartica*, we focused our study on five pathogen microorganisms that are responsible of severe human disease and usually involved in hospital acquired illness: *Escherichia coli*, *Pseudomonas aeruginosa* and *Staphylococcus aureus* for bacteria and *Candida albicans* and *Aspergillus fumigatus* for fungi. All strains were supplied by the ATCC (American Type Culture Collection).

The three bacteria were cultured on the non-selective medium Tryptic Soy Agar, whereas the two fungi were cultured on the selective medium Sabouraud, which is a selective medium for fungi and yeasts. These two media were provided by the manufacturer Grosseron, France, and prepared according to the instructions.

For each microorganism, starting cultures were prepared on Tryptic Soy Agar media for bacteria and Sabouraud media for fungi. From these starting cultures, microbial suspensions were created by picking two colonies from the starting cultures and adding them to 2 mL of sterile water. Finally, 200 µL of these microbial suspensions were used to inoculate the culture media used to test the antimicrobial activities.

Disc diffusion assay was used in order to test the antimicrobial activities of interaction between *P. tetrio* and *A. cathartica*. Sterile filter paper discs of Whatman no.1 (6 mm in diameter) were placed on the surface of the culture media with sterile forceps and gently pressed to ensure good contact with the surface. The different organic extracts were prepared as follows: 5 mg of each extract was mixed in 500 µL of Dimethyl sulfoxide (DMSO) and 10 µL of these mixtures were placed on antibiotic disks (100 µg/disk). The plates were incubated for 24 h to 48 h at 37 °C for bacteria and 30 °C for fungi. The zone of inhibition was calculated by measuring the diameter of the inhibition zone around the well (mm). All tests were conducted in duplicate.

4. Conclusions

In conclusion, we investigated the plant–herbivore interaction between the caterpillar *Pseudosphinx tetrio* and the flowering plant *A. cathartica*. In order to better understand this trophic relationship, several techniques were used, such as TLC, 1H NMR, HPLC-MS (analyzed using a 13 multivariate PCA) and an innovative approach using electrochemical methods (electrochemical ecology). The measured antimicrobial activities support the physicochemical tests. The results show a similar profile between the leaves of healthy and predated *A. cathartica* and the excretions of the caterpillars. The similar analytical profile between the leaves of *A. cathartica* and the excretions of *P. tetrio*, and the difference with the caterpillar bodies, suggests a selective excretion of compounds by the caterpillar (proposed hypothesis 2). These organic compounds found selectively in the excretions (rather than in the body) could explain the ability of *P. tetrio* to feed on this toxic Apocynaceae species.

Author Contributions: Conceptualization, G.C.-T. and M.S.; methodology, validation, formal analysis, investigation, L.M., M.M.L., M.M., M.V.C., D.P.V., M.V.P.-M., M.-N.S., L.P., A.D., Z.B., P.M. and G.C.-T. All authors have read and agreed to the published version of the manuscript.

Funding: The authors would like to acknowledge the financial support of the "AgroEcoDiv" project from the European Union Fund (FEDER) and "Région Guadeloupe". We would also like the Caribaea Initiative association for a two month study grant from Matignon. DPV would like to thank the OMICAS program: Optimización Multiescala In-silico de Cultivos Agrícolas Sostenibles (Infraestructura y validación en Arroz y Caña de Azúcar) Scientific Ecosystem belonging to the Colombia Científica Program, sponsored by The World Bank, The Ministry of Science, Technology and Innovation (MINCIENCIAS), ICETEX, the Colombian Ministry of Education and the Colombian Ministry of Commerce, Industry and Tourism, under GRANT ID: FP44842-217-2018, OMICAS Award ID: 792-61187.

Data Availability Statement: Not applicable.

Acknowledgments: Thanks are due to Tara Joy Massad (Gorongosa National Park, Mozambique) for her attentive lecture and fruitful discussion.

Conflicts of Interest: The authors declare no conflict of interest.

References

1. Cock, M.J.W. Hawk-moths (Sphingidae) of Trinidad, West Indies: An illustrated and annotated catalogue. *Living World J. Trinidad Tobago Field Nat. Club* **2018**, *2018*, 10–81.
2. Ríos, S.; Drechsel, U. Nuevos registros de Pseudosphinx tetrio (Linnaeus 1771) en el Paraguay (Lepidoptera: Sphingidae). *Parag. Biodivers.* **2017**, *4*, 60–65.
3. Santiago-Blay, J. Notes on *Pseudosphinx tetrio* (L.) (Sphingidae) in Puerto Rico. *J. Lepid. Soc.* **1985**, *39*, 208–214.
4. Dunford, J.; Barbara, K. Tetrio Sphinx, Giant Gray Sphinx, Frangipani Hornworm, *Pseudosphinx tetrio* (Linnaeus) (Insecta: Lepidoptera: Sphingidae)1. *EDIS* **2019**, *2005*, 5. [CrossRef]
5. Cock, M.J. *Pseudosphinx tetrio* (L.) (Lepidotera: Sphingidae) in Trinidad and Tobago. *Living World J. Trinidad Tobago Field Nat. Club* **2008**, *2000*, 49–52.
6. Meurgey, F.; Ramage, T. Challenging the Wallacean shortfall: A total assessment of insect diversity on Guadeloupe (French West Indies), a checklist and bibliography. *Insecta Mundi* **2020**, *786*, 1–183.
7. D'Aguilar, J. Catalogue raisonné des Insectes des Antilles françaises 1. Lépidoptèreq, Sphingidae. *Annales Des Epiphyties 1950* **1966**, *17*, 251–262.
8. Chalumeau, F.; Delplanque, A. Catalogue commenté des Ctenuchidae (Lepidoptera) des Antilles françaises. *Bull. De La Société Linnéenne De Lyon* **1978**, *47*, 176–187. [CrossRef]
9. Sautière, C. Insectes capturés aux abords et dans les limites de Parc national de la Guadeloupe. Cerambycidae, Scarabaeidae, Sphingidae. *Bull. De L'entomologie Tourangelle* **1999**, *20*, 13–34.
10. Meurgey, F. *Les Arthropodes Continentaux de Guadeloupe: Synthèse Bibliographique Pour un État des Lieux Des Connaissances*; Ropport SHNLH pour le Parc national de Guadeloupe; 2011.
11. MNHN & OFB. Fiche de Pseudosphinx tetrio (Linnaeus, 1771). *Inventaire Natl. Du Patrim. Nat. INPN* **2022**, *2003*. Available online: https://inpn.mnhn.fr/espece/cd_nom/641756 (accessed on 5 December 2022).
12. Warrell, D.A. Researching nature's venoms and poisons. *Trans. R. Soc. Trop. Med. Hyg.* **2009**, *103*, 860–866. [CrossRef] [PubMed]
13. De Souza Leão Veiga, A.F.; de Melo, J.P.R.; Moreira, M.D. Aspectos Morfológicos da Forma Imatura de Pseudosphinx tetrio (LINNAEUS, 1771) (LEPIDOP-TERA: SPHINGIDAE). *Jorn. De Ensino Pesqui. E Extensão-JEPEX 9. Recife* **2009**.
14. Haxaire, J.; Rasplus, J.-Y. Contribution à la Connaissance des Sphingidae de Guyane Française. 1e Partie. *Bull. De La Société Entomol. De Fr.* **1986**, *91*, 275–285. [CrossRef]
15. Haxaire, J.; Herbin, D. Les Lépidoptères Sphingidae de Bolivie. Ecologie et Systématique. 2ème Partie: Les Sous-Familles des Smerinthinae et Macroglossinae Pro Parte (1). *Rev. Assoc. Rousillon. Entomol* **2000**, *9*, 4–19.
16. Meerman, J.C. The Annual Outbreak of Colorful Caterpillars of the Frangipani Hawkmoth (Pseudosphinx tetrio). *Biol. Divers. Belize* **2009**. Available online: http://biological-diversity.info/pseudosphinx.htm (accessed on 8 December 2009).
17. Sloan, S.A.; Zimmerman, J.K.; Sabat, A.M. Phenology of Plumeria alba and its Herbivores in a Tropical Dry Forest. *Biotropica* **2007**, *39*, 195–201. [CrossRef]
18. Rajamanickam, K.; Sudha, S.S. In-Vitro Antimicrobial Activity and In-Vivo Toxicity of Moringa Oleifera and Allamanda Cathartica against Multiple Drug Resistant Clinical Pathogens. *Int. J. Pharma. Bio. Sci.* **2013**, *4*, B768–B775.
19. Khanam, M.N.; Anis, M.; Ahmad, S. Establishment of adventitious root cultures of Allamanda cathartica L. for the production of iridoid glycosides and its identification using HPTLC MS. *Ind. Crops Prod.* **2018**, *125*, 198–206. [CrossRef]
20. Wong, S.K.; Lim, Y.; Chan, E. Botany, uses, phytochemistry and pharmacology of selected Apocynaceae species: A review. *Pharmcognosy Commun.* **2013**, *3*, 2–11.
21. Chandreyi, G.; Labani, H.; Sudip, K.N.; Sayantan, S.; Alokila, D.; Swagata, B.; Maitrayee, B.; Pranabesh, G.; Sirshendu Chatterjee Allamanda cathartica Linn. Apocynaceae: A mini review. *Int. J. Herb. Med.* **2019**, *7*, 29–33.
22. Francis, J.K. *Wildland Shrubs of the United States and Its Territories: Thamnic Descriptions, Volume 1*; U.S. Department of Agriculture, Forest Service, International Institute of Tropical Forestry: San Juan, PR, USA, 2004; p. IITF-GTR-26. [CrossRef]
23. WFO. *Allamanda* L. Published on the Internet. 2022. Available online: http://www.worldfloraonline.org/taxon/wfo-4000001228 (accessed on 5 December 2022).
24. Pessoa, C.; Costa-Lotufo, L.V.; Leyva, A.; de Moraes, M.E.A.; de Moraes, M.O. Anticancer potential of Northeast Brazilian plants. In *Advances in Phytomedicine*; Elsevier: Amsterdam, The Netherlands, 2006; Volume 2, pp. 197–211. [CrossRef]

25. Rahman, A.H.M.; Akter, M. Taxonomy and Traditional Medicinal Uses of Apocynaceae (Dogbane) Family of Rajshahi District, Bangladesh. *Res. Rev. J. Bot. Sci.* **2015**, *4*, 5–13.
26. Nayak, S.; Nalabothu, P.; Sandiford, S.; Bhogadi, V.; Adogwa, A. Evaluation of wound healing activity of Allamanda cathartica. L. and Laurus nobilis. L. extracts on rats. *BMC Complement. Altern. Med.* **2006**, *6*, 12. [CrossRef] [PubMed]
27. Ghosh, C.; Banerjee, S. Floral extracts of allamanda blanchetii and allamanda cathartica are comparatively higher resource of anti-oxidants and polysaccharides than leaf and stem extracts. *Int. J. Curr. Pharm. Sci.* **2018**, *10*, 36–39. [CrossRef]
28. Hameed, A.; Nawaz, G.; Gulzar, T. Chemical composition, antioxidant activities and protein profiling of different parts of *Allamanda cathartica*. *Nat. Prod. Res.* **2014**, *28*, 2066–2071. [CrossRef]
29. Sarker, R.; Tasnuva, S.; Farhana, I.; Sharmin, R.C. In vitro antioxidant, total phenolic, membrane stabilizing and antimicrobial activity of Allamanda cathartica L.: A medicinal plant of Bangladesh. *J. Med. Plants Res.* **2014**, *8*, 63–67. [CrossRef]
30. Karunakaran, G.; Jagathambal, M.; Gusev, A.; Kolesnikov, E.; Mandal, A.R.; Kuznetsov, D. Allamanda cathartica flower's aqueous extract-mediated green synthesis of silver nanoparticles with excellent antioxidant and antibacterial potential for biomedical application. *MRS Commun.* **2016**, *6*, 41–46. [CrossRef]
31. Islam, M.R.; Ahamed, R.; Rahman, M.O.; Al-Amin, M.; Alam, K.D.; Lyzu, F. In Vitro Antimicrobial Activities of Four Medicinally Important Plants in Bangladesh. *Eur. J. Sci. Res.* **2010**, *39*, 199–206.
32. Sharmin, M.; Banya, P.D.; Paul, L.; Chowdhury, F.F.K.; Afrin, S.; Acharjee, M.; Rahman, T.; Noor, R. Study of microbial proliferation and the in vitro antibacterial traits of commonly available flowers in Dhaka Metropolis. *Asian Pac. J. Trop. Dis.* **2015**, *5*, 91–97. [CrossRef]
33. Souza, E.; Barcellos, V.D.A.; Sbaraini, N.; Reuwsaat, J.C.V.; Schneider, R.d.O.; da Silva, A.C.; Garcia, A.W.A.; von Poser, G.L.; Barbosa, E.G.; Lima, J.P.M.S.; et al. A Plumieridine-Rich Fraction From Allamanda polyantha Inhibits Chitinolytic Activity and Exhibits Antifungal Properties Against Cryptococcus neoformans. *Front. Microbiol.* **2020**, *11*, 2058. [CrossRef]
34. Conrad, O.A.; Dike, I.P.; Agbara, U. In vivo antioxidant assessment of two antimalarial plants–Allamamda cathartica and Bixa orellana. *Asian Pac. J. Trop. Biomed.* **2013**, *3*, 388–394. [CrossRef]
35. Boeing, T.; de Souza, P.; Bonomini, T.J.; Mariano, L.N.B.; Somensi, L.B.; Lucinda, R.M.; Malheiros, A.; da Silva, L.M.; Andrade, S.F.d. Antioxidant and anti-inflammatory effect of plumieride in dextran sulfate sodium-induced colitis in mice. *Biomed. Pharmacother.* **2018**, *99*, 697–703. [CrossRef] [PubMed]
36. Kupchan, S.M.; Dessertine, A.L.; Blaylock, B.T.; Bryan, R.F. Isolation and structural elucidation of allamandin, and antileukemic iridoid lactone from allamanda cathartica. *J. Org. Chem.* **1974**, *39*, 2477–2482. [CrossRef] [PubMed]
37. Wong, S.; Lim, Y.; Abdullah, N.; Nordin, F. Antiproliferative and phytochemical analyses of leaf extracts of ten Apocynaceae species. *Phcog. Res.* **2011**, *3*, 100. [CrossRef] [PubMed]
38. Akah, P.A.; Offiah, V.N. Gastrointestinal Effects of *Allamanda cathartica* Leaf Extracts. *Int. J. Pharmacogn.* **1992**, *30*, 213–217. [CrossRef]
39. Nithya, K.; Muthumary, J. Bioactive Metabolite Produced by Phomopsis sp., an Endophytic Fungus in Allamanda cathartica Linn. *Recent Res. Sci. Technol.* **2011**, *3*, 44–48.
40. Nahar, A.; Ashrafuzzaman, S.; Islam, M.N.; Alam, M.S. Studies on antidermatophytic effect of Allamanda cathertica. *Bangladesh J. Pharm.* **2010**, *5*, 5–7. [CrossRef]
41. Janzen, D.H. Patterns of Herbivory in a Tropical Deciduous Forest. *Biotropica* **1981**, *13*, 271–282. [CrossRef]
42. Petricevich, V.; Abarca-Vargas, R. Allamanda cathartica: A Review of the Phytochemistry, Pharmacology, Toxicology, and Biotechnology. *Molecules* **2019**, *24*, 1238. [CrossRef]
43. Hema, K.; Krishnaveni, R. Antibacterial and antifungal activities of Allamanda cathartica Linn. *Int. J. Pharma Bio. Sci.* **2014**, *5*, 588–593.
44. Chaithra, A.B.; Satish, S.; Karunakara, H. Therapeutic uses of Allamanda Cathartica Linn. With a note on its pharmacological actions: A Review. *Int. J. Pharma Chem. Res.* **2016**, *2*, 227–232.
45. Bihani, T. Plumeria rubra L.–A review on its ethnopharmacological, morphological, phytochemical, pharmacological and toxicological studies. *J. Ethnopharmacol.* **2021**, *264*, 113291. [CrossRef] [PubMed]
46. Khan, I.A.; Hussain, M.; Syed, S.K.; Saadullah, M.; Alqahtani, A.M.; Alqahtani, T.; Aldahish, A.A.; Asiri, S.; Zeng, L.-H. Pharmacological Justification for the Medicinal Use of Plumeria rubra Linn. in Cardiovascular Disorders. *Molecules* **2021**, *27*, 251. [CrossRef] [PubMed]
47. Shefali, A.; Kanchan, D.B.; Shailey, S.; Mamta, L.; Tanuja, U.S.; Deepak, K. A review on phytopharmacological activity of plumeria species. *Int. J. Eng. Sci. Math.* **2018**, *7*. Available online: http://www.ijma.us (accessed on 5 December 2022).
48. Siti Mohamed Isa; Ablat, A.; Mohamad, J. The Antioxidant and Xanthine Oxidase Inhibitory Activity of Plumeria rubra Flowers. *Molecules* **2018**, *23*, 400. [CrossRef]
49. Thakur, M.; Gupta, P.; Roy, A.; Chandrakar, S.; Sahu, J.; Gupta, U.P. Therapeutic potential of Plumeria rubra plant: A review. *Adv. J. Bioact. Mol.* **2020**, *1*, 40–47.
50. Forister, M.L.; Novotny, V.; Panorska, A.K.; Baje, L.; Basset, Y.; Butterill, P.T.; Cizek, L.; Coley, P.D.; Dem, F.; Diniz, I.R.; et al. The global distribution of diet breadth in insect herbivores. *Proc. Natl. Acad. Sci. USA* **2015**, *112*, 442–447. [CrossRef] [PubMed]
51. Suzuki, H.C.; Ozaki, K.; Makino, T.; Uchiyama, H.; Yajima, S.; Kawata, M. Evolution of Gustatory Receptor Gene Family Provides Insights into Adaptation to Diverse Host Plants in Nymphalid Butterflies. *Genome Biol. Evol.* **2018**, *10*, 1351–1362. [CrossRef]
52. Opitz, S.E.W.; Müller, C. Plant chemistry and insect sequestration. *Chemoecology* **2009**, *19*, 117–154. [CrossRef]

53. Eisner, T.; Meinwald, J. *Chemical Ecology: The Chemistry of Biotic Interaction*; National Academies Press: Washington, DC, USA, 1995; p. 4979. [CrossRef]
54. McCoy, V.E.; Gee, C.T.; Michalski, J.M.; Wings, O. Oldest fossil evidence of latex sabotaging behavior by herbivorous insects. *Rev. Palaeobot. Palynol.* **2022**, *300*, 104631. [CrossRef]
55. Ali, J.G.; Agrawal, A.A. Specialist versus generalist insect herbivores and plant defense. *Trends Plant Sci.* **2012**, *17*, 293–302. [CrossRef]
56. Ehrlich, P.R.; Raven, P.H. Butterflies and plants: A study in coevolution. *Evolution* **1964**, *18*, 586–608. [CrossRef]
57. Bernays, E.A.; Chapman, R.E. *Host-Plant Selection by Phytophagous Insects*; Springer: Boston, MA, USA, 1994. [CrossRef]
58. Jones, P.L.; Petschenka, G.; Flacht, L.; Agrawal, A.A. Cardenolide Intake, Sequestration, and Excretion by the Monarch Butterfly along Gradients of Plant Toxicity and Larval Ontogeny. *J. Chem. Ecol.* **2019**, *45*, 264–277. [CrossRef] [PubMed]
59. Engler-Chaouat, H.S.; Gilbert, L.E. De novo Synthesis vs. Sequestration: Negatively Correlated Metabolic Traits and the Evolution of Host Plant Specialization in Cyanogenic Butterflies. *J. Chem. Ecol.* **2007**, *33*, 25–42. [CrossRef] [PubMed]
60. Lampert, E.C.; Dyer, L.A.; Bowers, M.D. Chemical Defense Across Three Trophic Levels: Catalpa bignonioides, the Caterpillar Ceratomia catalpae, and its Endoparasitoid Cotesia congregata. *J. Chem. Ecol.* **2011**, *37*, 1063–1070. [CrossRef] [PubMed]
61. Winde, I.; Wittstock, U. Insect herbivore counteradaptations to the plant glucosinolate–myrosinase system. *Phytochemistry* **2011**, *72*, 1566–1575. [CrossRef]
62. Berenbaum, M.R.; Zangerl, A.R. Furanocoumarin metabolism in Papilio polyxenes: Biochemistry, genetic variability, and ecological significance. *Oecologia* **1993**, *95*, 370–375. [CrossRef]
63. Frankfater, C.; Schühly, W.; Fronczek, F.R.; Slattery, M. Processing of a Sesquiterpene Lactone by Papilio glaucus Caterpillars. *J. Chem. Ecol.* **2005**, *31*, 2541–2550. [CrossRef]
64. Barny, L.A.; Tasca, J.A.; Sanchez, H.A.; Smith, C.R.; Koptur, S.; Livshultz, T.; Minbiole, K.P.C. Chemotaxonomic investigation of Apocynaceae for retronecine-type pyrrolizidine alkaloids using HPLC-MS/MS. *Phytochemistry* **2021**, *185*, 112662. [CrossRef]
65. Prabhadevi, V.; Sahaya, S.S.; Johnson, M.; Venkatramani, B.; Janakiraman, N. Phytochemical studies on Allamanda cathartica L. using GC–MS. *Asian Pac. J. Trop. Biomed.* **2012**, *2*, S550–S554. [CrossRef]
66. Marques, J.; Alves, J.; Torres-Rêgo, M.; Furtado, A.; Siqueira, E.; Galinari, E.; Araújo, D.; Guerra, G.; Azevedo, E.; Fernandes-Pedrosa, M.; et al. Phytochemical Analysis by HPLC–HRESI-MS and Anti-Inflammatory Activity of Tabernaemontana catharinensis. *IJMS* **2018**, *19*, 636. [CrossRef]
67. Palmeira-Mello, M.V.; Souza, J.L.; Pérez, A.F.; Cavalcanti, A.D.S.; Kahn, S.A.; Passe-Coutrin, N.; Guerra, I.R.; Doménech-Carbó, A.; Jauregui-Haza, U.J.; de Souza, A.M.T.; et al. Insights of Tris(2-pyridylmethyl)amine as anti-tumor agent for osteosarcoma: Experimental and in silico studies. *J. Mol. Struct.* **2021**, *1228*, 129773. [CrossRef]
68. Rozoy, E.; Araya-Farias, M.; Simard, S.; Kitts, D.; Lessard, J.; Bazinet, L. Redox properties of catechins and enriched green tea extracts effectively preserve l-5-methyltetrahydrofolate: Assessment using cyclic voltammetry analysis. *Food Chem.* **2013**, *138*, 1982–1991. [CrossRef] [PubMed]
69. Cebrián-Torrejón, G.; Kablan, L.; Ferreira, M.E.; Rodríguez de la Cruz, D.; Doménech-Carbó, A.; Vera de Bilbao, N.; Rojas de Arias, A.; Figadère, B.; Poupon, E.; Fournet, A. Harvesting canthinones: Identification of the optimal seasonal point of harvest of *Zanthoxylum chiloperone* leaves as a source of 5-methoxycanthin-6-one. *Nat. Prod. Res.* **2015**, *29*, 2054–2058. [CrossRef] [PubMed]
70. Doménech-Carbó, A.; Cebrián-Torrejón, G.; Lopes-Souto, A.; Martins-de-Moraes, M.; Jorge-Kato, M.; Fechine-Tavares, J.; Barbosa-Filho, J.M. Electrochemical ecology: VIMP monitoring of plant defense against external stressors. *RSC Adv.* **2015**, *5*, 61006–61011. [CrossRef]
71. Doménech-Carbó, A.; Labuda, J.; Scholz, F. Electroanalytical chemistry for the analysis of solids: Characterization and classification (IUPAC Technical Report). *Pure Appl. Chem.* **2012**, *85*, 609–631. [CrossRef]
72. Singh, S.; Kaur, I.; Kariyat, R. The Multifunctional Roles of Polyphenols in Plant-Herbivore Interactions. *IJMS* **2021**, *22*, 1442. [CrossRef]
73. Mannan, M.; Alam, M.; Mustari, F.; Kudrat-E-Zahan, M.; Ali, R.; Haque, A.; Zaman, S.; Talukder, D. In vitro Antioxidant, Antimicrobial, Insecticidal and Cytotoxic Activities of the Medicinal Plants: Allamanda cathartica and Mimusops elengi. *EJMP* **2017**, *20*, 1–12. [CrossRef]
74. Speed, M.P.; Fenton, A.; Jones, M.G.; Ruxton, G.D.; Brockhurst, M.A. Coevolution can explain defensive secondary metabolite diversity in plants. *New Phytol.* **2015**, *208*, 1251–1263. [CrossRef]
75. Petschenka, G.; Agrawal, A.A. Milkweed butterfly resistance to plant toxins is linked to sequestration, not coping with a toxic diet. *Proc. R. Soc. B.* **2015**, *282*, 20151865. [CrossRef]
76. Duffey, S.S. Sequestration of Plant Natural Products by Insects. *Annu. Rev. Entomol.* **1980**, *25*, 447–477. [CrossRef]
77. Dyer, L.A. Tasty Generalists and Nasty Specialists? Antipredator Mechanisms in Tropical Lepidopteran Larvae. *Ecology* **1995**, *76*, 1483–1496. [CrossRef]
78. Karban, R. The ecology and evolution of induced resistance against herbivores. *Funct. Ecol.* **2011**, *25*, 339–347. [CrossRef]
79. Ramos, M.V.; Pereira, D.A.; Souza, D.P.; Silva, M.-L.S.; Alencar, L.M.R.; Sousa, J.S.; Queiroz, J.-F.N.; Freitas, C.D.T. Peptidases and peptidase inhibitors in gut of caterpillars and in the latex of their host plants. *Planta* **2015**, *241*, 167–178. [CrossRef] [PubMed]
80. Mushtaq, M.Y.; Choi, Y.H.; Verpoorte, R.; Wilson, E.G. Extraction for Metabolomics: Access to The Metabolome: Extraction for Metabolomics: Access to the Metabolome. *Phytochem. Anal.* **2014**, *25*, 291–306. [CrossRef]

81. Lin, C.Y.; Wu, H.; Tjeerdema, R.S.; Viant, M.R. Evaluation of metabolite extraction strategies from tissue samples using NMR metabolomics. *Metabolomics* **2007**, *3*, 55–67. [CrossRef]
82. Tautenhahn, R.; Patti, G.J.; Rinehart, D.; Siuzdak, G. XCMS Online: A Web-Based Platform to Process Untargeted Metabolomic Data. *Anal. Chem.* **2012**, *84*, 5035–5039. [CrossRef] [PubMed]
83. Mahieu, N.G.; Genenbacher, J.L.; Patti, G.J. A roadmap for the XCMS family of software solutions in metabolomics. *Curr. Opin. Chem. Biol.* **2016**, *30*, 87–93. [CrossRef] [PubMed]
84. Chong, J.; Soufan, O.; Li, C.; Caraus, I.; Li, S.; Bourque, G.; Wishart, D.S.; Xia, J. MetaboAnalyst 4.0: Towards more transparent and integrative metabolomics analysis. *Nucleic Acids Res.* **2018**, *46*, W486–W494. [CrossRef]
85. Jolliffe, I. Principal Component Analysis. In *Encyclopedia of Statistics in Behavioral Science*; Everitt, B.S., Howell, D.C., Eds.; John Wiley & Sons, Ltd.: Chichester, UK, 2005; p. bsa501. [CrossRef]
86. Wold, S.; Esbensen, K.; Geladi, P. Principal component analysis. *Chemom. Intell. Lab. Syst.* **1987**, *2*, 37–52. [CrossRef]
87. Rodrigues, D.; Muller, A.F.; Bonomini, T.; Klein-Júnior, L.; Lucinda-Silva, R.; Malheiros, A. Chemical variability of allamanda cathartica l. flowers assessed by multivariate data analysis. *Quím. Nova* **2020**, *43*, 201–205. [CrossRef]

plants

MDPI

Article

Allelochemicals from the Rhizosphere Soil of Potato (*Solanum tuberosum* L.) and Their Interactions with the Soilborne Pathogens

Aiyi Xin [1,2], Hui Jin [1], Xiaoyan Yang [1], Jinfeng Guan [3], Heping Hui [1], Haoyue Liu [1], Zengtuan Cui [4], Zhiheng Dun [4] and Bo Qin [1,*]

[1] CAS Key Laboratory of Chemistry of Northwestern Plant Resources and Key Laboratory for Natural Medicine of Gansu Province, Lanzhou Institute of Chemical Physics, Chinese Academy of Sciences (CAS), Lanzhou 730000, China; cpuxay@163.com (A.X.); comefine@licp.cas.cn (H.J.); yangxiaoy@licp.cas.cn (X.Y.); mysy105@163.com (H.H.); liuhaoyue0820@163.com (H.L.)

[2] School of Chemistry and Chemical Engineering, Mianyang Normal University, Mianyang 621000, China

[3] Institute for Food and Drug Control, Tongliao City, Inner Mongolia Autonomous Region, Tongliao 028000, China; nmtlgjf@sina.com

[4] Cultivated Land Quality Construction and Management Station of Gansu Province, Lanzhou 730030, China; konphy@126.com (Z.C.); ruoxi020xw@163.com (Z.D.)

* Correspondence: bqin@licp.cas.cn; Tel.: +86-931-4968371; Fax: +86-931-4968019

Abstract: To reveal the allelopathic effects of potato, seven compounds were isolated from the rhizosphere soil: 7-methoxycoumarin (**1**), palmitic acid (**2**), caffeic acid (**3**), chlorogenic acid (**4**), quercetin dehydrate (**5**), quercitrin (**6**), and rutin (**7**). Bioassays showed that compounds **1**, **2**, **4**, and **6** had inhibitory effects on the growth of *L. sativa* and tissue culture seedlings of potato. The existence of the allelochemicals was confirmed by HPLC, and their contents were quantified with a total concentration of 9.02 μg/g in the rhizosphere soil of replanted potato. Approaches on the interactions of the allelochemicals and pathogens of potato including *A. solani*, *B. cinerea*, *F. solani*, *F. oxysporum*, *C. coccodes*, and *V. dahlia* revealed that compound **1** had inhibitory effects but compounds **2**–**4** promoted the colony growth of the pathogens. These findings demonstrated that the autotoxic allelopathy and enhancement of the pathogens caused by the accumulation of the allelochemicals in the continuously cropped soil should be one of the main reasons for the replant problems of potato.

Keywords: potato (*Solanum tuberosum* L.); replant problems; allelochemicals; autotoxicity; antifungal effects

Citation: Xin, A.; Jin, H.; Yang, X.; Guan, J.; Hui, H.; Liu, H.; Cui, Z.; Dun, Z.; Qin, B. Allelochemicals from the Rhizosphere Soil of Potato (*Solanum tuberosum* L.) and Their Interactions with the Soilborne Pathogens. *Plants* **2022**, *11*, 1934. https://doi.org/10.3390/plants11151934

Academic Editors: Margot Schulz and Vincenzo Tabaglio

Received: 14 June 2022
Accepted: 14 July 2022
Published: 26 July 2022

1. Introduction

Tubers of potato (*Solanum tuberosum* L.) are the third most important food crops in terms of human consumption with a population of more than a billion people worldwide consuming them. The global total crop production exceeds 374 million metric tons each year, following wheat (*Triticum aestivum* L.), rice (*Oryza sativa* L.), and maize (*Zea mays* L.) according to the statistics data of the Food and Agriculture Organization of the United Nations (FAO, Rome, Italy) in 2017 [1,2]. Being rich in starch and crude protein, potato is used for both a staple food and a vegetable. Research indicated that potato can provide a great diversity of nutrients such as protein, lipids, minerals, vitamins, and dietary fiber [3]. Owing to the high content of essential amino acids, potato protein is considered as one of the most valuable non-animal proteins [4]. Meanwhile, it is a typical highly productive crop and has been used as a source of starch and alcohol for industrial uses; thus the production of potato has been regarded as a pillar industry for farmers to get rid of poverty and increase income in some major producing areas [5,6].

However, the crop production of potato has often been hampered by replant problems. The occurrence of replant failure not only affects the yield and quality of potato, but also reduces the safety of the product [7]. Therefore, it is urgent to declare the reasons causing the replant failure and explore an efficient way to achieve high and stable production of potato. Replant failure is a phenomenon in which several factors are involved; allelopathy (especially autotoxic allelopathy) and the enhancement of soilborne pathogens coupled with an imbalance in the soil microbial community caused by allelochemicals have been considered as essential factors on this issue [8–11]. A great variety of secondary metabolites produced by the crop and released into the soil environment through leaching, root exudation, and/or residue decomposition have been shown to interact with the replanted plant itself and other organisms [12–16]. Further knowledge indicated that soil-borne pathogens play a primary role in the occurrence of replant disease [17]. It is proposed that the activities of soil-borne pathogens would be improved indirectly by the allelochemicals which are released by the crop [18]. The allelochemicals are not only autotoxic compounds for the replanted plant, but also a precipitating factor for changing the composition of the microbial community to lead to the build-up of soil-borne pathogens [19,20]. The accumulation of phytotoxic substances and the decrease in bacterial diversity is attributed to replant failure which was occurred during continuous cropping [21]. Several common diseases of potato were caused by the fungi in major producing regions around the world, such as *Alternaria solani* that may result in early blight, *Botrytis cinerea* that may bring about gray mold, *Fusarium solani* that may increase the risk of dry rot, *Fusarium oxysporum* that may cause root rot, *Colletotrichum coccodes* that may lead to potato black dot, *Verticillium dahliae* that may give rise to potato early dying, etc.

It was reported that water extracts from different organs and root exudates of potato had potential phytotoxicity against potato itself and other plants [22–25]. However, the specific compounds which are responsible for the phytotoxicity still remain unknown, and the relative roles of autotoxic factors and pathogens of potato are unclear as well. Elucidating the chemical reasons for the replant problem is necessary to explor efficient approaches to eliminating the replant failure of potato. Thus, in this work, the allelochemicals from the rhizosphere soil of cultivated potato were isolated and identified for the first time. Further investigations on the interactions of the allelochemicals and pathogens were carried out as well to get more helpful information to understand the chemical basis of the replant failure of potato.

2. Results

2.1. Allelopathic Activity of the Crude Extract of the Rhizosphere Soil of Potato

The phytotoxic activity of the crude extract was evaluated with *Lactuca sativa* L. (Figure 1a) and potato tissue culture (Figure 1b) seedlings. The results revealed that the crude methanol extract significantly inhibited the growth of the treated seedlings and the effect was in a concentration-dependent manner. The crude extract exhibited a more remarkable inhibitory effect on the growth of roots than that on stems for *L. sativa* seedlings. As Figure 1a showed, the inhibition ratios of roots at the treated concentrations of 100, 200, and 400 μg/mL were 17.82%, 27.39%, and 65.96%, respectively. The growth of potato seedlings was observably inhibited by the crude extract in a similar manner. The inhibitory effect on the growth of the roots was more pronounced than that on the leaves and stems for the potato seedlings, which was in accordance with the growth of *L. sativa*. As shown in Figure 1b, at a concentration of 100 μg/mL, the inhibition ratios of the height of the potato seedlings, the fresh weight of leave and stems, and the fresh weight of the roots were 37.03%, 29.73%, and 50.17%, respectively. The inhibition ratios at the treated concentration of 200 μg/mL reached up to 54.25%, 36.68%, and 55.63%, respectively. At the maximum concentration (400 μg/mL), the inhibition ratios ran up to 61.26%, 48.20%, and 66.28%, respectively.

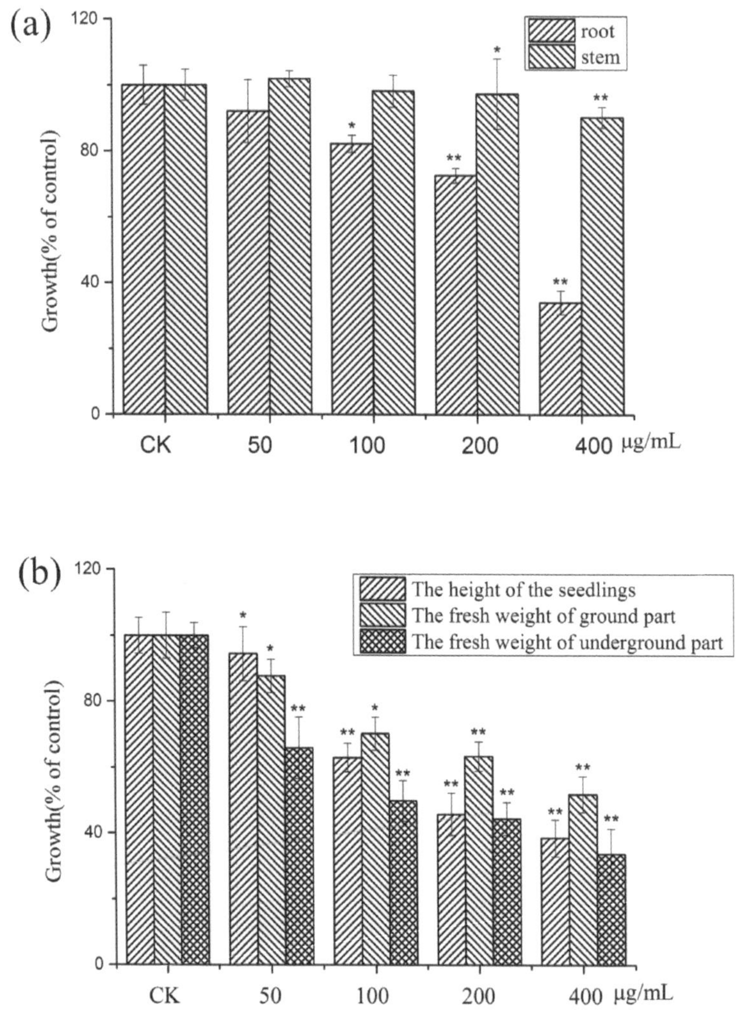

Figure 1. Phytotoxic effects of the crude extract of the rhizosphere soil of potato on seedlings of *L. sativa* (**a**) and potato (**b**) at concentrations of 50, 100, 200, and 400 μg/mL, respectively. Values are presented as a percentage of the mean compared to the control. Means significantly lower than the methanol controls are indicated by * (one way ANOVA; $p < 0.05$) or ** ($p < 0.01$). Error bars are one standard error of the mean. $n = 3$.

2.2. *Isolation of Allelochemicals from the Rhizosphere Soil of Potato*

Seven compounds were isolated from the extract of the rhizosphere soil of potato; they were identified as: 7-methoxycoumarin (**1**), palmitic acid (**2**), caffeic acid (**3**), chlorogenic acid (**4**), quercetin dehydrate (**5**), quercitrin (**6**), and rutin (**7**). Their structures were characterized by spectroscopic analysis in comparison with literature data. The structures of compounds **1**–**7** are shown in Figure 2.

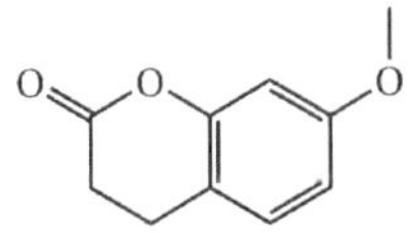

1

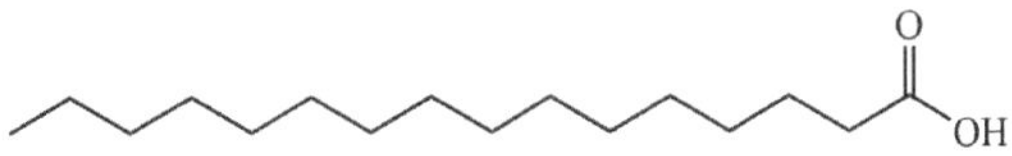

2

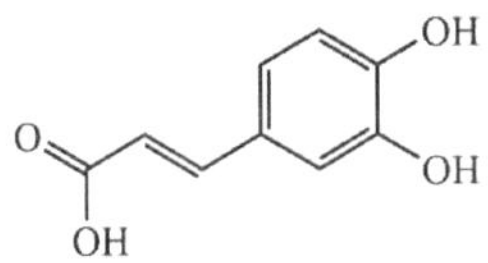

3

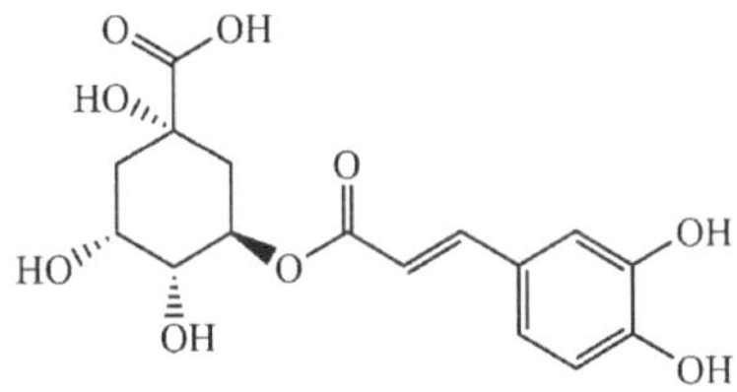

4

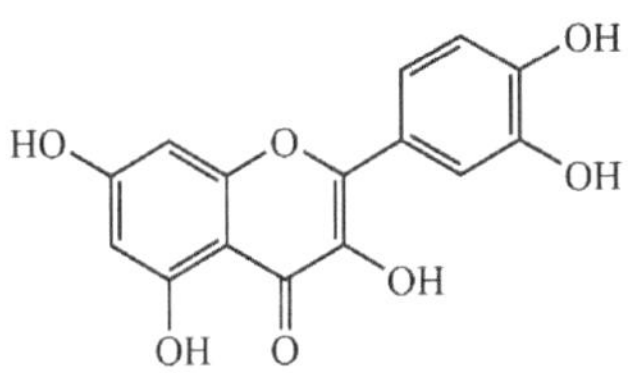

5

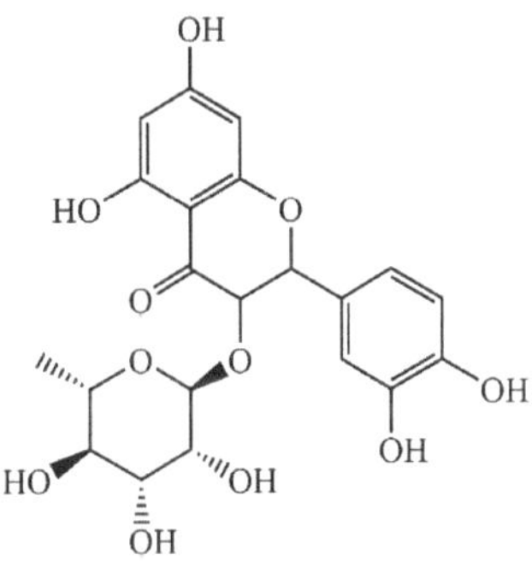

6

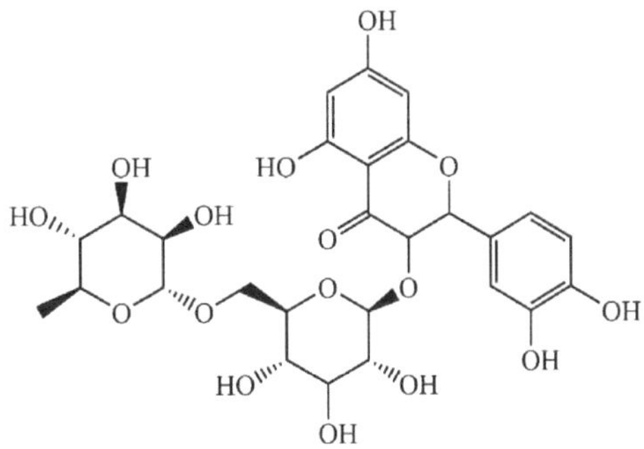

7

Figure 2. Structures of compounds **1**–**7** isolated from the rhizosphere soil of potato. (**1**. 7-methoxycoumarin, **2**. palmitic acid, **3**. caffeic acid, **4**. chlorogenic acid, **5**. quercetin dehydrate, **6**. quercitrin, and **7**. rutin).

Compound **1**: $C_{10}H_8O_3$

MS: $[M + Na]^+$ $C_{10}H_8NaO_3$, measured m/z 199.0363, calculated m/z 199.0371, err 3.9 ppm.

^{1}H NMR (400 MHz, CD_3OD) δ 7.65 (d, J = 9.5 Hz, 1H), 7.38 (d, J = 8.6 Hz, 1H), 6.85 (dd, J = 8.6, 2.3 Hz, 1H), 6.80 (d, J = 2.1 Hz, 1H), 6.25 (d, J = 9.5 Hz, 1H).

^{13}C NMR (100 MHz, CD_3OD) δ 162.57, 160.94, 155.63, 143.21, 128.54, 112.79, 112.30, 100.57, 55.52.

Compound **2:** $C_{16}H_{32}O_2$

MS: $[M + Na]^+$ $C_{16}H_{32}NaO_2$, measured m/z 279.2295, calculated m/z 279.2300, err 1.8 ppm.

^{1}H NMR (400 MHz, $CDCl_3$) δ 2.34 (t, J = 7.5 Hz, 2H), 1.66-1.60 (m, 2H), 1.26 (s, 24H), 0.88 (t, J = 6.7 Hz, 3H).

^{13}C NMR (100 MHz, $CDCl_3$) δ 180.57, 34.13, 31.94, 29.71, 29.70, 29.68, 29.67, 29.66, 29.61, 29.45, 29.38, 29.25, 29.08, 24.68, 22.70, 14.10.

Compound **3:** $C_9H_8O_4$

MS: $[M + Na]^+$ $C_9H_8NaO_4$, measured m/z 203.0310, calculated m/z 203.0315, err 2.5 ppm.

^{1}H NMR (400 MHz, CD_3OD) δ 7.53 (d, J = 15.9 Hz, 1H), 7.03 (d, J = 1.9 Hz, 1H), 6.92 (dd, J = 8.2, 1.9 Hz, 1H), 6.77 (d, J = 8.2 Hz, 1H), 6.21 (d, J = 15.9 Hz, 1H).

^{13}C NMR (100 MHz, CD_3OD) δ 169.68, 148.06, 145.69, 145.39, 126.41, 121.50, 115.11, 114.11, 113.71.

Compound **4:** $C_{16}H_{18}O_9$

MS: $[M + Na]^+$ $C_{16}H_{18}NaO_9$, measured m/z 377.0840, calculated m/z 377.0843, err 0.8 ppm.

^{1}H NMR (400 MHz, Acetone) δ 7.56 (d, J = 15.9 Hz, 1H), 7.16 (d, J = 1.9 Hz, 1H), 7.05 (dd, J = 8.2, 1.9 Hz, 1H), 6.88 (d, J = 8.2 Hz, 1H), 6.27 (d, J = 15.9 Hz, 1H), 5.42–5.36 (m, 1H), 4.25 (d, J = 3.3 Hz, 1H), 3.78 (d, J = 3.0 Hz, 1H), 2.29–2.22 (m, 1H), 2.16 (d, J = 3.1 Hz, 1H), 2.05 (d, J = 4.4 Hz, 2H).

^{13}C NMR (100 MHz, Acetone) δ 174.20, 166.24, 147.86, 145.42, 144.88, 126.79, 121.65, 115.51, 115.00, 114.34, 75.36, 72.64, 70.76, 70.47, 38.26, 37.03.

Compound **5:** $C_{15}H_{10}O_7$

MS: $[M + Na]^+$ $C_{15}H_{10}NaO_7$, measured m/z 325.0317, calculated m/z 325.0319, err 0.6 ppm.

^{1}H NMR (400 MHz, CD_3OD) δ 7.71 (d, J = 2.0 Hz, 1H), 7.61 (dd, J = 8.5, 2.1 Hz, 1H), 6.86 (d, J = 8.5 Hz, 1H), 6.36 (d, J = 2.0 Hz, 1H), 6.15 (d, J = 2.0 Hz, 1H).

^{13}C NMR (100 MHz, CD_3OD) δ 177.33, 165.57, 162.51, 158.22, 148.76, 147.97, 146.22, 137.24, 124.14, 121.66, 116.21, 115.97, 104.51, 99.21, 94.39.

Compound **6:** $C_{21}H_{20}O_{11}$

MS: $[M + Na]^+$ $C_{21}H_{20}NaO_{11}$, measured m/z 47.0908, calculated m/z 471.0898, err −2.1 ppm.

^{1}H NMR (400 MHz, CD_3OD) δ 7.34 (d, J = 1.9 Hz, 1H), 7.32 (dd, J = 8.3, 2.0 Hz, 1H), 6.92 (d, J = 8.3 Hz, 1H), 6.38 (d, J = 2.0 Hz, 1H), 6.21 (d, J = 2.0 Hz, 1H), 5.36 (s, 1H), 4.22 (d, J = 1.5 Hz, 1H), 3.75 (dd, J = 9.3, 3.3 Hz, 1H), 3.43 (dd, J = 9.5, 6.1 Hz, 1H), 3.35 (s, 1H), 0.95 (d, J = 6.1 Hz, 3H).

^{13}C NMR (100 MHz, CD_3OD) δ 179.45, 165.69, 163.03, 159.11, 158.33, 149.60, 146.22, 136.03, 122.76, 122.64, 116.71, 116.16, 105.69, 103.34, 99.60, 94.49, 73.04, 71.91, 71.83, 71.70, 17.44.

Compound **7:** $C_{27}H_{30}O_{16}$

MS: $[M + Na]^+$ $C_{27}H_{30}NaO_{16}$, measured m/z 633.1435, calculated m/z 633.1426, err −1.5 ppm.

^{1}H NMR (400 MHz, CD_3OD) δ 7.66 (s, 1H), 7.62 (d, J = 8.4 Hz, 1H), 6.86 (d, J = 8.4 Hz, 1H), 6.37 (d, J = 1.9 Hz, 1H), 6.18 (d, J = 1.9 Hz, 1H), 5.10 (d, J = 7.4 Hz, 1H), 4.52 (s, 1H), 3.80 (d, J = 10.5 Hz, 1H), 3.64 (d, J = 1.5 Hz, 1H), 3.54 (dd, J = 9.5, 3.3 Hz, 1H), 3.48 (d, J = 9.1 Hz, 1H), 3.45 (d, J = 5.1 Hz, 2H), 3.40 (d, J = 4.8 Hz, 1H), 3.36 (d, J = 4.5 Hz, 1H), 3.33 (s, 1H), 3.27 (d, J = 3.2 Hz, 1H), 1.12 (d, J = 6.2 Hz, 3H).

^{13}C NMR (100 MHz, CD_3OD) δ 177.98, 164.58, 161.53, 157.92, 157.07, 148.39, 144.41, 134.26, 129.05, 122.18, 121.72, 116.32, 114.66, 104.22, 103.37, 101.02, 98.55, 93.48, 76.78, 75.79, 74.34, 72.55, 70.70, 69.99, 68.31, 67.16, 16.50.

2.3. Allelopathic Activities of the Purified Compounds on L. sativa Seedlings

As shown in Figure 3, compounds **1**, **2**, **4**, and **6** inhibited the growth of the seedlings of *L. sativa*, while compounds **3**, **5**, and **7** had no effect. The inhibitory effect increased with the increase in treated concentrations. Compound **1** displayed a strong inhibition on the growth of *L. sativa* at all treated concentrations. The inhibition ratios to the roots and stems of *L. sativa* at the concentration of 200 μg/mL were 78.87% and 87.87%, respectively. Compounds **2**, **4**, and **6** showed strong inhibition on the growth of *L. sativa* at higher concentrations (≥50 μg/mL), while there were slight or no effects at low treated concentrations. Meanwhile, the inhibition ratios on roots were more remarkable than that on stems for *L. sativa* seedlings. At the maximum concentration (200 μg/mL), the inhibition ratios of roots reached to 64.29%, 35.50%, and 65.53%, respectively.

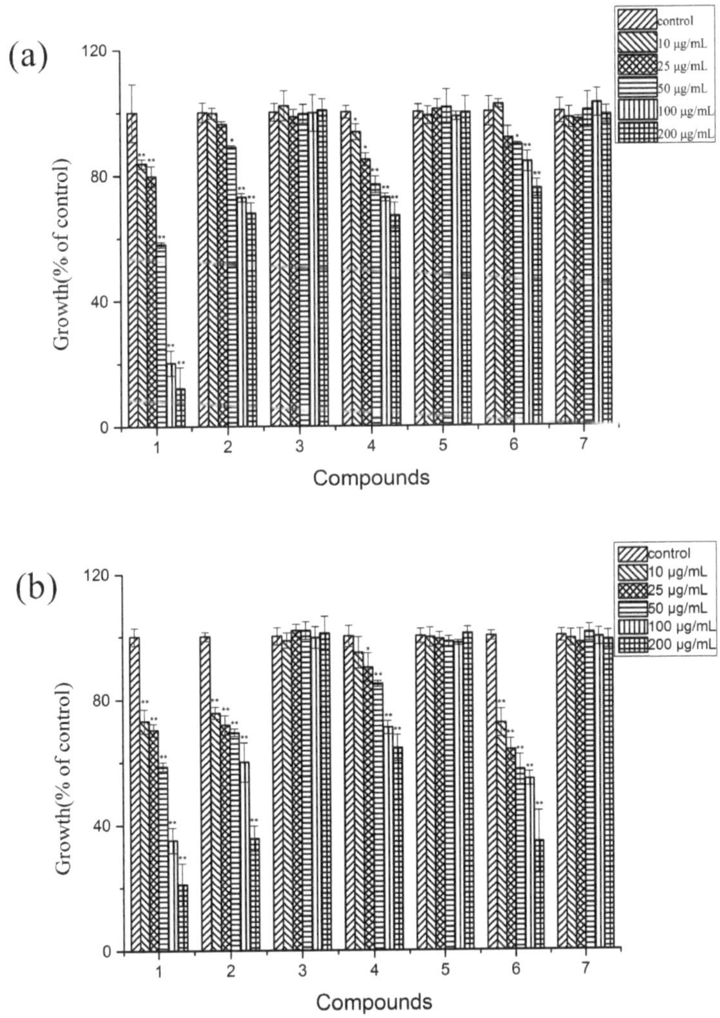

Figure 3. Phytotoxic effects of compounds **1–7** on root (**a**) and stems (**b**) lengths of *L. sativa* seedlings at concentrations of 10, 25, 50, 100, and 200 μg/mL, respectively. Values are presented as a percentage of the mean compared to the control. Means significantly lower than the DMSO controls are indicated by * (one way ANOVA; $p < 0.05$) or ** ($p < 0.01$). Error bars are one standard error of the mean. $n = 3$.

2.4. Autotoxic Activities of the Purified Compounds on Tissue Culture Seedlings of Potato

The autotoxic activity of the purified compounds was evaluated with tissue culture seedlings of potato. The results revealed that compounds **1**, **2**, **4**, and **6** had an inhibition effect on the growth of the potato seedlings, while compounds **3**, **5**, and **7** had no effect (Figure 4). Compound **1** showed an inhibitory effect on the growth of tissue culture seedlings of potato at the highest treated concentration (200 μg/mL), while there were no effects at lower concentrations. The inhibition ratios at the concentration of 200 μg/mL on the height of the seedlings, the fresh weight of the leaves and stems, and the fresh weight of the roots were 14.06%, 12.67%, and 20.17%, respectively. At higher treated concentrations (≥50 μg/mL), compounds **2**, **4**, and **6** displayed moderate inhibitions on the growth of potato tissue culture seedlings, while there were slight or no effects at lower treated concentrations. Similarly, the autotoxic effects on the fresh weight of the roots of tissue seedlings were more distinguished in a concentration-dependent manner. The inhibition ratios of the fresh weight of the leaves and stems at the maximum concentration (200 μg/mL) reached up to 42.07%, 34.41%, and 43.08%, respectively. The typical patterns of the effects of compounds **1**, **2**, **4**, and **6** on the growth length of *L. sativa* seedlings (a) and potato tissue culture seedlings (b) at concentrations of 0, 10, 25, 50, 100, and 200 μg/mL are shown in Figure 5.

2.5. Confirmation and Quantification of the Allelochemicals in the Rhizosphere Soil

High-performance liquid chromatography (HPLC) was performed to quantify the allelochemicals in the rhizosphere soil of replanted potato. The presence of the compounds in the soil was confirmed by comparing the retention times of the standard compounds **1**, **2**, **4**, and **6** under the same chromatographic conditions (Figure 6). At the same time, the standard curve of the content of compounds **1**, **2**, **4**, and **6** was determined at a wavelength of 240 nm (compounds **1**, **2**, and **4**) and 280 nm (compound **6**) (Table 1). The contents of compounds **1**, **2**, **4**, and **6** in the rhizosphere soil were calculated to be 1.46, 6.64, 0.35, and 0.57 μg/g, respectively.

Table 1. Regression equations, correlation coefficients, linear range, retention time, and contents for the four allelochemicals.

Compounds	Linear Equations [a]	R^2	Linear Range (ng/mL)	Retention Time (min)	Content (μg/g)
1	$y = 0.0093x - 343.9$	0.9995	2000.0–30,000.0	23.503	1.46
2	$y = 0.0133x + 123.7$	0.9998	2000.0–30,000.0	32.605	6.64
4	$y = 0.0102x + 212.15$	0.9999	500.0–30,000.0	10.687	0.35
6	$y = 0.0152x + 518.21$	0.9993	2000.0–30,000.0	13.900	0.57

[a] y represents the peak area, x represents the concentration (ng/mL).

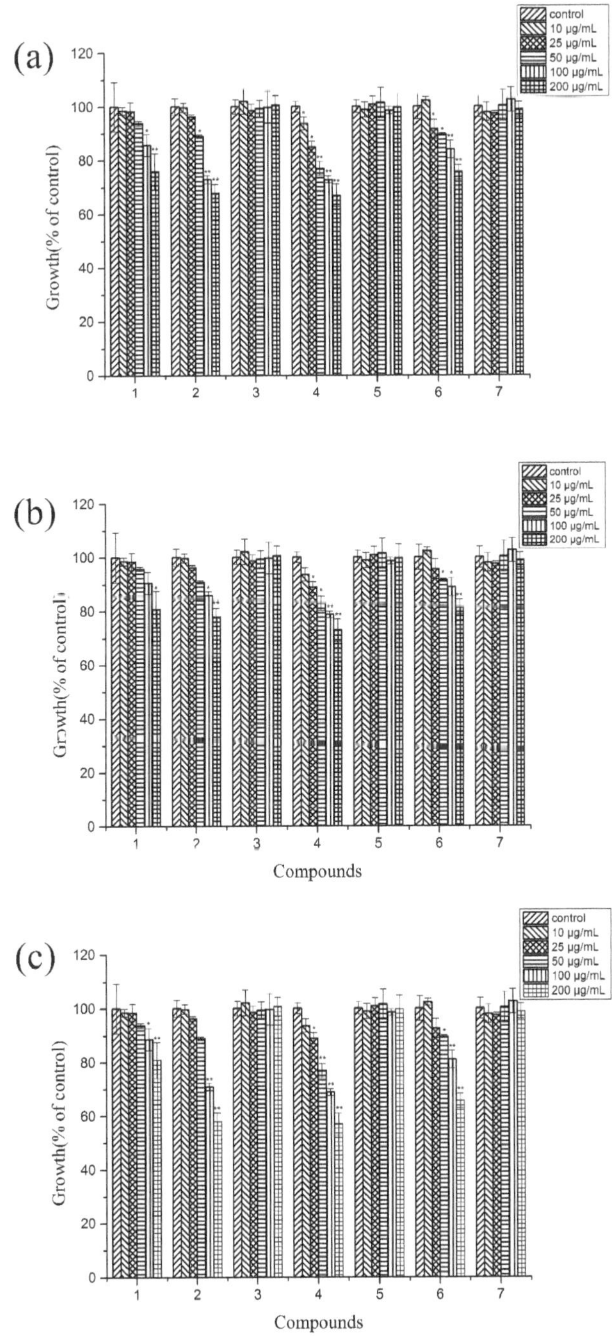

Figure 4. Autotoxic effects of compounds **1–7** on the height of the seedlings (**a**), the fresh weight of the leaves and stems (**b**), and the fresh weight of the roots (**c**) at concentrations of 10, 25, 50, 100, and 200 μg/mL, respectively. Values are presented as a percentage of the mean compared to the control. Means significantly lower than the DMSO controls are indicated by * (one way ANOVA; $p < 0.05$) or ** ($p < 0.01$). Error bars are one standard error of the mean. $n = 3$.

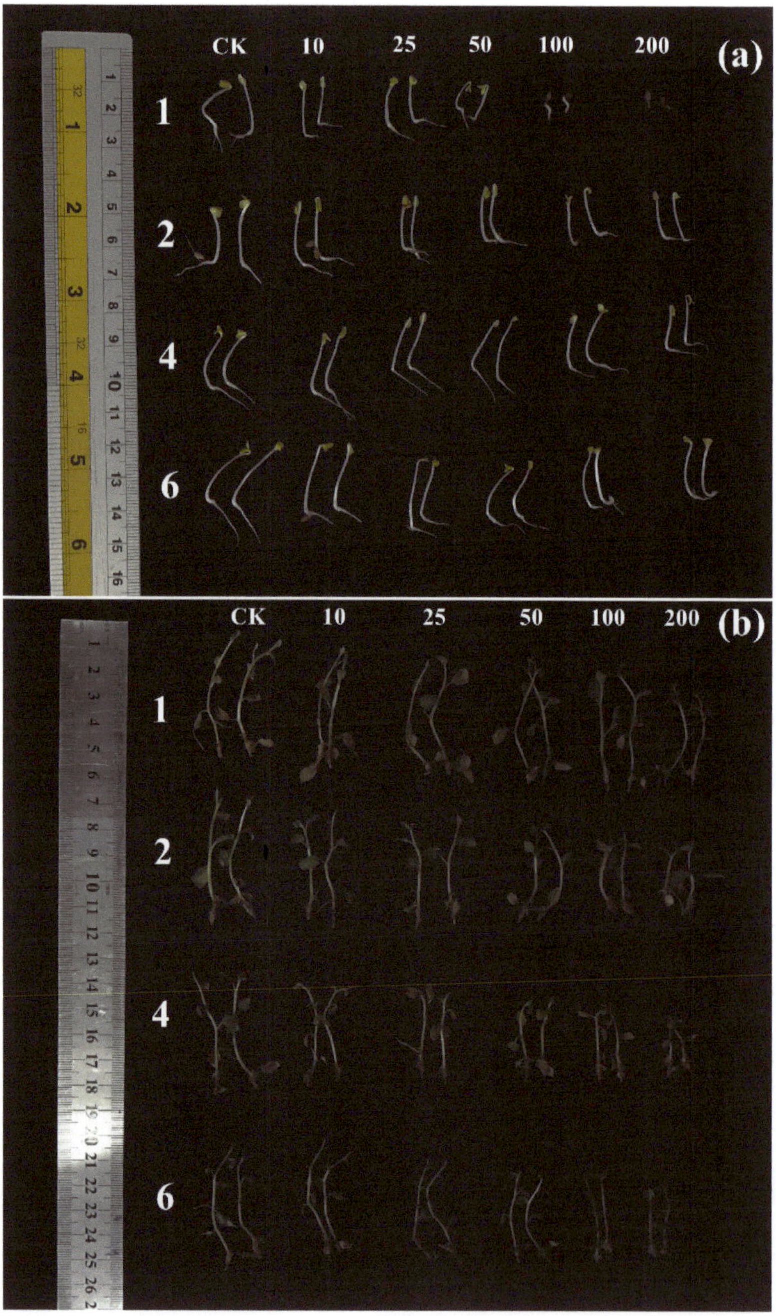

Figure 5. Typical pictures of the effects of compounds **1**, **2**, **4** and **6** on the growth of *L. sativa* seedlings (**a**) and potato tissue culture seedlings (**b**) at concentrations of 0, 10, 25, 50, 100, and 200 μg/mL.

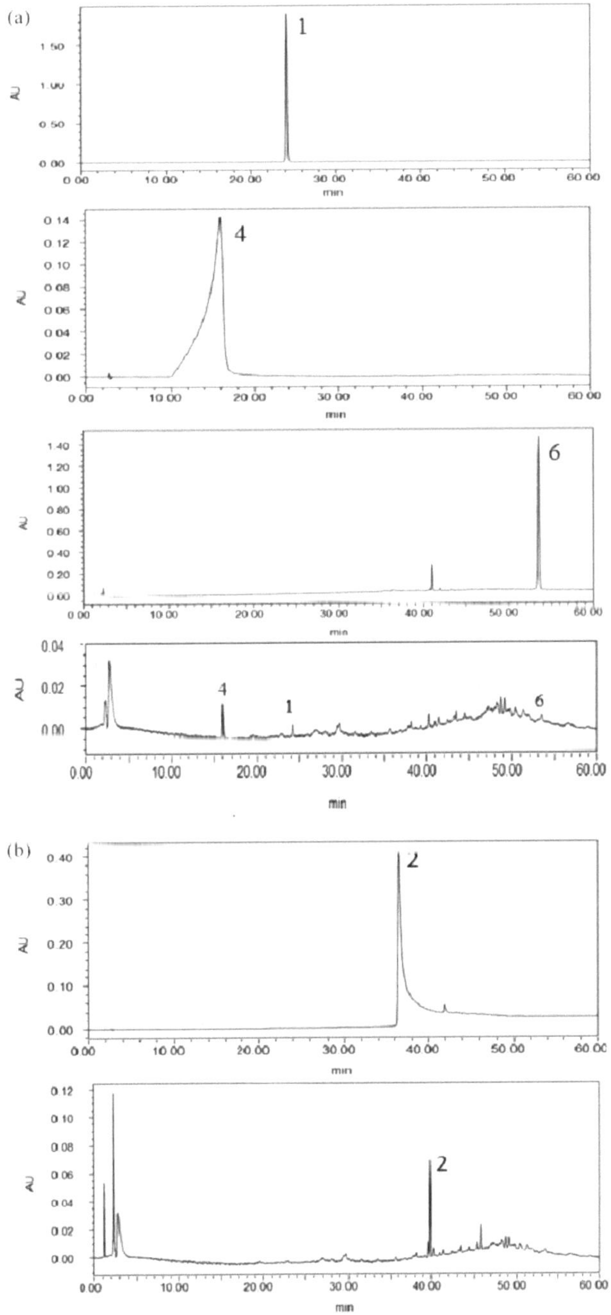

Figure 6. HPLC analysis of the allelochemicals in the rhizosphere soil of continuously cropped potato at 240 nm (**a**) and 320 nm (**b**), respectively; purified compounds **1**, **4**, and **6** were detected at 240 nm (**a**) and compound **2** was detected at 320 nm (**b**).

2.6. Antifungal Activity on Potato Pathogens of the Allelochemicals

Compounds **1–7** showed different antifungal activity at a treated concentration of 200 μg/mL against pathogens of potato, including *A. solani*, *B. cinerea*, *F. solani*, *F. oxysporum*, *C. coccodes* and *V. dahlia*, as shown in Figures 7 and 8. The results indicated that compounds **2–4** promoted the growth of the colonies of the six pathogens of potato; only compound **1** showed an inhibitory effect. The promotive indices of compound **2** on *A. solani*, *B. cinerea*, and *C. coccodes* at 200 μg/mL were 33.8%, 11.5%, and 7.7%, respectively. The promotive indices of compound **3** on *A. solani*, *B. cinerea*, *C. coccodes*, and *V. dahlia* at 200 μg/mL were 29.3%, 8.5%, 11.6%, and 7.2%, respectively. The promotive indices of compound **4** on *A. solani*, *B. cinerea*, *F. solani*, *F. oxysporum*, and *V. dahlia* at 200 μg/mL were 20.6%, 7.8%, 7.6%, 8.0%, and 14.9%, respectively.

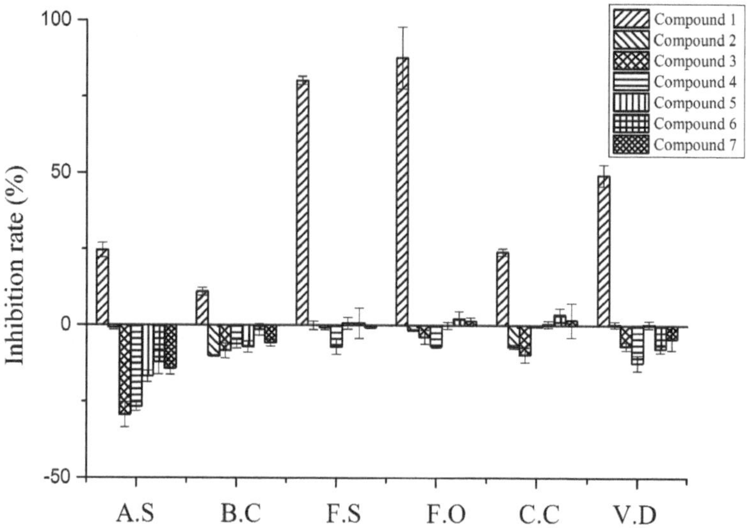

Figure 7. The effects of compounds **1–7** against the potato pathogens including *A. solani*, *B. cinerea*, *F. solani*, *F. oxysporum*, *C. coccodes*, and *V. dahlia* at a concentration of 200 μg/mL.

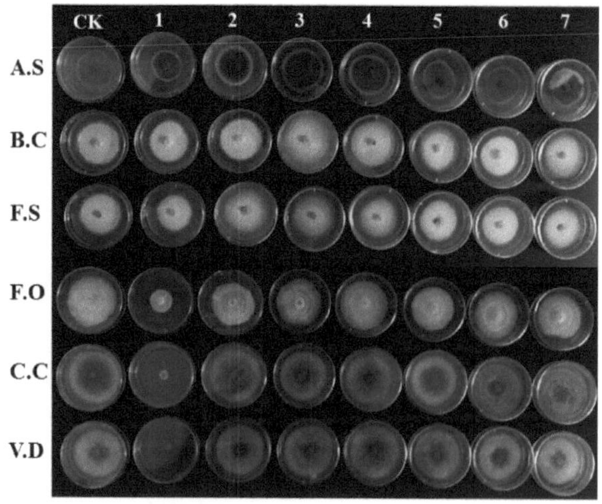

Figure 8. Typical pictures of the effects of compounds **1–7** against potato pathogens including *A. solani*, *B. cinerea*, *F. solani*, *F. oxysporum*, *C. coccodes*, and *V. dahlia* at a concentration of 200 μg/mL.

3. Discussion

Negative plant–soil feedback is one of the important factors of the replant problem, which plays an essential role in hindering the sustainable development of agriculture [26]. Investigations indicated that the accumulation of allelochemicals in the soil environment is a crucial factor for the replant failure of large amounts of plants. The growth of replanted plants may be inhibited by the reduction in the number of meristem cells, the reduction in elongation zones, and the accumulation of auxins caused by allelochemicals [27].

The bioassay results of the crude extract in the present study indicated that there were potential allelochemicals in the rhizosphere soil of potato that contributed to the inhibitory effect on the seedling growth of itself and other plants. Nevertheless, the autotoxic effect of the allelochemicals of potato has not been previously clarified.

Hence, in this work, seven compounds were isolated from the rhizosphere soil of potato and bioassays indicated that compounds **1**, **2**, **4**, and **6** had inhibitory effects on the growth of *L. sativa* and potato tissue culture seedlings. Compound **1** belongs to coumarin, which is considered to be one kind of allelochemical by Rice [28]. It was reported that coumarin inhibited the seed germination and plant growth (length and mass of shoots and roots) of several plants [29,30]. The results in this study revealed that it had an inhibitory effect on the growth of lettuce seedlings and tissue culture seedlings of potato at the maximum treatment concentration. Compound **2** is in the range of fatty acids; investigations demonstrate that it has an enormous capacity of phytotoxic activity [31]. The results in this work showed that it inhibited the growth of lettuce seedlings and tissue culture seedlings of potato at high treated concentrations, which are consistent with the previous study. Chlorogenic acid (compound **3**) is abundant in potato peels. It has a broad range of therapeutic properties, including antioxidant activity andantibacterial, anti-inflammatory, and anti-obesity properties [32]. However, phenolic acids are one kind of important allelochemical as well, which generally exist in their cultivated soils [33]. It was reported that chlorogenic acid inhibited the growth of lettuce seedlings at high concentrations, and ROS in lettuce could be accumulated under the stress of chlorogenic acid [34]. The results in the present study indicated that the growth of seedlings of lettuce and potato were suppressed at a high treated concentration of compound **3**. Compound **6** was regarded as a flavonoid; it is reported that flavonoids affect many physiological functions of plants, including the inhibition of auxin transport, phytochemical defense, an allelopathic effect, and an antioxidant effect [35–37]. In the present study, compound **6** only displayed an inhibitory effect on the growth of the *L. sativa* and potato at the maximum treated concentration. Analysis of the characterized compounds in the rhizosphere soil showed that compound **2** had the highest content with a value of 6.64 μg/g, and the total content of the four autotoxic allelochemicals in the rhizosphere soil reached 9.02 μg/g.

Soil microorganisms are critical elements in agricultural ecosystems, and the compositions of soil microbial communities are influenced through rhizodeposition [38]. By this way, beneficial microbes were promoted, but also soil-borne pathogens were built-up [39,40]. The imbalance of soil microbial community has been considered as one of the essential factors for replant problems. Many cultivated plants, which suffer soil sickness, are often connected with soil-borne pathogens [41]. Several research studies suggested that replant disease could be induced directly by the occurrence of pathogens and indirectly by the release of toxic substances which were decomposed by plants [42,43]. Currently, allelochemicals are regarded as a significant cause which influences the microbial community [44]. The accumulation of allelochemicals brings about the increase in harmful pathogens in rhizosphere soil, leads to an unbalance of the structure of the microbial community, and results in replant diseases during continuous cropping [45]. In this work, the antifungal assays showed that compound **1** had an inhibitory effect on the six tested pathogens of potato, but compounds **2–4** promoted the colony growth of the pathogens, suggesting that the release of the autotoxic allelochemicals induced the disequilibrium of the microbial community in the continuously cropped potato soil. The compositions of the allelochemicals affected the composition of the microbial community in the replanted soil and enhanced the feasibility of the outbreak of the soil-borne diseases of potato during continuous cultivation.

4. Materials and Methods

4.1. General Experimental Instruments and Reagents

MCI gel CHP20/P120 (Mitsubishi chemical corporation, Tokyo, Japan) and silica gel 60 (200–300 mesh, Qingdao Haiyang Chemical Co., Ltd., Qingdao, China) were used for Column Chromatography (CC). Thin-layer chromatography (TLC) and preparative thin-layer chromatography were carried out on silica gel F254 plates, and the spots were detected under UV detection or by heating after spraying with 5% H_2SO_4 in C_2H_5OH (v/v). By using a Bruker Avance III-400 spectrometer (Bruker AXS GmbH, Karlsruhe, Germany), ^{1}H and ^{13}C NMR spectra (400 and 100 MHz, respectively) were recorded. Electrospray ionization quadruple time -of-flight mass spectrometry (ESI-Q-TOF MS) was used to obtain the high-resolution mass data.

The High-Performance Liquid Chromatographic analysis was conducted on a Waters 1525 binary HPLC pump (Waters, Milford, MA, USA) with a Waters 2998 photodiode array detector, coupled with a Waters C18 column (250 mm × 4.6 mm, 5 μm).

4.2. Soil Samples

Soil samples around the roots of potato cultivar Long 7 (within 5 cm) were collected in five different areas of Lanzhou, Gansu Province, China, in October 2018. The collected soil samples were dried in the dark at room temperature and passed through a 1 mm screen sieve. Then, the soil samples were extracted three times (90 min each) in methanol with ultrasonic treatments, and the filtrate was concentrated in vacuum to dryness.

4.3. Isolation of Allelochemicals from the Rhizosphere Soil

The residue of the rhizosphere soil was dissolved in water, then extracted by petroleum ether, chloroform, ethyl acetate, and n-butanol saturated with water, to finally obtain five fractions (Fr. 1–5). Fr. 1 was fractionated by MCI gel column chromatography eluting with methanol/water (3:7, 1:1, 7:3 and 9:1, v/v), and finally eluted with methanol to get compound **1** (3.6 mg) and compound **2** (12.4 mg), respectively. Fr. 2 was separated by PTLC (petroleum ether/EtOAc, 10:1, v/v) to yield compound **3** (4.9 mg), compound **4** (2.3 mg), and Fr. 2–3. Fr. 2–3 was subjected on silica gel CC with petroleum ether/EtOAc (3:1) in mobile phase to give compound **5** (3.5 mg). Fr. 3 was submitted to silica gel CC and eluting with chloroform/methanol (10:1, v/v) to obtain compound **6** (2.5 mg). Fr. 4 was eluted with chloroform/methanol (3:1, v/v) on silica gel column chromatography to obtain compound **7** (6.3 mg).

4.4. Plant Material and Growth Conditions

Seeds of *Lactuca sativa* L. were purchased from Gansu Academy of Agricultural Sciences. After soaked with 10% sodium hypochlorite for 10 min, the test seeds were washed with distilled sterile water 3 times, then germinated on filter paper in the dark at 25 °C for 1 day.

Healthy potato cultivar Long 7 with strong growth potential and no virus carrying was used for the germination treatment. After germination, the lateral or top buds were cut into a beaker with a scalpel and rinsed with flowing water for 6 h. These materials were soaked with 75% alcohol for 45 s. Then, the buds were sanitized in 0.1% mercuric chloride for 10 min, and washed with sterile water 5 times in order to thoroughly remove residual Hg^+ from explant material. The medium for the tissue culture of potatoes was modified on MS medium containing MS medium, agar (7 g/L), sucrose (30 g/L), and 1.5 mg/L naphthaleneacetic acid (NAA). Finally, the tissue culture seedlings (about 1–2 cm length) were cut and inserted into the medium, and placed in a culture chamber for light culture (in the light for 14 h and in the dark for 10 h per day) at 25 °C.

4.5. Bioassays

L. sativa was selected as the receptor plant, and the allelopathic activities of the crude extract and compounds **1–7** were evaluated by a plate culture method. The crude extract and compounds **1–7** were dissolved in DMSO and diluted with distilled sterile water, with final concentrations of 50, 100, 200, and 400 μg/mL for the crude extract and 10, 25, 50, 100, and 200 μg/mL for compounds **1–7**, respectively. The concentration of DMSO should be ensured not to exceed 1% (v/v). The same ratio of DMSO was added into distilled sterile water as a control. After germination, the test seedlings were transferred to 6-well plates (Corning, Inc., Corning, NY, USA). 0.5 mL of the solutions was added to each of the wells, and three replicates (six seedlings for each replicate) were set for each treatment. Then, the six-well plates were placed in a constant temperature and humidity chamber in the dark at 25 °C for 2 days. The lengths of the roots and stems were measured and the relative length of the control was calculated.

Tissue culture seedlings of potato were used to evaluate the autotoxic activities of the crude extract and compounds **1–7**. The crude extract and compounds **1–7** were dissolved in DMSO and added into the medium, with final concentrations of 50, 100, 200, and 400 μg/mL for the crude extract and 10, 25, 50, 100, and 200 μg/mL for compounds **1–7**, respectively. The same ratio of DMSO was added into the medium as a control. The tissue culture seedlings with the same growing trend were transferred to the prepared treatment and control medium, and three replicates (six seedlings for each replicate) were set for each treatment. Then, the seedlings were cultured in the same conditions as above for 10 days. The height of the seedlings, the fresh weight of the leaves and stems, and the fresh weight of the roots were measured and the relative ratio of the control was calculated.

4.6. High-Performance Liquid Chromatography (HPLC) Analysis

The existence of the allelochemicals in the rhizosphere soil of potato was confirmed by high-performance liquid chromatography (HPLC), and the contents of the allelochemicals were quantified simultaneously. The optimum gradient of the mobile phase was: 20–100% A (A: methanol; B: water) at 0–30 min and 100% (A) at 30–45 min. The detection wavelength was selected as 240 nm for compounds **1**, **2**, and **4**, and 280 nm for compound **6**. The flow rate was 1 mL/min with an injection volume of 20 μL, and the column temperature was 35 °C. The existence of these compounds in the soil was confirmed by comparing the retention times of the standard compounds **1**, **2**, **4**, and **6** under the same chromatographic conditions. By using the linear regression method, the standard curves were achieved. The contents of allelochemicals in the rhizosphere soil of potato could be calculated by using the peak area and linear equations. The results are presented in Table 1.

4.7. Antifungal Activity

The antifungal activity of compounds **1–7** was evaluated against *Alternaria solani* Sorauer, *Botrytis cinerea* Persoon, *Fusarium solani* (Mart.) Sacc, *Fusarium oxysporum* Schlecht, *Colletotrichum coccodes* (Wall) Hughes, and *Verticillium dahlia* by a mycelium growth rate test with some modifications [46] using PDA medium. All the plant pathogens were from the CAS Key Laboratory of Chemistry of Northwestern Plant Resources, Lanzhou Institute of Chemical Physics, Chinese Academy of Sciences, Lanzhou, China. The strains were identified by Hui Jin and Xiaoyan Yang from Lanzhou Institute of Chemical Physics, Chinese Academy of Sciences, Lanzhou, China. Compounds **1–7** were dissolved in DMSO and added into the PDA medium with a final concentration of 200 μg/mL; the concentration of DMSO should be ensured not exceed 1% (v/v). 10 mL of the treatment and control medium were poured into each sterilized Petri dish (60 mm in diameter), and the fungi mycelia disk with a 5.0 mm diameter was placed in the center of the Petri dish and incubated at 25 °C. When the mycelium of the fungi reached the edges of the control plate, the mycelia diameter of each treatment was measured.

4.8. Statistical Analysis

All data were subject to an analysis of variance by SPSS 18. The significant differences between the treatment groups and the control group were calculated by the Fisher's least significant difference (LSD) test and one-way analysis of variance (ANOVA). Statistical significance was accepted at * $p < 0.05$, ** $p < 0.01$. The relative ratio (percent) was determined by the formula [treated group/control group] $\times$ 100.

5. Conclusions

In summary, the isolated compounds from the rhizosphere soil of potato had a close relation to the allelopathy and autotoxicity. The accumulation of the allelochemicals could induce an imbalance of the microbial community in the potato replanted soil and result in replant failure during continuous cropping. These findings are helpful for understanding the allelopathic factors and mechanism that lead to replant failure in the cultivation of potato.

Author Contributions: Conceptualization, A.X., X.Y. and B.Q.; validation, J.G., Z.C. and Z.D.; formal analysis, A.X.; investigation, A.X. and X.Y.; resources, X.Y. and H.J.; data curation, A.X.; writing—original draft preparation, A.X.; writing—review and editing, H.J. and B.Q.; visualization, A.X., H.H. and H.L.; supervision, B.Q.; project administration, A.X. and H.J.; funding acquisition, B.Q. All authors have read and agreed to the published version of the manuscript.

Funding: This research was funded by the National Natural Science Foundation of China (22076195) and Cooperation Program to Gansu Province of Lanzhou Branch of the Chinese Academy of Sciences (2021).

Institutional Review Board Statement: Not applicable.

Informed Consent Statement: Not applicable.

Data Availability Statement: Not applicable.

Conflicts of Interest: The authors declare no conflict of interest.

Abbreviations

ANOVA—analysis of variance; CC—column chromatography; GC−MS—gas chromatography−mass spectrometry; LSD—least significant difference; MCI—gel, middle chromatogram isolated gel; PTLC—preparative thin-layer chromatography; TLC—Thin-layer chromatography.

References

1. Lopez, A.; Arazuri, S.; Garcia, I.; Mangado, J.; Jaren, C. A review of the application of near-infrared spectroscopy for the analysis of potatoes. *J. Agric. Food Chem.* **2013**, *61*, 5413–5424. [CrossRef] [PubMed]
2. Food and Agriculture Organization of the United Nations. Available online: https://www.fao.org/about/en (accessed on 1 January 2022).
3. Camire, M.E.; Kubow, S.; Donnelly, D.J. Potatoes and human health. *Crit. Rev. Food Sci. Nutr.* **2009**, *49*, 823–840. [CrossRef] [PubMed]
4. Kondo, Y.; Sakuma, R.; Ichisawa, M.; Ishihara, K.; Kubo, M.; Handa, S.; Mugita, H.; Maruyama, N.; Koga, H.; Ishigami, A. Potato chip intake increases ascorbic acid levels and decreases reactive oxygen species in SMP30/GNL knockout mouse tissues. *J. Agric. Food Chem.* **2014**, *62*, 9286–9295. [CrossRef] [PubMed]
5. Orawetz, T.; Malinova, I.; Orzechowski, S.; Fettke, J. Reduction of the plastidial phosphorylase in potato (*Solanum tuberosum* L.) reveals impact on storage starch structure during growth at low temperature. *Plant Physiol. Biochem.* **2016**, *100*, 141–149. [CrossRef]
6. Sulc, M.; Kotikova, Z.; Paznocht, L.; Pivec, V.; Hamouz, K.; Lachman, J. Changes in anthocyanidin levels during the maturation of color-fleshed potato (*Solanum tuberosum* L.) tubers. *Food Chem.* **2017**, *237*, 981–988. [CrossRef]
7. Qin, Y.; Ma, K.; Liu, P. Effect of potato continuous cropping on genetic diversity of soil microorganisms. *Chin. J. Eco-Agric.* **2015**, *23*, 225–232.
8. Cesarano, G.; Zotti, M.; Antignani, V.; Marra, R.; Scala, F.; Bonanomi, G. Soil sickness and negative plant-soil feedback: A reappraisal of hypotheses. *J. Plant Pathol.* **2017**, *99*, 545–570.

9. Manici, L.M.; Kelderer, M.; Franke-Whittle, I.H.; Ruhmer, T.; Baab, G.; Nicoletti, F.; Caputo, F.; Topp, A.; Insam, H.; Naef, A. Relationship between root-endophytic microbial communities and replant disease in specialized apple growing areas in Europe. *Appl. Soil Ecol.* **2013**, *72*, 207–214. [CrossRef]
10. Kardol, P.; Cornips, N.J.; van Kempen, M.M.L.; Bakx-Schotman, J.M.T.; van der Putten, W.H. Microbe mediated plant soil feedback causes historical contingency effects in plant community assembly. *Ecol. Monogr.* **2007**, *77*, 147–162. [CrossRef]
11. Van de Voorde, T.F.J.; Ruijten, M.; van der Putten, W.H.; Bezemer, T.M. Can the negative plant-soil feedback of *Jacobaea vulgaris* be explained by autotoxicity? *Basic Appl. Ecol.* **2012**, *13*, 533–541. [CrossRef]
12. Liu, J.K.; Yan, Z.Q.; Li, X.Z.; Jin, H.; Yang, X.Y.; Xie, M.; Su, A.X.; Qin, B. Characterization of allelochemicals from the rhizosphere soil of *Pinellia ternate* (Thnub.) and their inhibition activity on protective enzymes. *Appl. Soil Ecol.* **2018**, *125*, 301–306. [CrossRef]
13. Guo, K.; He, X.F.; Yan, Z.Q.; Li, X.Z.; Ren, X.; Pan, L.; Qin, B. Allelochemicals from the Rhizosphere Soil of Cultivated *Astragalus hoantchy* L. *J. Agric. Food Chem.* **2016**, *64*, 3345–3352. [CrossRef] [PubMed]
14. Ren, X.; Yan, Z.Q.; He, X.F.; Li, X.Z.; Qin, B. Allelochemicals from rhizosphere soils of *Glycyrrhiza uralensis* Fisch: Discovery of the autotoxic compounds of a traditional herbal medicine. *Ind. Crops Prod.* **2017**, *97*, 302–307. [CrossRef]
15. Xie, M.; Yan, Z.Q.; Ren, X.; Li, X.Z.; Qin, B. Codonopilate A, a triterpenyl ester as main autotoxin in cultivated soil of *Codonopsis pilosula* (Franch.) Nannf. *J. Agric. Food Chem.* **2017**, *65*, 2032–2038. [CrossRef] [PubMed]
16. Hasegawa, T.; Kato, Y.; Okabe, A.; Itoi, C.; Ooshiro, A.; Kawaide, H.; Natsume, M. Effect of secondary metabolites of tomato (*Solanum lycopersicum*) on chemotaxis of *ralstonia solanacearum*, pathogen of bacterial wilt disease. *J. Agric. Food Chem.* **2019**, *67*, 1807–1813. [CrossRef] [PubMed]
17. Manici, L.M.; Ciavatta, C.; Kelderer, M.; Erschbaumer, G. Replant problems in south tyrol: Role of fungal pathogens and microbial population in conventional and organic apple orchards. *Plant Soil.* **2003**, *256*, 315–324. [CrossRef]
18. Reddy, M.A.; Patrick, Z.A. Colonization of tobacco seedling roots by fluorescent pseudomonad suppressive to black root rot caused by *Thielaviopsis basicola*. *Crop Protection.* **1992**, *11*, 148–154. [CrossRef]
19. Bonanomi, G.; de Filippis, F.; Cesarano, G.; La Storia, A.; Ercolini, D.; Scala, F. Organic farming induces changes in soil microbiota that affect agro-ecosystem functions. *Soil Biol. Biochem.* **2016**, *103*, 327–336. [CrossRef]
20. Xia, Z.C.; Kong, C.H.; Chen, L.C.; Wang, S.L. Allelochemical-mediated soil microbial community in long-term monospecific Chinese fir forest plantations. *Appl. Soil Ecol.* **2015**, *96*, 52–59. [CrossRef]
21. Dong, L.L.; Xu, J.; Li, Y.; Fang, H.L.; Niu, W.H.; Li, X.W.; Zhang, Y.J.; Ding, W.L.; Chen, S.L. Manipulation of microbial community in the rhizosphere alleviates the replanting issues in *Panax ginseng*. *Soil Biol. Biochem.* **2018**, *125*, 64–74. [CrossRef]
22. Soltys-Kalina, D.; Murawska, Z.; Strzelczyk Zyta, D.; Wasilewicz-Flis, I.; Marczewski, W. Phytotoxic potential of cultivated and wild potato species (*Solanum sp.*): Role of glycoalkaloids, phenolics and flavonoids in phytotoxicity against mustard (*Sinapis alba* L.). *Acta Physiol. Plant.* **2019**, *41*, 41–55. [CrossRef]
23. Wan, N.X.; Yuan, J.C.; He, W.; Long, W.J.; Zhang, Q.; Zhou, S.M.; Zheng, S.L. Autotoxicity of water extracts from different organs of potato. *J. Zhejiang Univ.* **2016**, *42*, 411–418.
24. Zhang, W.M.; Qiu, H.Z.; Zhang, C.H.; Liu, X.; Gao, Y.A.; Shen, Q.R. Identification and autotoxicity of root exudates of continuous cropping potato at different growth stages. *Chin. J. Eco-Agric.* **2015**, *23*, 215–224.
25. Guo, C.P.; Zi, S.H.; Ou-Yang, C.R.; Wu, B.Z. Effect of aboveground aqueous extracts of maize and potato on the maize growth. *J. Maize Sci.* **2016**, *24*, 79–84.
26. Gamiz, B.; Facenda, G.; Celis, R. Nanoengineered sorbents to increase the persistence of the allelochemical carvone in the rhizosphere. *J. Agric. Food Chem.* **2019**, *67*, 589–596. [CrossRef]
27. Zhang, W.; Lu, L.Y.; Hu, L.Y.; Cao, W.; Sun, K.; Sun, Q.B.; Siddikee, A.; Shi, R.H.; Dai, C.C. Evidences for the Involvement of auxin, ethylene and ROS signaling during allelochemical benzoic acid-mediated primary root inhibition of Arabidopsis. *Plant Cell Physiol.* **2018**, *59*, 1889–1904. [CrossRef]
28. Rice, E.L. Allelopathy-Update. *Bot. Rev.* **1979**, *45*, 15–109. [CrossRef]
29. Niro, E.; Marzaioli, R.; de Crescenzo, S.; D'Abrosca, B.; Castaldi, S.; Esposito, A.; Fiorentino, A.; Rutigliano, F.A. Effects of the allelochemical coumarin on plants and soil microbial community. *Soil Biol. Biochem.* **2016**, *95*, 30–39. [CrossRef]
30. Abenavoli, M.R.; Cacco, G.; Sorgona, A.; Marabottini, R.; Paolacci, A.R.; Ciaffi, M.; Badiani, M. The inhibitory effects of coumarin on the germination of durum wheat (*Triticum turgidum* ssp. durum, cv. Simeto) seeds. *J. Chem. Ecol.* **2006**, *32*, 489–506. [CrossRef]
31. Mushtaq, W.; Ain, Q.; Siddiqui, M.B.; Hakeem, K.R. Cytotoxic allelochemicals induce ultrastructural modifications in *Cassia tora* L. and mitotic changes in *Allium cepa* L.: A weed versus weed allelopathy approach. *Protoplasma* **2019**, *256*, 857–871. [CrossRef]
32. Naveed, M.; Hejazi, V.; Abbas, M.; Kamboh, A.A.; Khan, G.J.; Shumzaid, M.; Ahmad, F.; Babazadeh, D.; Xia, F.F.; Modarresi-Ghazani, F.; et al. Chlorogenic acid (CGA): A pharmacological review and call for further research. *Biomed. Pharmacother.* **2018**, *97*, 67–74. [CrossRef] [PubMed]
33. Fu, Y.H.; Quan, W.X.; Li, C.C.; Qian, C.Y.; Tang, F.H.; Chen, X.J. Allelopathic effects of phenolic acids on seedling growth and photosynthesis in *Rhododendron delavayi* Franch. *Photosynthetica* **2019**, *57*, 377–387. [CrossRef]
34. Wang, D.D.; Yang, J.S.; Yan, Z.Q.; Sun, Y.H.; Cui, H.Y.; Ding, L.; Qin, B. Allelopathic effect and mechanism of chlorogenic acid on the growth of lettuce. *Acta Bot. Boreali-Occident. Sin.* **2014**, *34*, 957–962.
35. Yin, R.H.; Han, K.; Heller, W.; Albert, A.; Dobrev, P.I.; Zazimalova, E.; Schaffner, A.R. Kaempferol 3-O-rhamnoside-7-O-rhamnoside is an endogenous flavonol inhibitor of polar auxin transport in Arabidopsis shoots. *New Phytol.* **2014**, *20*, 466–475. [CrossRef]
36. Yang, C.Q.; Yang, W.Y.; Liu, J. Advances on chemical ecology of plant flavonoids. *Nat. Prod. Res. Dev.* **2018**, *30*, 2009–2016.

37. Brunetti, C.; Di Ferdinando, M.; Fini, A.; Pollastri, S.; Tattini, M. Flavonoids as antioxidants and developmental regulators: Relative significance in plants and humans. *Int. J. Mol. Sci.* **2013**, *14*, 3540–3555. [CrossRef] [PubMed]
38. Paterson, E.; Gebbing, T.; Abel, C.; Sim, A.; Telfer, G. Rhizodeposition shapes rhizosphere microbial community structure in organic soil. *New Phytol.* **2007**, *173*, 600–610. [CrossRef]
39. Hayat, R.; Ali, S.; Amara, U.; Khalid, R.; Ahmed, I. Soil beneficial bacteria and their role in plant growth promotion: A review. *Ann. Microbiol.* **2010**, *60*, 579–598. [CrossRef]
40. Huang, L.F.; Song, L.X.; Xia, X.J.; Mao, W.H.; Shi, K.; Zhou, Y.H.; Yu, J.Q. Plant-soil feedbacks and soil sickness: From mechanisms to application in agriculture. *J. Chem. Ecol.* **2013**, *39*, 232–242. [CrossRef]
41. Mazzola, M.; Manici, L.M. Apple replant disease: Role of microbial ecology in cause and control. *Annu. Rev. Phytopathol.* **2012**, *50*, 45–65. [CrossRef]
42. Yang, J.I.; Ruegger, P.M.; McKenry, M.V.; Becker, J.O.; Borneman, J. Correlations between root-associated microorganisms and peach replant disease symptoms in a california soil. *PLoS ONE* **2012**, *7*, e46420. [CrossRef] [PubMed]
43. Bonanomi, G.; D'Ascoli, R.; Antignani, V.; Capodilupo, M.; Cozzolino, L.; Marzaioli, R.; Puopolo, G.; Rutigliano, F.A.; Scelza, R.; Scotti, R.; et al. Assessing soil quality under intensive cultivation and tree orchards in southern Italy. *Appl. Soil Ecol.* **2011**, *47*, 184–194. [CrossRef]
44. Bertin, C.; Yang, X.H.; Weston, L.A. The role of root exudates and allelochemicals in the rhizosphere. *Plant Soil.* **2003**, *256*, 67–83. [CrossRef]
45. Lin, S.; Dai, L.Q.; Chen, T.; Li, Z.F.; Zhang, Z.Y.; Lin, W.X. Screening and identification of harmful and beneficial microorganisms associated with replanting disease in rhizosphere soil of *Pseudostellariae heterophylla*. *Int. J. Agric. Biol.* **2015**, *17*, 458–466. [CrossRef]
46. Wei, L.J.; Tan, W.Q.; Wang, G.; Li, Q.; Dong, F.; Guo, Z.Y. The antioxidant and antifungal activity of chitosan derivatives bearing Schiff bases and quaternary ammonium salts. *Carbohydr. Polym.* **2019**, *226*, 115256. [CrossRef] [PubMed]

plants

MDPI

Review

Novel Cognitions in Allelopathy: Implications from the "Horizontal Natural Product Transfer"

Laura Lewerenz [1], Sara Abouzeid [2], Mahdi Yahyazadeh [3], Tahani Hijazin [4] and Dirk Selmar [1,*]

1 Institut für Pflanzenbiologie, Technische Universität Braunschweig, Mendelssohnstraße 4, D-38106 Braunschweig, Germany
2 Pharmacognosy Department, Faculty of Pharmacy, Mansoura University, Mansoura 35516, Egypt
3 Research Institute of Forests and Rangelands, Agricultural Research, Education and Extension Organization (AREEO), Tehran P.O. Box 13185-116, Iran
4 Biology Department, Faculty of Science, Mutah University, P.O. Box 7, Mutah 61710, Jordan
* Correspondence: d.selmar@tu-bs.de

Abstract: Whereas the translocation of allelochemicals between plants is well established, a related general transfer of genuine specialized metabolites has not been considered so far. The elucidation of the so-called "Horizontal Natural Product Transfer" revealed that alkaloids, such as nicotine and pyrrolizidine alkaloids, which are leached out from decomposing alkaloid-containing plants (donor plants), are indeed taken up by the roots of plants growing in the vicinity (acceptor plants). Further studies demonstrated that phenolic compounds, such as coumarins or stilbenes, are also taken up by acceptor plants. Contemporary analyses from co-cultivation experiments outlined that natural products are not exclusively transferred from dead and rotting donor plant materials, but also from vital plants. In analogy to xenobiotics, the imported specialized metabolites might also be modified within the acceptor plants. As known from the uptake of xenobiotics, the import of specialized metabolites is also generally due to a simple diffusion of the substances across the biomembranes and does not require a carrier. The uptake depends in stricto sensu on the physicochemical properties of the certain compound. This article presents a current overview of the phenomenon of "Horizontal Natural Product Transfer" and discusses its relevance for our understanding of allelopathic interactions. The knowledge that specialized metabolites might in general be readily translocated from one plant into others should significantly contribute to our understanding of plant–plant interactions and—in particular—to the evolution of typical allelopathic effects, such as inhibition of growth and germination of potential competitors.

Keywords: horizontal natural product transfer; allelopathy; alkaloids; uptake

Citation: Lewerenz, L.; Abouzeid, S.; Yahyazadeh, M.; Hijazin, T.; Selmar, D. Novel Cognitions in Allelopathy: Implications from the "Horizontal Natural Product Transfer". *Plants* **2022**, *11*, 3264. https://doi.org/10.3390/plants11233264

Academic Editors: Margot Schulz, Vincenzo Tabaglio and Per Kudsk

Received: 5 October 2022
Accepted: 22 November 2022
Published: 28 November 2022

1. Background

By their roots, plants take up plenty of inorganic nutrients from the soil, as well as many other compounds. Concerning biochemical ecology, in particular, the uptake of allelochemicals, i.e., active substances, which inhibit germination or growth of putative competitors, is of special interest [1]. In general, these substances are released from donor plants into the soil and exhibit their inhibitory effect on the plants growing in their vicinity [2,3] (see also other chapters of this Special Issue). To execute their inhibitory effects, the allelochemicals have to be taken up by the acceptor plants [1,4]. A further import of natural products is related to Systemic Acquired Resistance (SAR) of plants: it was shown that salicylic acid, which provokes SAR, is also taken up by roots [5,6]. In this context, it has to be emphasized that the uptake of ionic compounds such as salicylic acid, as well as of nutrients such as nitrate, sulfate or metal ions, requires specific transporters [7–9]. As a consequence, by regulating the carrier proteins' number and activity, plants are capable of controlling the import of the related substances by their roots. As opposed to this, the situation is entirely different when considering the uptake of xenobiotics, i.e., systemic herbicides, fungicides [10], or veterinary medicines [11]. Because of their partially hydrophobic character, these substances are able to diffuse smoothly through membranes [12–14]. Thus, these substances are taken up by the roots passively (see below), and, in consequence, the plants cannot control or even prevent the import of these substances. Although these nexuses were well established, corresponding contemplations about common specialized metabolites have not been considered at all. However, this attitude has plainly changed within the last few years due to comprehensive investigations to identify the potential sources of various contaminations of plant-derived commodities.

Triggered by alarming reports of the European Food Safety Authority (EFSA) on contaminations of herbal medicines and phytopharmaceuticals by various toxic alkaloids, i.e., nicotine [15], pyrrolizidine alkaloids [16,17], or tropane alkaloids [18], several studies on the source of these contaminations were conducted. It became apparent that plants take up nicotine [19] and pyrrolizidine alkaloids [20] from the soil. Subsequent field trials demonstrated that nicotine, which is leached out from discarded cigarette butts, is taken up massively by acceptor plants. The related nicotine concentrations in the commodities were around 1 µg/g d.w. [21], thus exceeding the official maximum residue limit (0.01 µg/g d.w.) about a hundred-fold. In the same manner, pyrrolizidine alkaloids (PAs), which are leached out from rotting PA-containing weeds into the soil, e.g., *Senecio jacobaea*, are also taken up by the roots of the acceptor plants and translocated into their leaves [20,22]. Accordingly, such transfer of alkaloids—at least in part—accounts for the numerous and widespread alkaloidal contaminations of spice and medicinal plants reported comprehensively [15,16,23]. Based on these coherences, the so-called "Horizontal Natural Product Transfer" was outlined [24], a so far unrecognized phenomenon which has been studied intensively during the last few years (for review, see [25–27]). It became evident that apart from alkaloids, other natural products, such as coumarins [28], stilbenes [29], or betalains [30], are also taken up by plant roots. In this context, it has to be mentioned that in some cases, the natural products whose import into acceptor plants was verified, i.e., the stilbene resveratrol [29] or the coumarins [27], are known to exhibit allelopathic effects [31,32]. These examples vividly demonstrate that the line between "simple natural products" and "allelochemicals", which are leached out, is quite diffuse. Up to now, this issue has not been taken into consideration when evolutionary aspects of allelopathic effects are discussed.

2. Horizontal Natural Product Transfer—Basic Principles and Cognitions

2.1. Uptake of Chemicals from the Soil

A vast number of studies on the uptake of xenobiotics showed that by far, most of these substances can smoothly diffuse across membranes [12–14]. The ability for such passive membrane transfer can roughly be deduced from the so-called K_{OW} value [10], which corresponds to the distribution coefficient of a certain substance between octanol and water. In most related publications, its decadal logarithm, denoted as pK_{OW}, is outlined [33]. Frequently, this parameter is also denominated as log$P$ [34]. Comprehensive research displayed that substances exhibiting log$P$ values between −1 and 3 are inherently able to diffuse easily through biomembranes [33,35]. It is self-evident that these coherences, which pertain to the uptake of xenobiotics, also apply to all other organic substances. This phenomenon was approved by Hurtado et al. [36], who demonstrated that organic compounds exhibiting appropriate log$P$ values are generally taken up by plants. In consequence, there is no doubt that also all plant-derived natural products exhibiting an appropriate log$P$ value are passively taken up by plants when they are present in the soil.

Nevertheless, an additional factor must be considered when contemplating the membrane permeability of alkaloids, i.e., the pH-dependent protonation of these alkaline specialized metabolites [37]. Whereas the free bases of most alkaloids pass biomembranes smoothly, the protonated forms cannot, since the positive charge strongly increases the hydrophilic character and thus prevents the penetration of biomembranes. This manifestation nicely coincides with the quite negative log$P$ values of charged alkaloid salts. This implies that—apart from the log$P$ of the various alkaloids—the pH of the medium also strongly determines their membrane permeability and thereby massively influences the extent of the uptake of alkaloids from the soil [38,39]. In acidic soils, due to the high degree of protonation, most alkaloids are unable to cross the plasmalemma of the root cells and accordingly, they are not taken up [10]. In contrast, in soils exhibiting enhanced pH values, protonation is pushed back and the alkaloids exist as free bases, which smoothly penetrate the plasmalemma of the root cells. provoking their strong uptake [39].

2.2. Horizontal Natural Product Transfer between Vital Plants

When contemplating the basic coherences of the horizontal natural product transfer, the question arises whether—in analogy to the relocation of natural products from rotting plant material—a corresponding transfer also might occur between living plants. This seems to be obvious because it is well known that allelochemicals are also released into the environment by various mechanisms [40]: they might either be leached out of decomposing plant materials, or they could actively be exuded from vital plants by their roots [2,3] or by their leaves [40,41]. In order to ascertain a putative release of common and typical natural products into the soil and a subsequent transfer into acceptor plants, co-culture experiments have been conducted: *Senecio jacobaea* plants containing high concentrations of pyrrolizidine alkaloids (Pas) were cultivated in the same pot together with parsley plants. After two months of co-cultivation, the plants were harvested and analyzed. Amazingly, in all parsley plants, which had been co-cultivated in the same pot together with a *Senecio* plant, significant concentrations of Pas were present; on average more than 200 μg/kg d.w. [22]. To rule out that the transfer of alkaloids is due to a too intensive contact between the roots of the donor and acceptor plants, a further study employing various distances between the plants was required. Accordingly, in a corresponding field trial, different acceptor plants were co-cultivated with *S. jacobaea*. It turned out that, in all plants grown in the field together with *S. jacobaea*, significant amounts of PAs were also present [22]. Thus, the alkaloids, which previously had been synthesized in the donor plants, are now present in the plants growing in the vicinity. Accordingly, there is no doubt that the PAs indeed had been transferred from vital donor plants via the soil into acceptor plants growing nearby (Figure 1). In further approaches, it could be verified that non-alkaloidal specialized metabolites are also transferred via the soil from donor to acceptor plants, i.e., coumarins [28], stilbenes [29], aristolochic acids [42,43], or betalains [25]. Meanwhile, several studies outlined a corresponding transfer of specialized metabolites from vital donor plants to acceptor plants growing in their vicinity (Table 1). However, up to now, there is no clue concerning the mode of this transfer.

First and foremost, it could be assumed that the described PA-transfer between living plants is due to the shedding of *Senecio* leaves; when the leaves decay, the Pas could be leached out from the rotting plant materials. In principle, this possibility cannot be excluded, but during the pot experiments described above, no leaves had been shed [26]. Thus, at least in these experiments, a transfer via shed leaves could be ruled out. Nonetheless, when referring to the mentioned field experiment, such an option cannot be excluded. These coherences also account for most of the reports on the transfer of specialized metabolites mentioned in Table 1. When considering an entire vegetation period, it has to be regarded that a senescence-induced shedding always takes place. Thus, this option certainly might be relevant, at least to some extent.

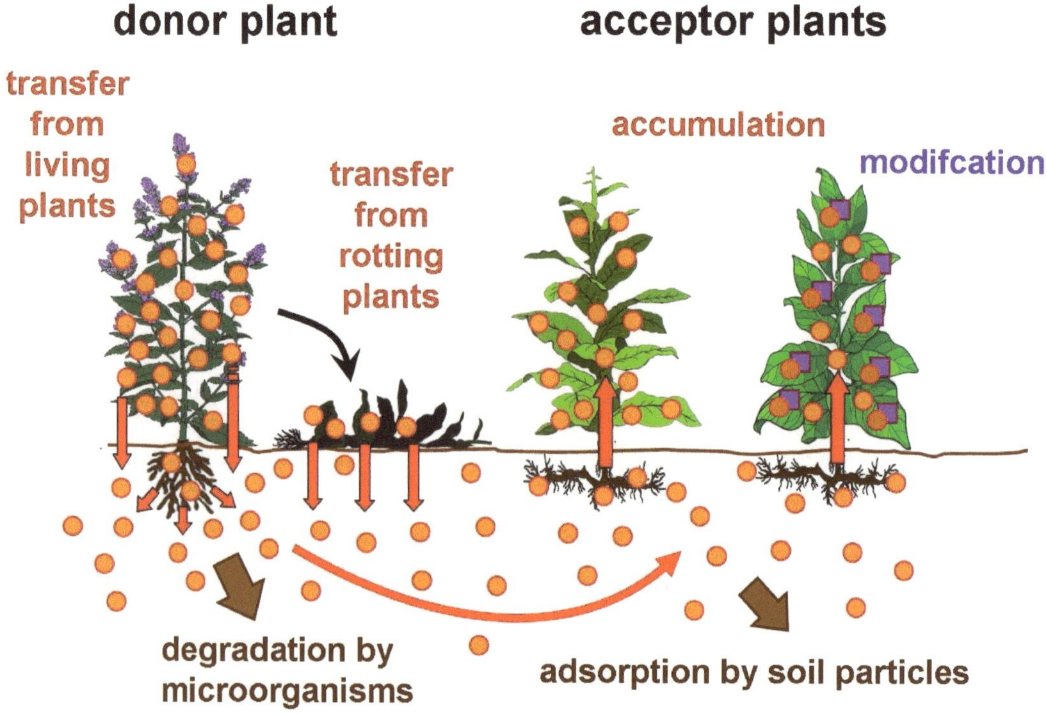

Figure 1. Horizontal natural product transfer. The scheme displayed by Selmar et al. [26] is supplemented with the processes occurring in the rhizosphere.

Table 1. Transfer of alkaloids via soil between vital donor and acceptor plants.

Donor Plant	Type of Alkaloid	Acceptor Plant	Authors
Chromolaena odorata	Pyrrolizidine alkaloids	*Zea mays*	[44]
Secale cereale	Benzoxazinoids	*Vicia villosa*	[45]
Senecio jacobaea	Pyrrolizidine alkaloids	*Matricaria chamomilla*	[20,22]
Senecio jacobaea	Pyrrolizidine alkaloids	*Melissa officinalis*	[20,22]
Senecio jacobaea	Pyrrolizidine alkaloids	*Mentha × piperita*	[20,22]
Senecio jacobaea	Pyrrolizidine alkaloids	*Petroselinum crispum*	[20,22,25]
Senecio jacobaea	Pyrrolizidine alkaloids	*Tropaeolum majus*	[22]
Solanaceae	Atropine	*Triticum aestivum*	[46]

Another option for the transfer of the alkaloids could be due to their bleeding out from minor leaf injuries, which might be caused by herbivore or pathogen attacks. Yet, the plants used in the co-culture experiments had been described as healthy and no marks of herbivory were visible [22]. Thus, it seems unlikely that the observed PA-transfer is related to plain bleeding of the alkaloids from injured leaves. This conclusion is underlined by the massive differences in the PA-spectra of donor and acceptor plants [22]. In case of wound-induced bleeding, all the different PAs accumulated in the donor plant would be leached out and taken up by the acceptor plants to more or less the same extent. The observed differences must be due to a process, whose extent differs for the various alkaloids. In consequence, simple bleeding of the alkaloids from injured leaves can be ruled out. The same deduction is also valid for wound-induced bleeding from putatively injured roots. Thus, we have to suppose that the PAs present in the acceptor plants indeed had been released from vital, healthy and uninjured *Senecio* donor plants.

When bearing in mind the well-established active exudation of allelochemicals [2,3,40], it seems to be obvious that PAs, as well as other alkaloids, are also exuded by the donor plants, either by their roots or by their leaves. When discussing such a release, it readily becomes apparent that in principle there are two possibilities: the alkaloids could either be diffused passively out of the cells, or they could actively be exuded by the roots or shoots. In this context, the finding of García-Jorgensen et al. [47] becomes relevant, which outlined that the carcinogenic phytotoxin ptaquiloside, which is accumulated in the fern *Pteridium aquilinum*, indeed passively diffuses out of its leaves. Consequently, all alkaloids and other specialized metabolites exhibiting appropriate log$P$ values might also simply diffuse passively out of leaves or roots into the soil, and accordingly will be taken up passively by the roots of plants growing in the vicinity. Yet, in case of polyphenolic compounds, an uptake might be prevented due to their oxidation to yield quinones, which efficiently bind to proteins, forming insoluble tanning products. The observed differences in the PA spectra of donor and acceptor plants might result from disparities in the log$P$- and pK_a-values of the various PAs (see below) and thus, in their ability to diffuse through biomembranes. However, in this context, it has to be considered that in plant cells alkaloids are efficiently trapped in their vacuoles. Already half a century ago, Matile [48] outlined his famous "ion trap mechanism": within the neutral cytosol, most alkaloids are present as free bases and—provided that their log$P$-values are in the range between -1 and 3 (see above), they are able to diffuse through biomembranes, such as the tonoplast. However, due to the acidic conditions in the vacuoles, the alkaloids become protonated. Consequently, they are no longer able to pass the tonoplast and accordingly are trapped and accumulated in the vacuoles. Thus, a passive release out of the cells is quite unlikely. Yet, these coherences only apply to alkaloids revealing a relatively high pKa, ensuring a sufficient protonation in the vacuole [39]. In contrast, alkaloids, which reveal a very low pKa, e.g., caffeine (pKa = 0.7) are not protonated in the acidic vacuole. Since they do not exhibit a positive charge that typically prevents their membrane permeability, these alkaloids are not retained within this compartment. Nevertheless, due to an effective complex formation with chlorogenic acids, caffeine—to a certain extent—is also retained in the vacuole [49], since—in the same manner as protonated alkaloids—the caffeine-chlorogenic acids-complex also cannot pass the tonoplast passively and thus, it is retained in the vacuole.

Most of the literature dealing with the release of alkaloids from vital plant parts is based on organ or cell culture experiments, e.g., harmine and harmaline are exuded from root culture cells of *Oxalis tuberose* [50], the indole alkaloid ajmalicine is secreted from hairy root cultures of *Catharanthus roseus* [51], and nicotine is present in the medium of root cultures from *Nicotiana tabacum* [52]. Based on the coherences outlined above, the occurrence of alkaloids in the culture medium might just be a hint for an active exudation. Nonetheless, Toppel et al. [53] demonstrated that in the culture medium of *Senecio* root cells, senkirkine was exclusively present, although the cultured cells contained a wide variety of PAs. This discrepancy can only be explained by the fact that senkirkine is actively and specifically exuded from the cells into the culture medium [53]. It is evident that such exudation requires specific carriers, which catalyze the transfer of the protonated alkaloids across the tonoplast and thus out of the vacuole. Indeed, numerous alkaloid transporters are described in the literature [54–56], for review see [57]. Unfortunately, in most of these reports, the ability of alkaloids to easily pass biomembranes is entirely ignored, and the authors are not aware that the transporters are required for the membrane transfer of protonated alkaloids rather than for the free bases [27].

Apart from the data on the release of alkaloids into the medium of organ and cell cultures, various data on the liberation of alkaloids from genuine plants into the soil are available. Baumann and Gabriel [58] showed that the roots of coffee seedlings exude caffeine, and Schulz et al. [59] demonstrated that dihydroxybenzoxazinone is released from the roots of *Agropyron repens* into the soil. Analogously, *Oxalis tuberosa* roots liberate carboline alkaloids, i.e., harmine and harmaline, into the soil [60]. Moreover, lupine alkaloids (quinolizidine alkaloids) had been detected in soils well-vegetated by narrow-leaf and yellow lupines [61]. Similarly, Hama and Strobel [62] found PAs in soils, overgrown with PA-containing *Senecio* plants. Apart from the occurrence of various alkaloids in soils, the assumption that alkaloids are released by the roots of various alkaloidal plants is supported by various indirect results, i.e., by detecting them in acceptor plants growing in their vicinity (Table 1). However, when contemplating the mode of release of these alkaloids, any solid proof on the mechanism of the release is lacking. In principle, these alkaloids might be eluted from shed leaves or leached out from injured leaves or roots, respectively. Moreover, they might be released from vital tissues, either by passive diffusion or an active exudation by the means of transporters. Indeed, with respect to the data mentioned above, up to now, no option can ultimately be excluded. Much more research is required to soundly unveil the actual path of alkaloid release into the soil.

2.3. *Translocation of Chemicals within the Acceptor Plants*

The alkaloids, which had been taken up by the roots of the acceptor plants, are subsequently present in their leaves [19,20]. Thus, the imported specialized metabolites have to be translocated within the acceptor plants. In principle, there are only two options for such allocation, either via xylem or phloem. In the case of xenobiotics, it is well documented that their transport from root to shoot is accomplished via the xylem [33,63]. Accordingly, it seems to be obvious that analogously also the natural products taken up by acceptor plants are translocated within the xylem. This theory was confirmed by the finding that imported alkaloids accumulated to a high extent in old leaves, whereas the alkaloid concentrations in young leaves were quite lower [19,20]. This different distribution is due to expansion-related differences in the transpiration rates, which in turn drives the xylem transport, and thus is responsible for the different extent of allocation of substances via the xylem. Furthermore, no imported alkaloids had been detected in the flowers of the acceptor plants [20], which represent a major physiological sink. Thus, a source-sink-translocation via phloem can be excluded. Finally, the implication of a xylem-based allocation of imported natural products is verified by the finding that, within the acceptor plants, these substances are also present in guttation droplets [20], whose composition corresponds to that of the xylem sap [64]. From these coherences, it has to be deduced that the imported natural compounds, in particular nicotine and PAs, indeed are translocated within the acceptor plants in the xylem.

When comparing these insights with the transport of PAs in genuine alkaloidal plants, such as *Senecio*, a strong discrepancy becomes evident: In the group of Hartmann, it was soundly verified that the PAs synthesized in the roots of *Senecio* plants are transferred as PA-*N*-oxides via phloem into the physiological sinks, such as flowers and seeds [65–67]. Based on the physicochemical properties of the alkaloids and their corresponding *N*-oxides, this putative contradiction could be elucidated [37]: as outlined above, the free bases of alkaloids readily diffuse through biomembranes. Accordingly, due to the ion trap mechanism already explained for the accumulation of protonated alkaloids within the vacuoles (see above), they are also displaced from the alkaline phloem and trapped within the acidic xylem. In consequence, they are allocated within the xylem into the transpiring leaves—a fact that is fully in accordance with the findings displayed for the alkaloids taken up from the soil. In contrast, the strongly enhanced hydrophilic properties of alkaloid-*N*-oxides or quaternary alkaloids, respectively, are not able to diffuse through biomembranes. Consequently, they would be retained in the alkaline phloem, and thereby enable the option for a long-distance transport to the physiological sinks via phloem as reported by the Hartmann group [65–67].

Since the specialized metabolites are passively taken up by the roots of the acceptor plants, the velocity and the extent of their import should be quite similar in the various species. The only factor which determines the uptake is the corresponding difference in concentration between the soil and the cytosol of the rhizodermis cells. Nevertheless, this gradient between "inside and outside" depends on various additional factors. Apart from the pH, which strongly influences the share of protonated and thus membrane-permeable alkaloids (see above), the most important factor corresponds to the decrease in the concentration of imported compounds within the root cells and thus, how the concentration gradient between "inside and outside" is maintained. In principle, there are two possibilities, either their allocation into other organs or their modification (see below). The velocity of a xylem-based allocation is mainly determined by the rate of transpiration [68]. Indeed, this property varies greatly among different plant species and is therefore responsible for the observed differences in the uptake of a certain natural substance imported into different plants.

Additionally, the gradient of a certain substance between "inside and outside" might also be impacted by a corresponding decrease in its concentration in the soil, either by its adsorption to various soil particles or by a related microbial decomposition or modification, respectively.

In addition to these insights, the transfer of natural products from living donor plants might also open new doors to unveil the beneficial impact of some hitherto not fully understood processes, i.e., co-cultivation of certain vegetables and crop rotations. The findings related to the exchange of natural products among vital plants might provide new approaches for the explanations of these ambiguous phenomena.

2.4. Modification of the Imported Compounds in the Acceptor Plants

Astonishingly, the contents of alkaloids imported into the acceptor plants decreased over time: a massive decline in concentration was reported for the content of nicotine in peppermint plants [19], as well as for the PAs in various spice plants [20]. Since a complete decomposition of these compounds to CO_2 and water seemed to be very unlikely, it was argued that the imported compounds had been modified within the acceptor plants. In this context, it is appropriate to outline that xenobiotics, which are taken up from the soil, are also known to be frequently modified within the acceptor plants (Figure 1). The most abundant modifications comprise oxidations, hydroxylations, and conjugations with glucose [69,70]. Based on these cognitions, the so-called green liver concept—a hypothesis that proposed that these reactions are part of a deliberate detoxification system for xenobiotics—was developed [70,71]. In consequence, it seems to be natural that the time-dependent decrease in the alkaloid content in the acceptor plants might also be due to the modification of the imported natural compounds. The confirmation of this conjecture was elaborated by advanced analyses of PAs present in the acceptor plants: standard quantification of PAs is generally performed based on standard LC-MS methods, quantifying and summing up the individual contents of 27 genuine PAs [20,72]. Thus, in these studies, solely the original already known PAs, which had previously been present in the donor plant, were estimated, whereas their putative derivatives could not be determined. Yet, in a further approach to quantify the PAs in the acceptor plants, an alternative method—denoted as "sum parameter method"—was employed. For this approach, the PAs are degraded to their basic skeleton, the so-called necine base, which subsequently is quantified by HPLC-ESI-MS [73]. In consequence, apart from the genuine Pas, all putative derivatives will also be determined. The related results were really astonishing: in contrast to the putative decline of the alkaloids over time—as estimated by applying the standard method—a further massive increase in the alkaloids content was verified [30]. In parsley, two weeks after the mulching with PA-containing *Senecio* plant material, more than three-quarters of the imported alkaloids had already been modified within the acceptor plants [30], and thus became invisible to the standard LC-MS quantification. Unfortunately, up to now, the structure of these PA derivatives could not be elucidated. There is a massive demand for further research, since the content of alkaloids in contaminated plant-derived commodities is obviously far higher than that previously stated. This especially accounts for PAs to soundly evaluate the actual health risk.

In order to elucidate the basics of the modification of imported natural products in the acceptor plants, fundamental studies employing umbelliferone as a model substance had been performed. In contrast to most alkaloids, due to their high fluorescence, coumarins as well as their putative derivatives could easily be identified. The studies dealing with the uptake of umbelliferone into various acceptor plants clearly unveiled that the metabolic fate of imported natural products strongly depends on the plant species [28]: in various acceptor plants like flax, radish, or pea, the umbelliferone imported by their roots is just accumulated in their leaves. In contrast, in garden cress and barley, it is hydroxylated to yield esculetin. However, in garden cress, the generated esculetin is glucosylated to esculin, whereas in barley, it is methylated to yield scopoletin [28]. Based on the employment of naproxen, an efficient inhibitor of P450 enzymes, it was demonstrated that the oxidative conversion of umbelliferone is catalyzed by cytochrome P450 enzymes [74]. In contrast, the oxidation of harmaline taken up by barley seedlings to yield harmine is not catalyzed by a P450 enzyme [75].

From the coherences mentioned above—in particular that the extent and mode of modification of the imported products strongly vary between plant species—the occurrence of a general detoxification system, hypothesized as "green liver concept" [69,71], has to be doubted. In contrast, the imported substances seem to be accidentally modified by enzymes involved in the inherent specialized metabolism, which are present in some plants and absent in others [27].

3. Horizontal Natural Product Transfer and Allelopathy

Allelochemicals released from numerous plants inhibit potential competitors for limited light, water, or nutrient resources [76]. For a number of these allelopathic compounds, their underlying physiological mechanism is related to the inhibition of a wide array of corresponding metabolic processes: allelochemicals can affect photosynthesis by reducing PSII efficiency, might impair nutrient uptake and ATP synthesis, or harm cell cycle and gene expression or the phytohormone metabolism [1,77]. Due to these inhibitory effects, they are already employed in agricultural systems as useful tools for controlling weeds [78,79], inhibiting pests [80,81], and fighting diseases [2,82]. In consequence, the usage of synthetical and potentially toxic plant-protecting agents can significantly be reduced. Indeed, the absorption of some allelochemicals by crop plants might also cause health risks as mentioned for the uptake of PAs. However, in general, the related concentrations are far too low to entail serious consequences for human health.

Apart from that, many plants are able to cope with inhibitory allelochemicals. Whereas seed germination of cucumber, alfalfa, garden cress, or tomato is strongly inhibited by juglone, the germination of seeds from wheat, barley, watermelon, corn, and radish is not affected; and the seedling growth of muskmelon is even enhanced by juglone [83]. Obviously, the sensitivity of seeds for allelochemicals varies strongly between different plant species. As outlined above, the corresponding mechanisms seem to be quite diverse and may pertain to various levels of metabolic processes. In this sense, several primary responses to the presence of allelochemicals are described. One of the most important mechanisms of the inhibitory effect of juglone is due to the generation of reactive oxygen species (ROS) [84]. However, the application of juglone may also enhance the ROS-scavenging systems and thereby reduce radical-related damages [84]. Furthermore, detoxifying enzymes and transporters, which inactivate or eliminate the toxic compounds, can be induced by the allelochemicals. In maize, a glutathione transferase, which modifies juglone, is up-regulated in response to the presence of the allelochemical. As a consequence of this derivatization, its toxic properties are strongly diminished [85]. A comprehensive study on the derivatization of allelochemicals by Schulz et al. [59] outlined that the mode and extent of the detoxification process may vary between different plant species. The authors showed that many acceptor plants detoxify benzoxazinones, such as DIBOA and DIMBO, by glucosylation, whereas others detoxify them by the generation of related carbamates. Furthermore, in several other plant species, some further metabolites of the benzoxazinones had been detected [86]. Apart from varying sensitivities of acceptor plants for allelochemicals based on differences in the detoxification mechanism, the different sensitivities might also be due to quite other causations. In this sense, the actual inhibitory effects of a certain allelochemical could be more or less pronounced in different plant species. In this sense,

Ding et al. [87], unveiled the molecular mechanism, being responsible for the massive differences in the toxicity of cinnamic acid, which is released as an allelochemical from cucumber roots. Even in the roots of *Cucumis sativus*, the cinnamic acid significantly decreases the activity of membrane H^+-ATPase, whereas no corresponding inhibition is observed in the roots of fig-leaf gourd (*Cucurbita ficifolia*). Accordingly, it can be argued that observed differences in the sensitivity for a certain allelochemical are indeed related to differences in its direct impact on metabolic events.

These examples outline that more than one factor or mode of action might be responsible for the observed differences in sensitivity. These coherences become even more complex when considering that further environmental factors, e.g., stress, could alter the sensitivity for a certain allelochemical. In this context, the observation of Amari et al. [88] became evident, who showed that the sensitivity for juglone is strongly diminished when reactions involved in the SAR (systemic acquired resistance) are induced. Furthermore, it has to be considered that allelopathy comprises a wide array of interaction between plants by influencing their competition either by the direct effects of allelochemicals, e.g., by inhibiting the germination and growth of potential competitors or by indirect impacts, e.g., by affecting the microbiome, the nutrient availability, and the character of the soil [89].

At this juncture, the question arises whether the observed differences in the sensitivity of the various plant species indeed originate unfailingly from evolutionary adaptations or if some of them are due to random and accidental detoxification processes catalyzed by intrinsic promiscuous enzymes. Depending on the plant species, the specialized metabolism and thus, the composition of immanent enzymes, massively differs. In consequence—in analogy to the proposed random modification of xenobiotics [27]—in some plants the imported allelochemicals might also be modified "accidentally", whereas in others, they are not. However, in contrast to allelopathic interactions which represent quite ancient features, plants had been confronted with xenobiotics, i.e., herbicides or fungicides only for one and a half century. The timespan until today is far too low for the evolutionary generation of a comprehensive detoxification system, which is aimed to "deliberately" detoxify these xenobiotics, i.e., as proposed by the so-called green liver concept [69,71]. Nonetheless, the processes generating the well-known resistances against herbicide can be realized in a very short time, since only one or few mutations are necessary to thwart the toxic effect, i.e., mutations in genes encoding enzyme proteins, non-enzyme proteins, or factors controlling cell division [90]. In consequence, it could be argued that the postulated green liver concept for the detoxification of xenobiotics [69,71] does not represent an adaptation of plants coping with synthetic herbicides or fungicides, but that these detoxification systems had obviously evolved in response to toxic allelochemicals. In consequence, the evolution of the various adaptations of plants to cope with allelochemicals of plants growing in their vicinity ought to be considered as important and principal factors for the deployment and sustaining of a certain plant community. Unfortunately, these insights have not been considered so far, either in plant sociology or in vegetation science.

4. Conclusions

The coherences outlined in this treatise display that the horizontal natural product transfer, which represents a quite general mechanism pertaining to all plants, might be the basis for both the evolution of allelochemicals and subsequently, of the corresponding detoxification mechanisms, which in turn currently also include coping with xenobiotics. Moreover, the insights into the differences in the potential detoxification of allelochemicals may open new doors for the understanding of "why and how" the composition of the various plant communities has been evolved and preserved.

Author Contributions: All authors contributed on the conceptuation and writing in an eaqual manner. All authors have read and agreed to the published version of the manuscript.

Funding: The research received no funding.

Data Availability Statement: Not applicable.

Conflicts of Interest: The authors declare no conflict of interest.

References

1. Inderjit, K.; Duke, S.O. Ecophysiological aspects of allelopathy. *Planta* **2003**, *17*, 529–539.
2. Bertin, C.; Yang, X.; West, L.A. The role of root exudates and allelochemicals in the rhizosphere. *Plant Soil* **2003**, *256*, 67–83. [CrossRef]
3. Kalinova, J.; Vrchotova, N.; Triska, J. Exudation of Allelopathic Substances in Buckwheat (*Fagopyrum esculentum* Moench). *J. Agri. Food Chem.* **2007**, *55*, 6453–6459. [CrossRef] [PubMed]
4. Willis, R.J. The Historical Bases of the Concept of Allelopathy. *J. Hist. Biol.* **1985**, *18*, 71–102. [CrossRef]
5. De Meyer, G.; Capieau, K.; Audenaert, K.; Buchala, A.; Métraux, J.P.; Höfte, M. Nanogram Amounts of salicylic acid produced by the Rhizobacterium *Pseudomonas aeruginosa* 7NSK2 activate the systemic acquired resistance pathway in bean. *Mol. Plant Microbe Interact.* **1999**, *12*, 450–458. [CrossRef] [PubMed]
6. Manthe, B.; Schulz, M.; Schnabl, H. Effects of salicylic acid on growth and stomatal movements of *Vicia faba* L.: Evidence for salicylic acid metabolization. *J. Chem. Ecol.* **1992**, *18*, 1525–1539. [CrossRef] [PubMed]
7. Forde, B.G. Review: Nitrate transporters in plants: Structure, function and regulation. *Biochim. Biophys. Acta* **2000**, *1465*, 219–235. [CrossRef] [PubMed]
8. Buchner, P.; Takahashi, H.; Hawkesford, M.J. Plant sulphate transporters: Co-ordination of uptake, intracellular and long-distance transport. *J. Exp. Bot.* **2004**, *55*, 1765–1773. [CrossRef] [PubMed]
9. Kobayashi, T.; Nishizawa, N.K. Iron uptake, translocation, and regulation in higher plants. *Annu. Rev. Plant Biol.* **2012**, *63*, 131–152. [CrossRef] [PubMed]
10. Trapp, S.; Legind, C.N. Uptake of organic contaminants from soil into vegetables and fruits. In *Dealing with Contaminated Sites*; Swartjes, F.A., Ed.; Springer: Heidelberg, The Netherlands, 2011.
11. Boxall, A.B.A.; Johnson, P.; Smith, E.J.; Sinclair, C.J.; Stutt, E.; Levy, L.S. Uptake of veterinary medicines from soils into plants. *J. Agri. Food Chem.* **2006**, *54*, 2288–2297. [CrossRef]
12. Inoue, J.; Chamberlain, K.; Bromilow, R.H. Physicochemical factors affecting the uptake by roots and translocation to shoots of amine bases in barley. *Pest. Manag. Sci.* **1998**, *54*, 8–21. [CrossRef]
13. Nwoko, C.O. Trends in phytoremediation of toxic elemental and organic pollutants. *Afr. J. Biotechnol.* **2010**, *9*, 6010–6016.
14. Sibout, R.; Höfte, H. Plant Cell Biology: The ABC of Monolignol Transport. *Curr. Biol.* **2012**, *22*, R533–R535. [CrossRef] [PubMed]
15. EFSA—European Food Safety Authority. Setting of temporary MRLs for nicotine in tea, herbal infusions, spices, rose hips and fresh herbs. *EFSA J.* **2011**, *9*, 2098. [CrossRef]
16. EFSA—European Food Safety Authority. Scientific Opinion on Pyrrolizidine alkaloids in food and feed. *EFSA J.* **2011**, *9*, 2406. [CrossRef]
17. EFSA—European Food Safety Authority. Risks for human health related to the presence of pyrrolizidine alkaloids in honey, tea, herbal infusions and food supplements. *EFSA J.* **2017**, *15*, 4908. [CrossRef]
18. EFSA—European Food Safety Authority. Human acute exposure assessment to tropane alkaloids. *EFSA J.* **2018**, *16*, 5160. [CrossRef]
19. Selmar, D.; Engelhardt, U.H.; Hänsel, S.; Thräne, C.; Nowak, M.; Kleinwächter, M. Nicotine uptake by peppermint plants as a possible source of nicotine in plant-derived products. *Agron. Sustain. Dev.* **2015**, *35*, 1185–1190. [CrossRef]
20. Nowak, M.; Wittke, C.; Lederer, I.; Klier, B.; Kleinwächter, M.; Selmar, D. Interspecific transfer of pyrrolizidine alkaloids: An unconsidered source of contaminations of phytopharmaceuticals and plant derived commodities. *Food Chem.* **2016**, *213*, 163–168. [CrossRef] [PubMed]
21. Selmar, D.; Radwan, A.; Abdalla, N.; Taha, H.; Wittke, C.; El-Henawy, A.; Alshaal, T.; Amer, M.; Nowak, M.; El-Ramady, H. Uptake of Nicotine from Discarded Cigarette Butts—A so far unconsidered path of contamination of plant derived commodities. *Environ. Pollut.* **2018**, *238*, 972–976. [CrossRef] [PubMed]
22. Selmar, D.; Wittke, C.; Beck-von Wolffersdorff, I.; Klier, B.; Lewerenz, L.; Kleinwächter, M.; Nowak, M. Transfer of pyrrolizidine alkaloids between living plants: A disregarded source of contaminations. *Environ. Pollut.* **2019**, *248*, 456–461. [CrossRef] [PubMed]
23. Mulder, P.P.J.; Sánchez, P.L.; These, A.; Preiss-Weigert, A.; Castellari, M. Occurrence of Pyrrolizidine Alkaloids in food. *EFSA Support. Publ.* **2015**, *12*, 859E. [CrossRef]
24. Selmar, D.; Radwan, A.; Nowak, M. Horizontal Natural Product Transfer: A So Far Unconsidered Source of Contamination of Plant-Derived Commodities. *J. Environ. Anal. Toxicol.* **2015**, *5*, 4. [CrossRef]
25. Nowak, M.; Yahyazadeh, M.; Lewerenz, L.; Selmar, D. Horizontal Natural Product Transfer: A so far unconsidered source of contamination of medicinal. In *Medicinal Plants and Environmental Challenges*; Ghorbanpour, M.M., Varma, A., Eds.; Springer International Publishing: Cham, Switzerland, 2017; pp. 215–226.
26. Selmar, D.; Radwan, A.; Hijazin, T.; Abouzeid, S.; Yahyazadeh, M.; Lewerenz, L.; Kleinwächter, M.; Nowak, M. Horizontal Natural Product Transfer: Intriguing Insights into a Newly Discovered Phenomenon. *J. Agri. Food Chem.* **2019**, *67*, 8740–8745. [CrossRef]
27. Hijazin, T.; Lewerenz, L.; Yahyazadeh, M.; Selmar, D. Horizontal Natural Product Transfer: A Phenomenon Which Is Responsible for the Widespread Alkaloidal Contaminations of Herbal Products. In *Environmental Challenges and Medicinal Plants: Environmental Challenges and Solutions*; Aftab, T., Ed.; Springer Nature: Cham, Switzerland, 2022; pp. 183–201.

28. Hijazin, T.; Radwan, A.; Abouzeid, S.; Dräger, G.; Selmar, D. Uptake and modification of umbelliferone by various seedlings. *Phytochemistry* **2019**, *157*, 194–199. [CrossRef]
29. Abouzeid, S.; Beutling, U.; Selmar, D. Stress-induced modification of indole alkaloids: Phytomodificines as a new category of specialized metabolites. *Phytochemistry* **2019**, *159*, 102–107. [CrossRef]
30. Selmar, D.; Abouzeid, S.; Radwan, A.; Hijazin, T.; Yahyazadeh, M.; Lewerenz, L.; Nowak, M.; Kleinwächter, M. Horizontal natural product transfer—A novel attribution in allelopathy. In *Reference Series in Phytochemistry. Co-Evolution of Secondary Metabolite*; Mérillon, J.-M., Ramawat, K.G., Eds.; Springer International Publishing: Cham, Switzerland, 2020; pp. 429–439.
31. Fan, P.; Hostettmann, K.; Lou, H. Allelochemicals of the invasive neophyte *Polygonum cuspidatum* Sieb. & Zucc. (Polygonaceae). *Chemoecology* **2010**, *20*, 223–227. [CrossRef]
32. Pan, L.; Li, X.Z.; Yan, Z.Q.; Guo, H.R.; Qin, B. Phytotoxicity of umbelliferone and its analogs: Structure-activity relationships and action mechanisms. *Plant Physiol. Biochem.* **2015**, *97*, 272–279. [CrossRef]
33. Trapp, S. Modelling uptake into roots and subsequent translocation of neutral and ionizable organic compounds. *Pest Manag. Sci.* **2000**, *56*, 767–778. [CrossRef]
34. Cronin, M.T.D.; Livingstone, J. Calculation of physiochemical properties. In *Predicting Chemical Toxicity and Fate*; Cronin, T.D., Livingstone, J., Eds.; CRC Press: Boca Raton, FL, USA, 2004; pp. 31–40.
35. Limmer, M.A.; Burken, J.G. Plant Translocation of Organic Compounds. Molecularand Physicochemical Predictors. *Environ. Sci. Technol. Lett.* **2014**, *1*, 156–161. [CrossRef]
36. Hurtado, C.; Domínguez, C.; Pérez-Babace, L.; Cãnameras, N.; Comas, J.; Bayona, J.M. Estimate of uptake and translocation of emerging organic contaminants from irrigation water concentration in lettuce grow under controlled conditions". *J. Hazard. Mat.* **2016**, *305*, 139–148. [CrossRef]
37. Nowak, M.; Selmar, D. Cellular distribution of alkaloids and their translocation via phloem and xylem: The importance of compartment pH. *Plant Biol.* **2016**, *18*, 879–882. [CrossRef] [PubMed]
38. Yahyazadeh, M.; Nowak, M.; Kıma, H.; Selmar, D. Horizontal natural product transfer: A potential source of alkaloidal contaminants in phytopharmaceuticals. *Phytomedicine* **2017**, *34*, 21–25. [CrossRef]
39. Hijazin, T.; Radwan, A.; Lewerenz, L.; Abouzeid, S.; Selmar, D. The uptake of alkaloids by plants from the soil is determined by rhizosphere pH. *Rhizosphere* **2020**, *15*, 100234. [CrossRef]
40. Nakano, H.; Nakajima, E.; Fujii, Y.; Yamada, K.; Shigemori, H.; Hasegawa, K. Leaching of the allelopathic substance, L-tryptophan from the foliage of mesquite (*Prosopis juliflora* DC.) plants by water spraying. *Plant Growth Regul.* **2003**, *40*, 49–52. [CrossRef]
41. Tukey, H.B. The leaching of substances from plants. *Annu. Rev. Plant Biol.* **1970**, *21*, 305–324. [CrossRef]
42. Pavlović, N.M.; Maksimović, V.; Maksimović, J.D.; Orem, W.H.; Tatu, C.A.; Lerch, H.E.; Bunnell, J.E.; Kostić, E.N.; Szilagyi, D.N.; Paunescu, V. Possible health impacts of naturally occurring uptake of aristolochic acids by maize and cucumber roots: Links to the etiology of endemic (Balkan) nephropathy. *Environ. Geochem. Health* **2013**, *35*, 5–226. [CrossRef] [PubMed]
43. Li, W.; Hu, Q.; Chan, W. Uptake and accumulation of nephrotoxic and carcinogenic aristolochic acids in food crops grown in *Aristolochia clematitis*-contaminated soil and water. *J. Agri. Food Chem.* **2016**, *64*, 107–112. [CrossRef]
44. Letsyo, E.; Adams, Z.S.; Dzikunoo, J.; Asante-Donyinah, D. Uptake and accumulation of pyrrolizidine alkaloids in the tissues of maize (*Zea mays* L.) plants from the soil of a 4-year-old *Chromolaena odorata* dominated fallow farmland. *Chemosphere* **2021**, *270*, 128669. [CrossRef]
45. Hazrati, H.; Fomsgaard, I.S.; Kudsk, P. Root-Exuded Benzoxazinoids: Uptake and translocation in neighboring plants. *J. Agri. Food Chem.* **2020**, *68*, 10609–10617. [CrossRef]
46. Jandrić, Z.; Rathor, M.; Ghhem-Kieth, S.; Adu-Gyamfi, J.; Mayr, L.; Resch, C.; Bado, S.; Švarc-Gajić, J.; Cannavan, A. Uptake of 14C-atropine and/or its transformation products from soil by wheat (*Triticum aestivum* var Kronjet) and their translocation to shoots. *J. Environ. Sci. Health B* **2013**, *48*, 1034–1104. [CrossRef] [PubMed]
47. García-Jorgensen, D.B.; Hansen, H.C.B.; Abrahamsen, P.; Diamantopoulos, E. A novel model concept for modelling the leaching of natural toxins: Results for the case of ptaquiloside. *Environ. Sci. Process. Impacts* **2020**, *22*, 1768–1779. [CrossRef] [PubMed]
48. Matile, P. Localization of alkaloids and mechanism of their accumulation in vacuoles of *Chelidonium majus* laticifers. *Nova Acta Leopold.* **1976**, *7*, 139–156.
49. Waldhauser, S.S.M.; Baumann, T.W. Compartmentation of caffeine and related purine alkaloids depends exclusively on the physical chemistry of their vacuolar complex formation with chlorogenic acids. *Phytochemistry* **1996**, *42*, 985–996. [CrossRef]
50. Bais, H.P.; Vepachedu, R.; Vivanco, J.M. Root specific elicitation and exudation of fluorescent β-carbolines in transformed root cultures of *Oxalis tuberosa*. *Plant Physiol. Biochem.* **2003**, *41*, 345–353. [CrossRef]
51. Ruiz-May, E.; Galaz-Ávalos, R.M.; Loyola-Vargas, V.M. Differential secretion and accumulation of terpene indole alkaloids in hairy roots of *Catharanthus roseus* treated with methyl jasmonate. *Mol. Biotechnol.* **2009**, *41*, 278–285. [CrossRef] [PubMed]
52. Zhao, B.; Agblevor, F.A.; Ritesh, K.C.; Jelesko, J.G. Enhanced production of the alkaloid nicotine in hairy root cultures of *Nicotiana tabacum*. *Plant Cell Tissue Organ Cult.* **2013**, *113*, 121–129. [CrossRef]
53. Toppel, G.; Witte, L.; Riebesehl, B.; Borstel, K.V.; Hartmann, T. Alkaloid patterns and biosynthetic capacity of root cultures from some pyrrolizidine alkaloid producing *Senecio* species. *Plant Cell Rep.* **1987**, *6*, 466–469. [CrossRef] [PubMed]
54. Shitan, N.; Bazin, I.; Dan, K.; Obata, K.; Kigawa, K.; Ueda, K.; Sato, F.; Forestier, C.; Yazaki, K. Involvement of CjMDR1, a plant MDR-type ABC protein, in alkaloid transport in *Coptis japonica*. *Proc. Natl. Acad. Sci. USA* **2003**, *100*, 751–756. [CrossRef] [PubMed]

55. Otani, M.; Shitan, N.; Sakai, K.; Martinoia, E.; Sato, F.; Yazaki, K. Characterization of vacuolar transport of the endogenous alkaloid berberine in *Coptis japonica*. *Plant Physiol.* **2005**, *138*, 1939–1946. [CrossRef] [PubMed]
56. Morita, M.; Shitan, N.; Sawada, K.; Van Montagu, M.C.E.; Inzé, D.; Rischer, H.; Goossens, A.; Oksman-Caldentey, K.-M.; Moriyama, Y.; Yazaki, K. Vacuolar transport of nicotine is mediated by a multidrug and toxic compound extrusion (MATE) transporter in *Nicotiana tabacum*. *Proc. Natl. Acad. Sci. USA* **2009**, *106*, 2447–2452. [CrossRef] [PubMed]
57. Yazaki, K.; Sugiyama, A.; Morita, M.; Shitan, N. Secondary transport as an efficient membrane transport mechanism for plant secondary metabolites. *Phytochem. Rev.* **2008**, *7*, 513–524. [CrossRef]
58. Baumann, T.W.; Gabriel, H. Metabolism and excretion of caffeine during germination of *Coffea arabica* L. *Plant Cell Physiol.* **1984**, *25*, 1431–1436. [CrossRef]
59. Schulz, M.; Knop, M.; Kant, S.; Sicker, D.; Voloshchuk, N.; Gryganski, A. Detoxification of allelochemicals- the case of bezoxazolin-2(3H)-one(BOA). In *Allelopathy: A Physiological Process with Ecological Implications*; Reigosa, R.M.J., Pedrol, N., González, L., Eds.; Springer: Dordrecht, The Netherlands, 2006; pp. 157–170.
60. Bais, H.P.; Park, S.W.; Stermitz, F.R.; Halligan, K.M.; Vivanco, J.M. Exudation of fluorescent β-carbolines from *Oxalis tuberosa* L. roots. *Phytochemistry* **2002**, *61*, 539–543. [CrossRef] [PubMed]
61. Hama, J.; Strobel, B.W. Natural alkaloids from narrow-leaf and yellow lupins transfer to soil and soil solution in agricultural fields. *Environ. Sci. Eur.* **2020**, *32*, 126. [CrossRef]
62. Hama, J.; Strobel, B.W. Occurrence of pyrrolizidine alkaloids in ragwort plants, soils and surface waters at the field scale in grassland. *Sci. Total Environ.* **2021**, *755*, 142822. [CrossRef] [PubMed]
63. Hsu, F.C.; Marxmiller, R.L.; Yang, A.Y.S. Study of root uptake and xylem translocation of cinmethylin and related compounds in detopped soybean roots using a pressure chamber technique. *Plant Physiol.* **1990**, *93*, 1573–1578. [CrossRef]
64. Singh, S. Guttation: Mechanism, Momentum and Modulation. *Bot. Rev.* **2016**, *82*, 149–182. [CrossRef]
65. Ehmke, A.; Borstel, K.V.; Hartmann, T. Alkaloid N-oxides as transport and vacuolar storage compounds of pyrrolizidine alkaloids in *Senecio vulgaris* L. *Planta* **1988**, *176*, 83–90. [CrossRef] [PubMed]
66. Hartmann, T.; Dierich, B. Chemical diversity and variation of pyrrolizidine alkaloids of the senecionine type: Biological need or coincidence? *Planta* **1989**, *206*, 443–451. [CrossRef]
67. Witte, L.; Ehmke, A.; Hartmann, T. Interspecific flow of pyrrolizidine alkaloids: From plants via aphids to ladybirds. *Naturwissenschaften* **1990**, *77*, 540–543. [CrossRef]
68. Riedell, W.E.; Schumacher, T.E. Transport of water and nutrients in plants. In *Agricultural Sciences*; Lal, R., Ed.; EOLSS Publications: Abu Dhabi, United Arab Emirates, 2009; Volume 1, pp. 371–387.
69. Sandermann, H. Higher plant metabolism of xenobiotics: The ′green liver′ concept. *Pharmacogenetics* **1994**, *44*, 225–241. [CrossRef] [PubMed]
70. Schäffner, A.; Messener, B.; Langebartels, C.; Sandermann, H. Genes and enzymes for in-planta phytoremediation of air, water and soil. *Acta Biotechnol.* **2002**, *22*, 141–151. [CrossRef]
71. Burken, J.G. Uptake and metabolism of organic compounds: Green-liver model. In *Phytoremediation: Transformation and Control of Contaminants*; McCutcheon, S.C., Schnoor, J.L., Eds.; Wiley: New York, NY, USA, 2003. [CrossRef]
72. Ronczka, S.; These, A.; Bodi, D.; Preiß-Weigert, A. International collaborative study for the determination of pyrrolizidine alkaloids in honey and herbal tea by SPE-LC-MS/MS. *BfR Wiss.* **2015**, *1*, 1–91.
73. Cramer, L.; Schiebel, H.-M.; Ernst, L.; Beuerle, T. Pyrrolizidine alkaloids in the food chain. Development, validation, and application of a new HPLC-ESI-MS/MS sum parameter method. *J. Agri. Food Chem.* **2013**, *61*, 11382–11391. [CrossRef] [PubMed]
74. Miners, J.O.; Coulter, S.; Tukey, R.H.; Veronese, M.E.; Birkett, D.J. Cytochromes P450, 1A2, and 2C9 are responsible for the human hepatic *O*-demethylation of *R*- and *S*-naproxen. *Biochem. Pharmacol.* **1996**, *51*, 1003–1008. [CrossRef] [PubMed]
75. Lewerenz, L.; Hijazin, T.; Abouzeid, S.; Hänsch, R.; Selmar, D. Pilot study on the uptake and modification of harmaline in acceptor plants: An innovative approach to visualize the interspecific transfer of natural products. *Phytochemistry* **2020**, *174*, 102–107. [CrossRef] [PubMed]
76. Bais, H.P.; Vepachedu, R.; Gilroy, S.; Callaway, R.M.; Vivanco, J.M. Allelopathy and exotic plant invasion: From molecules and genes to species interactions. *Science* **2003**, *301*, 1377–1380. [CrossRef] [PubMed]
77. Blum, U. Relationships between phenolic acid concentrations, transpiration, water utilization, leaf area expansion, and uptake of phenolic acids: Nutrient culture studies. *J. Chem. Ecol.* **2005**, *31*, 1907–1932. [CrossRef] [PubMed]
78. Boydston, R.A.; Morra, M.J.; Borek, V.; Clayton, L.; Vaughn, S.F. Onion and weed response to mustard (*Sinapis alba*) seed meal. *Weed Sci.* **2011**, *59*, 546–552. [CrossRef]
79. Tavella, L.B.; Silva, P.S.L.; Monteiro, A.L.; de Oliveira, V.R.; de Oliveira-Fernandes, P.L. *Gliricidia sepium* intercropping for weed management in immature corn ear production. *Cien. Agron.* **2017**, *48*, 50–656. [CrossRef]
80. Avato, P.; D'Addabbo, T.; Leonetti, P.; Argentieri, M.P. Nematicidal potential of Brassicaceae. *Phytochem. Rev.* **2013**, *12*, 791–802. [CrossRef]
81. Singh, A.; Weisser, W.W.; Hanna, R.; Houmgny, R.; Zytynska, S.E. Reduce pests, enhance production: Benefits of intercropping at high densities for okra farmers in Cameroon. *Pest Manag. Sci.* **2017**, *73*, 2017–2027. [CrossRef] [PubMed]
82. Abdel-Monaim, M.F.; Abo-Elyousr, K.A.M. Effect of preceding and intercropping crops on suppression of lentil damping-off and root rot disease in New Valley—Egypt. *Crop Prot.* **2012**, *32*, 41–46. [CrossRef]

3. Kocaçalikan, I.; Terzi, I. Allelopathic effects of walnut leaf extracts and juglone on seed germination and seedling growth. *J. Hort. Sci. Biotechnol.* **2001**, *76*, 436–440. [CrossRef]
84 Hadrami, A.; Kone, D.; Lepoivre, P. Effect of Juglone on Active Oxygen Species and Antioxidant Enzymes in Susceptible and Partially Resistant Banana Cultivars to Black Leaf Streak Disease. *Eur. J. Plant Pathol.* **2005**, *113*, 241–254. [CrossRef]
85. Sytykiewicz, H. Expression patterns of glutathione transferase gene (GstI) in maize seedlings under juglone-induced oxidative stress. *Int. J. Mol. Sci.* **2011**, *12*, 7982–7995. [CrossRef]
86. Schulz, M.; Wieland, I. Variation in metabolism of BOA among species in various field communities—Biochemical evidence for co-evolutionary processes in plant communities? *Chemoecology* **1999**, *9*, 133–141. [CrossRef]
87. Ding, J.; Sun, Y.; Xiao, C.L.; Shi, K.; Zhou, Y.H.; Yu, J.Q. Physiological basis of different allelopathic reactions of cucumber and figleaf gourd plants to cinnamic acid. *J. Exp. Bot.* **2007**, *58*, 3765–3773. [CrossRef] [PubMed]
88. Amari, L.D.G.E.; Cherif, M.; Kouakou, T.H.; Camara, B.; Konné, D. Salicylic acid and acibenzolar-s-methyl induced resistance against toxic effect of juglone, a toxin of mycosphaerella fijiensis causal agent of banana black leaf streak disease. *J. Adv. Agric.* **2015**, *3*, 204–217. [CrossRef]
89. Wardle, D.A.; Karban, R.; Callaway, R.M. The ecosystem and evolutionary contexts of allelopathy. *Trends Ecol. Evol.* **2011**, *26*, 655–662. [CrossRef]
90. Powles, S.B.; Yu, Q. Evolution in action: Plants resistant to herbicides. *Ann. Rev. Plant Biol.* **2010**, *61*, 317–347. [CrossRef] [PubMed]

plants

MDPI

Review

Cyperus (*Cyperus esculentus* L.): A Review of Its Compositions, Medical Efficacy, Antibacterial Activity and Allelopathic Potentials

Shengai Zhang [1,2], Peizhi Li [3], Zunmiao Wei [1], Yan Cheng [1], Jiayao Liu [1], Yanmin Yang [2], Yuyan Wang [1,2] and Zhongsheng Mu [1,*]

1 Institute of Economic Plants, Jilin Academy of Agricultural Sciences, Gongzhuling 136105, China; wodani999@163.com (S.Z.); miaomiao0825@126.com (Z.W.); chengyan199910@163.com (Y.C.); jiayaoliu@yeah.net (J.L.); wyy6677@126.com (Y.W.)
2 Binzhou Vocational College, Binzhou 256600, China; yanmin107@163.com
3 Jia Sixie Agricultural College, Weifang University of Science and Technology, Weifang 262700, China; llppzz123@126.com
* Correspondence: muzs@163.com

Abstract: Cyperus (*Cyperus esculentus* L.) is an edible perennial grass-like plant, which propagates exclusively with underground tubers. Its tubers are rich in starch (20–30%), fat (25–35%), sugar (10–20%), protein (10–15%) and dietary fiber (8–9%). In addition, the tubers also contain alkaloids, organic acids, vitamins (C and E), steroids, terpenoids and other active components. The contents of oleic acid and linoleic acid in Cyperus oil are very high, which have important medicinal value and health-promoting properties. Most of the extracts from the tubers, stems and leaves of Cyperus have allelopathic potential and antibacterial, antioxidant and insecticidal activities. In recent years, the planting area of Cyperus has increased significantly all over the world, especially in China and some other countries. This paper presents the current status of Cyperus and the recent trend in research in this area. Published reports on its nutritional contents, active ingredients, medicinal efficacy, antibacterial activity and allelopathic potential were also reviewed.

Keywords: Cyperus; *Cyperus esculentus* L.; bioactive; medicinal efficacy; antibacterial activity; allelopathic potential; phytoinsecticide

Citation: Zhang, S.; Li, P.; Wei, Z.; Cheng, Y.; Liu, J.; Yang, Y.; Wang, Y.; Mu, Z. Cyperus (*Cyperus esculentus* L.): A Review of Its Compositions, Medical Efficacy, Antibacterial Activity and Allelopathic Potentials. *Plants* **2022**, *11*, 1127. https://doi.org/10.3390/plants11091127

Academic Editors: Margot Schulz and Vincenzo Tabaglio

Received: 22 March 2022
Accepted: 11 April 2022
Published: 21 April 2022

1. Introduction

Cyperus (*Cyperus esculentus* L.) is a perennial herb developed from rhizomes with sweet nut-like tubers, but cultivated species are often annual. It belongs to the genus Cyperus in the family Cyperaceae, and it propagates entirely with underground tubers in soil [1] (Figures 1 and 2). Cyperus is also known by various other names, such as yellow nutsedge, tiger nut, chufa, iron water chestnut, underground chestnut, underground walnut, earth almond, ginseng fruit, ginseng bean, etc. [2,3]. Cyperus is rich in fat, protein, sugar, other nutrients and a variety of active substances. It is a new crop with high quality, high yield and high value [4]. The tuberous rhizomes of Cyperus have been used as food by hunter-gatherers and agricultural societies for millennia.

Varieties and selections of the plant are widely grown in Southern Europe, North Africa and West Africa. Their popular uses as foods, beverage and medicine suggest that these edible tubers may have functional food potential. The ancient Egyptians first recognized the importance of this plant, planting and using it for culinary and medicinal purposes [5]. Later, the application of Cyperus became more and more extensive, from foods and medicine to its use in bioenergy [6]. In addition, there are a variety of active components in Cyperus, such as volatile oil, organic acids, alkaloids, phenols, terpenoids, anthraquinone, steroids, etc., which have antibacterial, antioxidant and antitumor activities. It was reported that the consumption of Cyperus could help prevent heart disease and thrombosis, improve blood circulation and lower the risk of colon cancer [7]. At the same time, a variety of active components in Cyperus have strong allelopathic effects, which have a great impact on intercropping, interplanting and crop rotation. However, due to its grass-like appearance and the poor research on Cyperus, for a long time in the past, some countries had always treated it as a weed, so it had not been fully utilized and investigated [8]. In recent years, it has been widely cultivated in Spain, Africa, Australia, South and North America. The annual value of Cyperus production in Spain has risen to 3.3 million euros, and its production area has also increased significantly in China, Italy, Bulgaria, etc. At present, there is still a lack of understanding in the active components, antibacterial activity and allelopathic potential of Cyperus, which affects the development, utilization and cultivation of Cyperus. Therefore, this paper reviews the chemical composition, medical efficacy and allelopathic potential of Cyperus to provide the references for the in-depth research, further development and utilization of Cyperus.

Figure 1. Cyperus (*Cyperus esculentus* L.) in growth stage.

Figure 2. Cyperus (*Cyperus esculentus* L.) tubers at maturity.

2. Compositions

Cyperus is a new, high-quality oil crop. It is rich in oil, starch and sugar, and it also contains protein, dietary fiber, vitamin C, vitamin E and a variety of minerals (e.g., phosphorus, potassium, magnesium, sodium, calcium, etc.) (Table 1). From Table 1, (i) the content of lipid in Cyperus is higher than that of soybean. It can be used as a new oil crop and has great development potential; (ii) the content of starch in Cyperus is significantly higher than that of common oil crops, up to 23.2–29.9%; (iii) the contents of starch and carbohydrate in Cyperus are significantly higher than that of common oil crops, which indicates that Cyperus is a food of raw materials with high calories.

Table 1. Main composition and content of cultivated Cyperus (*Cyperus esculentus* L.) tubers compared to these of key oil crops.

Product	Cyperus (%)	Peanut (%)	Soybean (%)	Rape (%)	Sunflower (%)	Sesame (%)
Scientific name	*Cyperus esculentus*	*Arachis hypogaea*	*Glycine max*	*Brassica napus*	*Helianthus annuus*	*Sesamum indicum*
Starch	23.2–29.9	4.00	2.00–6.53	0.80–1.18	12.0–18.0	–
Lipid	20.1–34.5	39.2–49.8	16.0–22.0	37.5–46.3	49.9–51.2	44.9–57.9
Protein	5.04–8.00	20.3–26.7	35.0–44.6	23.1–32.4	15.4–23.9	18.4–30.4
Crude Fiber	8.06–8.91	3.30–8.50	5.23–15.5	5.7–9.6	4.90–6.10	7.45–12.6
Carbohydrate	43.3	7.48–22.1	22.0–30.0	15.9–24.8	13.0–16.7	14.4–31.5
Total sugar	15.4–23.4	3.54–6.53	5.06–7.70	8.00–12.9	16.0–23.0	8.44–12.3
Ash	1.60–2.19	2.02–2.27	4.35–4.60	4.10–5.30	3.20–5.40	4.77–6.18
Moisture	6.13–7.27	6.20–7.30	10.2–11.1	8.05	3.00–3.82	3.06–6.15
References	[9–12]	[13–15]	[16–18]	[19,20]	[14,21,22]	[23,24]

Note: "–" indicates no report.

In addition, Cyperus also contains a variety of active substances (e.g., alkaloid, saponin, phytosterol, flavone, terpenoid, tannin, etc.) [25]. These active substances are important pharmaceutical components. Most flavonoids have strong biological activities, such as an antioxidant effect, and can be used in pharmacological research (Table 2).

Table 2. Main active substances of cultivated Cyperus (*Cyperus esculentus* L.) tubers.

Ingredient	Concentration	Reference
Resins	+++	[7,26]
Cyanogenic glycosides	+	[7,26]
Alkaloids	+++	[7]
Glycosides	0	
Flavonoids	+++	[27,28]
Saponins	+	[7]
Tannins	+	[7]
Sterols	+++	[7]
Cardiac glycosides	0	
β-sitosterol	+++	[26]
δ^5-avenasterol	+	[26]
Vitamin E	+	[26]

Note: '+++' indicates very high concentration, '++' indicates medium concentration, '+' indicates trace concentration and '0' indicates uncertainty.

Due to the differences in climate, soil and cultivation conditions, the bioactive content in Cyperus varies. It was reported that, under normal conditions, the crude fat content in Cyperus tuber (dry tuber) is 25–35%, crude protein 10–15%, starch 20–30%, sugar 10–20%, vitamin E 0.8–1.4%, and the flavonoids in stems and leaves could reach 13.2 mg/g [11,27]. Chen et al. found that the contents of fat, starch, sugar, protein and water in the tubers of Cyperus were 26.5%, 23.2%, 23.4%, 8.0% and 7.0%, respectively, and other active components (e.g., organic acids, terpenoids, steroids, etc.) were also detected [29].

Liu showed that the basic components of Cyperus were 26.5% fat, 23.2% starch, 23.4% sugar, 8.0% protein and 7.0% water. At the same time, she also detected organic acids, vitamins, steroids and terpenes with health care effects, in which the content of vitamin E was 0.15% and the content of sterols was 0.53% [30]. The test results of Chen et al. and Liu are very similar. It can be concluded that the contents of fat, starch and sugar in Cyperus are indeed very high. In addition, Yan et al. detected the main components in the stems and leaves of Cyperus. The results showed that the stems and leaves of Cyperus contained 2.53% crude fat, 7.08% crude protein, 47.1% crude fiber, 10.5% ash and 7.75% sugar (in glucose) [26]. It also contained peptides, phenols, organic acids and alkaloids. These active ingredients play an important role in the fields of medicine and ecology [31]. Yan et al. conducted a qualitative analysis on the active components of Cyperus. The results showed that Cyperus contained organic acids, alkaloids, amino acids, peptides, polysaccharides, volatile oils, phenols, terpenes, anthraquinones, steroids and other active components [26]. Chukwuma et al. determined the phytochemical composition of Cyperus tubers. The results showed the presence of resins, alkaloids, cyanogenic glycosides, tannins, sterols and saponins were observed in the raw tubers, while only sterols, resins and alkaloids were observed in the roasted tuber. Meanwhile, analysis of the composition yielded saponins (0.88 ± 0.02 mg/100 g), tannins (9.50 ± 0.46 mg/100 g), oxalates (0.25 ± 0.65 mg/100 g), phytate (1.97 ± 0.81 mg/100 g) and cyanogenic glycosides (1.80 ± 0.69 mg/100 g) [7].

Cyperus (*Cyperus esculentus* L.) oil is golden in color, clear and transparent, and has a strong aroma. It is a rare raw material resource in the field of food, health products and cosmetics [32] (Table 3). From Table 3, the oleic acid content of Cyperus oil is significantly higher than that of common vegetable oils, and it does not contain erucic acid. Therefore, it is a healthy edible oil.

Table 3. Main fatty acid composition and content of cultivated Cyperus (*Cyperus esculentus* L.) oils compared to those of common vegetable oils (Crude oil).

Product	Cyperus (%)	Peanut (%)	Soybean (%)	Rape (%)	Sunflower (%)	Sesame (%)
Scientific name	*Cyperus esculentus*	*Arachis hypogaea*	*Glycine max*	*Brassica napus*	*Helianthus annuus*	*Sesamum indicum*
Palmitic acid (C16:0)	11.7–14.2	9.7–13.0	6.0–9.3	5.0	5.5	8.2
Stearic acid (C18:0)	1.1–4.9	2.7–5.0	3.0–5.5	3.6	2.9	1.9
Oleic acid (C18:1)	67.7–76.6	41.0–47.1	24.8–36.0	14.0–19.0	15.0–28.9	35.0–49.4
Linoleic acid (C18:2)	8.8–11.5	37.6	52.0–65.0	12.0–25.4	62.0–70.0	37.7–49.1
Linolenic acid (C18:3)	0.2–1.9	–	2.0–6.8	7.3–10.0	0.3	0.4
Arachidic acid (C20:0)	0.4–6.1	1.8–8.0	0.1–0.4	0.4–1.0	0.2	0.4–1.2
Erucic acid	–	–	0.23	31.0-55.0	–	–
References	[25,33,34]	[35,36]	[35,36]	[35,36]	[35,36]	[35,36]

Note: "–" indicates no report.

Cyperus oil contains 64–73.3% oleic acid and 11–15.5% linoleic acid, which is similar to olive oil (55–83% oleic acid and 3.5–21% linoleic acid) and is significantly higher than the oil contents in soybean, peanut and rapeseed oils [2,37]. It was found that Cyperus oil also contains active components such as rapeseed sterol, stigmasterol, 7-cholesterol, B-sitosterol, 24 methylene-Δ7-cholesterol, 5-oatsterol and cycloatroxanol. These plant components are important raw materials for the pharmaceutical industry [38,39].

3. Medicinal Research

3.1. Medicinal Efficacy

It is reported that the tuber of Cyperus (*Cyperus esculentus* L.) possesses an irregular fruit shape, rough surface, pungent smell, bitter taste and cool property. It has important medicinal value [25]. In India, Cyperus tuber is the oldest medicinal material, which can be used to treat critical and serious diseases (e.g., chronic gastritis, lymphatic tuberculosis, burns and scalds, coronary heart disease, acute cholecystitis, acute intracerebral hemorrhage, etc.) [40]. According to the *Xinhua outline of Materia Medica*, the tuber of Cyperus contains a kind of spicy, sweet and warm nature. It has the functions of soothing the liver, aerating and strengthening the spleen and stomach, and it can also prevent many diseases (e.g., obesity, diabetes, gastrointestinal diseases, etc.). The active ingredients of Cyperus can be used to treat chest tightness, dyspepsia, food stasis, female menstrual disorders, and so on. In Indian medical records, Cyperus has similar medicinal effects, too [41,42]. Another study showed that Cyperus contains an active component, anthraquinone (C14H8O2), which has the effects of anti-tumor, anti-oxidation and sterilization [43]. Achoribo et al. found, in their experiment, that the oleic acid contained in Cyperus oil can inhibit the overexpression of "Her-2/neu", thereby promoting the apoptosis of breast cancer cells by using "Her-2/neu" oncogene amplification [44].

Cyperus also has the effects of protecting nerves, the stomach and kidneys, strengthening yang and treating anemia. Wiebe et al. undertook a series of studies in order to indicate the central nervous system activity of the extract (Cyperol) from Cyperus (*Cyperus esculentus* L.). The results showed that Cyperol could induce changes in brain electrical activity (EEG), demonstrating that active ingredients from the extract were absorbed and were bioavailable. In summary, the results of the study indicate calming effects without sedation. The extract (Cyperol) does indeed have functional food potential and can be used to maintain a calm state of mind while dealing with cognitively demanding works [45]. Jing et al. studied the intervention effect of orientin in Cyperus on nerve injury in rats. The results showed that the orientin could significantly improve the neurological function of rats [46]. Elshamy et al. studied the effects of different concentration extracts from Cyperus on gastric ulcers induced by ethanol in rats. The results showed that the protective effect of the extract on gastric epithelial cells increased with the increase in concentration [47]. Imo Chinedu et al. examined the effects of ethanolic extracts from Cyperus (*Cyperus esculentus* L.) on selected indices of kidney function in male albino rats. The results showed the extracts had no apparent toxic effect on the kidney function of the experimental rats. The results also showed the extracts may help to reduce the retention of urea, thereby strengthening kidney functions. This plant material is therefore encouraged for use in general nutrition [48]. Nto et al. studied the intervention effect of Cyperus extracts on testicular dysfunction in Wistar rats. The results showed that Cyperus extracts significantly increased the number and volume of the epithelial cells. This finding indicates that Cyperus extracts can improve the vitality of testis in Wistar rats [49]. Jagpal studied the effects of Cyperus oils on reducing cell sickling in sickle cells anemia in vitro. The results showed that the Cyperus oil treatments resulted in an increase in the antioxidant presence of sickle cell samples when tested in vitro, as well as a morphological decrease in sickle cells and increase in spherical erythrocytes, thereby alleviating the symptoms of anemia [50]. In addition, Cyperus tubers (*Cyperus esculentus* L.) contain various compounds with intestinal health-promoting properties. Moral Anter David et al. studied the effect of Cyperus on *S. enteritidis* agglutination, oxidative stress and lactobacillus plantarum growth. The results revealed that, compared to controls, Cyperus partially restored "TER" in *S. enteritidis* infected cultures and promoted the growth of lactobacillus plantarum [51].

3.2. Health-Promoting Properties

Cyperus (*Cyperus esculentus* L.) is rich in fat, and the edible oil (Cyperus oil) can be extracted from it. Cyperus oil has a mellow taste and stable chemical properties, and it is clear and transparent, not easy to corrupt and very easy to be absorbed and utilized by the human body. It is a very high-quality edible oil [52]. The main components of Cyperus oil are unsaturated fatty acids (oleic acid and linoleic acid). Linoleic acid is an essential fatty acid for the human body. It is an important substance for the synthesis of prostaglandins. It can regulate physiological functions of the human body, promote the growth and development of the human body, promote the catabolism of cholesterol in the body, and it can prevent cardiovascular diseases. Cyperus oil is also rich in minerals, but the content of sodium is very low and does not contain cholesterol, which is very close to olive oil, hazelnut oil and avocado oil [53]. It is also reported that the content of sterols in Cyperus oil is about 1000 $\mu g \cdot g^{-1}$, which is significantly higher than that in olive oil (about 100 $\mu g \cdot g^{-1}$). These sterols in Cyperus oil could reduce the absorption of cholesterol and reduce the risk of cardiovascular disease [30]. Bamishaiye et al. studied the effects of Cyperus oil and soybean oil on tissue metabolism in rats. The results showed that, compared with feeding soybean oil, the treatments of feeding Cyperus oil significantly reduced the content of total cholesterol in rats [54]. Moon et al. studied the effect of Cyperus oil meal on lipid metabolism in mice. The results showed that Cyperus oil meal significantly reduced the increase of adipocyte volume, adipose tissue weight and body weight caused by high-fat diet, and it significantly reduced the contents of total cholesterol and triglyceride in serum and triglyceride in liver. The results showed that Cyperus is beneficial in preventing hyperlipidemia caused by diet [55]. Salem M L et al. studied the effects of Cyperus tubers on mice with atherosclerotic disease. The results showed that feeding Cyperus tubers to mice significantly alleviated the symptoms of atherosclerosis in mice [56]. According to research by Oluwajuyitan and others, Cyperus is a food that lowers the glycemic index, and it is suitable for diabetics [57]. Another study showed that Cyperus is a pollution-free and nutritious food raw material with the function of protecting the liver [58]. Onuoha et al. studied the protective effect of Cyperus milk on liver injury induced by acetaminophen (APAP). The results showed that Cyperus milk could significantly improve liver injury induced by acetaminophen (APAP) [59]. Hassanein et al. studied the protective effect of Cyperus essential oil on the hepatotoxicity induced by paracetamol. The results showed that the concentration of the essential oil at 12.5 µg/mL has a protective effect on the hepatotoxicity induced by paracetamol [60].

3.3. Antibacterial Activity

It is reported that Cyperus (*Cyperus esculentus* L.) contains a variety of active components, and most of its extracts have antibacterial effect [41]. Prakash et al. prepared different extracts of Cyperus, such as acetone, 50% ethanol, chloroform and petroleum ether to evaluate for their antibacterial activity against several human pathogens (*Escherichia coli, Staphylococcus aureus, Salmonella* sp, *Klebsiella pneumoniae, Proteus vulgaris, Pseudomonas aeruginosa* and *Citrobacter freundii*) by using the disc diffusion method. The acetone extract showed the highest inhibitory activity against *S. aureus, ILpneumoniae* and *P. vulgaris*. Furthermore, 50% ethanolic extract showed the maximum activity against *Escherichia coli, S. aureus* and *Salmonella* sp. Chloroform extract maximally inhibited the growth of *S. aureus*, whereas petroleum ether extract showed positive results against *Salmonella* sp., respectively. All extracts were sensitive to *C. freundii* [61]. Abdel et al. evaluated the antibacterial activity of Cyperus (*Cyperus esculentus* L.) oil via cup plate agar diffusion assay against six standard human pathogens (Gram positive: *Staphylococcus aureus* and *Bacillus subtilis*; Gram negative: *Escherichia coli, Pseudomonasa aeruginosa, the fungi Candida albicans* and *Aspergillus niger*). The oil showed different antimicrobial responses against testing organisms. It showed significant activity against the fungus *Candida albicans* and partial activity against *Staphylococcus aureus* [62]. Salem et al. studied the effect of Cyperus tubers on the reduction in atherosclerotic lesions in mice. The results indicated that the ingredients of Cyperus tubers exhibit anti-inflammatory properties upon inflammation and immunostimulatory effects in immunocompetent hosts [56].

3.4. Antioxidant Activity

Some studies showed that Cyperus (*Cyperus esculentus* L.) contains secondary metabolites of flavonoids, which showed antioxidant activity and an anticoagulant effect [63]. Jing et al. studied the antioxidant activity of the extracts from Cyperus. The results showed that the activities of superoxide dismutase (SOD) in liver tissue and serum of mice were significantly increased by different concentrations (high, medium and low) [64]. Willis et al. studied the antioxidant activities of water and methanol extracts from Cyperus. The results showed that both extracts had scavenging effects on free radicals [65]. Badejo et al. mixed Cyperus milk, ginger juice and water (7:1:2) to make a Cyperus mixed beverage, and they studied its effect of scavenging free radical. The results showed that the scavenging rate of free radical (DPPH) was 32.53% [66]. Another study found that the germinated tubers of Cyperus could further improve antioxidant activity [67,68]. Siqun Jing et al. evaluated the antioxidant activity and antibacterial property of the flavonoids from Cyperus (*Cyperus esculentus* L.) leaves. In all the assays, Cyperus leaves flavonoids had pronounced antioxidant activity in vivo that could significantly elevate the content of superoxide dismutase (SOD). Cyperus leaves flavonoids inhibited the growth of both Gram-positive and Gram-negative bacteria. The results indicate that the flavonoids from Cyperus leaves can be taken as a natural antioxidant and bacteriostatic substance in the food and pharmaceutical industry [69].

4. Allelopathic Potential

4.1. Impacts on Crops

It is reported that Cyperus has a strong allelopathic effect (inhibition) on many crops [70]. Dirk et al. studied the allelopathic effect of extracts and plant residues of Cyperus (*Cyperus esculentus* L.) on the growth of corn (*Zea mays* L.) and soybean (*Glycine max* L.) and found that tuber residues reduced the dry weight of corn and soybean more than foliage residues. As the concentration increased, the growth decreased, affecting soybean more than corn. Soybean growth was significantly reduced by the addition of tuber extracts. Growth inhibition was greatest when the tuber residue was in contact with the corn or soybean seeds [71]. This shows that extracts and residues of Cyperus have an allelopathic effect on corn and soybean. Reinhardt et al. studied the allelopathic effects of Cyperus (*Cyperus esculentus* L.) tubers on ectomycorrhiza (*Boletus*), lettuce (*Lactuca sativa*) and corn (*Zea mays*), and they further evaluated the allelopathic potential of Cyperus. The results showed that the aqueous extracts from Cyperus tubers significantly reduced the colony diameter of the ectomycorrhiza (*Boletus*) on agar medium. At 2% concentration, the aqueous extracts inhibited the germination of lettuce (*Lactuca sativa*) seeds significantly. The emergence of corn (*Zea mays* L.) was retarded in soil where Cyperus tubers were planted 28 days before planting the crop, irrespective of whether the Cyperus tubers continued to grow or were physically removed at the time the corn was sown. This shows that the root exudates or residues of Cyperus have an inhibitory effect on corn seed germination [72]. Later, H.P. Oamen et al. studied the allelopathic effects of aqueous extracts of *Cyperus esculentus* L. on the germination and seedling growth of *Cajanus cajan* L., *Phaseolus vulgaris* (brown and white varieties) and *Sphenotylis stenocarpa*. The results showed that the extracts have significant positive allelopathic effects on plumule and radicle lengths at $p < 0.05$. The number of root hairs for *C. cajan* was reduced at higher concentrations [73]. Papadi Asimina et al. studied the allelopathic potential of Cyperus (*Cyperus esculentus* L.) on the germination of the seeds of six commonly grown crops (*Zea mays* L., *Citrullus lanatus Thunb*, *Abelmoschus esculentus* (L), *Vigna unguiculata* (L), *Glycine max* (L) and *Arachis hypogaea* (L.).

The results showed that the aqueous extracts from Cyperus reduced the germination counts of the seeds by 10 to 100%. Cyperus shoot extracts were the most phytotoxic. In addition, the decomposing mulches showed varied but less inhibitory effects on the seeds with a trend toward increasing the inhibitory effect with an increasing mulch level and decreasing seed size [70]. Singh C.M. et al. studied the allelopathic effects of Cyperus (*Cyperus esculentus* L.) on the seed germination and seeding growth of wheat. The results showed that the extracts from Cyperus had inhibitory effects on the germination of wheat seeds. The time taken for germination prolonged; root, shoot and leaf developments were seriously inhibited. The findings of this study clearly indicate that before using the Cyperus for manuring in wheat fields, it would be safe to carry out a detailed field experiment to ascertain the allelopathic effects [74]. JJ Sani et al. evaluated the allelopathic effects of extracts from the aerial and underground organs of Cyperus (*Cyperus esculentus* L.) with three levels of extract (5, 10 and 20 percent) on rapeseed germination and growth. The results showed that the extract of Cyperus reduced rapeseed germination and growth. With increasing concentration of extracts, germination rate, germination percent, vigor index, root and shoot length were significantly reduced [75]. M.G. Patterson et al. conducted a series of experiments, from 1976 to 978, to determine the competitive relationship of Cyperus (*Cyperus esculentus* L.) with cotton. In the experiments, Cyperus was left undisturbed or removed from plots to give periods of competition of 2, 4, 6, 8, 10 and 25 weeks. The results showed that the seed cotton yield was reduced in 2 out of the 3 years in a full season (25 weeks) of competition but was unaffected by shorter periods of competition in all years. The main stem diameter of cotton was also reduced 2 out of the 3 years when competing the full season, while the main stem height was reduced in only 1 out of the 3 years in a full season competition [76].

4.2. *Phytoinsecticide*

Avicor et al. investigated the larvicidal potential of crude aqueous extracts from two Cyperus (*Cyperus esculentus* L.) varieties (black and yellow) on the mosquitoes *Aedes aegypti* (L.) and Culex *quinquefasciatus* (Say). Differential larvicidal responses were observed in the test mosquitoes, and extracts of black dried Cyperus were more larvicidal than yellow dried Cyperus. The results indicated the potential of Cyperus, particularly black dried Cyperus, as a resource of mosquito insecticide [77]. Based on the research results in this area, Cyperus (*Cyperus esculentus* L.) has great potential in developing phytoinsecticide and reducing people's dependence on synthetic insecticides.

5. Future Areas of Research

The following studies need be conducted in the future: (i) Preliminary indications suggest that the extract (Cyperol) from Cyperus (*Cyperus esculentus* L.) may have anxiolytic potential [45], which should be explored in clinical studies. (ii) The extraction and identification of allelochemicals from Cyperus. (iii) The allelopathic effects of extracts and decomposed substances from Cyperus on main crops (e.g., soybeans, corn, wheat, etc.) and its autotoxicity. (iv) The allelopathic effects of extracts and substances from decomposed Cyperus on soil microflora and activity. (v) The development of new allelopathic varieties of Cyperus. (vi) The larvicidal potential and identification of the extracts from Cyperus. Furthermore, (vii) studies on the medicinal efficacy and antibacterial potential of Cyperus should also be conducted in the future.

6. Conclusions

Some studies have been carried out on the chemical components, active substances and allelopathy of Cyperus (*Cyperus esculentus* L.), and it has been found that it has medicinal efficacy and antibacterial potential, but these studies are not detailed and in-depth enough. Therefore, it is necessary to study and evaluate its active components and antibacterial and insecticidal potential in more detail, in order to provide a scientific basis for the further development and utilization of Cyperus. In addition, Cyperus has strong allelopathy to many crops, and it inhibits the growth of adjacent crops. At the same time, Cyperus (*Cyperus esculentus* L.) has great potential in developing phytoinsecticide. Research on the allelopathy of Cyperus has a bright future in reducing our reliance on synthetic pesticides and promoting the sustainable development of Cyperus cropping.

Author Contributions: Conceptualization, S.Z. and Z.M.; methodology, P.L.; software, Z.W.; validation, Y.W.; formal analysis, P.L.; investigation, Y.C.; resources, J.L.; data curation, Y.Y.; writing—original draft preparation, S.Z. and Z.W.; writing—review and editing, Y.W. and Z.M. All authors have read and agreed to the published version of the manuscript.

Funding: Innovation of germplasm resources and breeding of new varieties of Cyperus (*Cyperus esculentus* L.) based on biotechnology (CXGC2021TD013).

Institutional Review Board Statement: Not applicable.

Informed Consent Statement: Not applicable.

Data Availability Statement: Not applicable.

Acknowledgments: This work was financially supported by the Agricultural Science and Technology Innovation project of Jilin Province "Innovation of germplasm resources and breeding of new varieties of Cyperus (*Cyperus esculentus* L.) based on biotechnology (CXGC2021TD013)". The authors thank Yuyan Wang and Zhongsheng Mu for their contributions to the preparation and publication of the paper. Compliance with ethical standards.

Conflicts of Interest: The authors declare no conflict of interest.

References

1. Michael, S. Defelice. Yellow nutsedge *Cyperus esculentus* L.: Snack food of the gods. *Weed Technol.* **2002**, *16*, 901–907.
2. Oderinde, R.A.; Tairu, O.A. Evaluation of the properties of yellow nut sedge (*Cyperus esculentus*) tuber oil. *Food Chem.* **1988**, *28*, 233–237. [CrossRef]
3. Yang, Z. Characteristics and Research Progress of *Cyperus esculentus* L. *North. Hortic.* **2017**, *17*, 192–201. (In Chinese)
4. Coskuner, Y.; Ercan, R.; Karababa, E.; Nazlican, A.N. Physical and chemical properties of chufa (*Cyperus esculentus* L.) tubers grown in the C, ukurova region of Turkey. *J. Sci. Food Agric.* **2002**, *82*, 625–631. [CrossRef]
5. Negbi, M. A sweetmeat plant, a perfume plant and their weedy relatives: A chapter in the history of *Cyperus esculentus* L. and *C. rotundus* L. *Econ. Bot.* **1992**, *46*, 64–71. [CrossRef]
6. Li, G.; Yang, L.; Zhu, Z.; Cai, B.; Kang, X. Study on Preparation of Biodiesel from Cyperus Oil. *China Grease* **2012**, *37*, 59–62.
7. Chukwuma, E.R.; Obioma, N.; Cristopher, O.I. The phytochemical composition and some biochemical effects of Nigerian tiger nut (*Cyperus esculentus* L.) tuber. *Pak. J. Nutr.* **2010**, *9*, 709–715. [CrossRef]
8. Femke, T.; de Vries, C. (*Cyperus esculentus, Cyperaceae*): A weedy cultivar or a cultivated weed. *Econ. Bot.* **1991**, *45*, 27–37.
9. Fu, L.; Li, G.; Liu, Y. Analysis and Study on Components of *Cyperus esculentus* Used as Soil, Grain and Feed Crop. *J. Zhengzhou Grain College* **1996**, *7*, 88–90.
10. Mokady, S.H.; Dolev, A. Nutritional evaluation of tubers of *Cyperus esculentus* L. *J. Sci. Food Agric.* **1970**, *21*, 211–214. [CrossRef]
11. Yu, H.; Jing, S. Research Progress on chemical constituents and application of Cyperus. *Food Ind.* **2015**, *36*, 242–245.
12. Yu, Y.; Lu, X.; Zhang, T.; Zhao, C.; Guan, S.; Pu, Y.; Gao, F. Tiger Nut (*Cyperus esculentus* L.): Nutrition, Processing, Function and Applications. *Foods* **2022**, *11*, 601. [CrossRef] [PubMed]
13. Taiyuan, S.; Miao, Y.; Yanqiu, H. Analysis on the Nutrients and Characteristic of Liaoning Peanut Varieties. *Food Res. Dev.* **2017**, *11*, 142–147.
14. Pickett, T.A. Composition of developing peanut seed. *Plant Physiol.* **1950**, *25*, 210–224. [CrossRef]
15. Asare, P.A.; Kpankpari, R.; Adu, M.O.; Afutu, E.; Adewumi, A.S. Phenotypic Characterization of Tiger Nuts (*Cyperus esculentus* L.) from Major Growing Areas in Ghana. *Sci. World J.* **2020**, *1*, 1–11. [CrossRef] [PubMed]
16. Aochen, L. Main Nutrientional Components and Values of Soybean. *Mod. Agric. Sci. Technol.* **2020**, *23*, 213–218.
17. Zhang, Y.; Hu, R.; Lin, G. Preliminary study on the content of soluble sugar and starch in soybean seeds. *Fujian J. Agric. Sci.* **2018**, *33*, 604–607.
18. Guillon, F.; Champ, M.M.-J. Carbohydrate fractions of legumes: Uses in human nutrition and potential for health. *Br. J. Nutr.* **2002**, *88*, 293–306. [CrossRef]
19. Zhenqian, Z.; Biqing, W.; Guohuai, W.; Chunyun, G.; Sheyuan, C. Study on physiological and biochemical characteristics of early maturing rape Varieties. *Crop Res.* **2014**, *28*, 349–353.
20. Yajun, Y.; Xiangyu, Z.; Yu, Z. Study on the changes of main nutrients in rapeseed during ripening. *Food Ind. Technol.* **2015**, *36*, 339–342.

21. Tipsina, H.H.; Yanxia, L. Comparative Analysis on the Composition and Function of Pine Nuts and Several Oil Crops. *J. Jilin Agric. Sci. Technol. Univ.* **2015**, *24*, 4–8.
22. Li, P.-J.; Li, B.-J.; Nan, W.; Mi, Y.; Liao, F.; Li, G. Discriminatory analysis on the morphology and major components of kernels of the edible and oil-type sunflowers. *J. Sichuan Univ. (Nat. Sci. Ed.)* **2017**, *54*, 203–208.
23. Wang, X.; Cui, Y.; Liu, B.; Ma, S. Correlation analysis of various components in sesame. *China Oils Fats* **2015**, *40*, 99–103.
24. Li, T.; Miao, H.; Wang, X.; Wang, D. Changes of main nutritional components and in vitro antioxidant activity of polysaccharide extracts in sesame leaves at different growth periods. *J. Henan Univ. Technol. (Nat. Sci. Ed.)* **2020**, *41*, 42–49.
25. Guo, T.; Wan, C.; Huang, F.; Wei, C.; Hu, Z. Research Progress on main nutritional components and physiological functions of tiger nut (*Cyperus esculentus* L.). *Chin. J. Oil Crop Sci.* **2021**, *43*, 1174–1180.
26. Yan, X.; Ren, Z.; Xiang, H.; Jing, S. Qualitative analysis of active components in Cyperus. *China Brew.* **2010**, *8*, 150–151.
27. Buweizuohere, A.; Deping, W.; Siqun, J.; Yixian, T. Research progress on bioactivity of total flavonoids in *Cyperus esculentus* L. leaves. *Cereals Oils* **2021**, *34*, 21–26.
28. Zhai, H.; Wang, D.; Jing, S. Effects of *Cyperus esculentus* total flavonoids on milk fresh-keeping. *Food Ind. Technol.* **2004**, *25*, 69–70.
29. Chen, X.; Chen, D.; Liu, L. *Cyperus esculentus* whole component analysis. *Food Sci. Technol.* **2009**, *34*, 165–168.
30. Liu, L. Study on the Active Ingredients of *Cyperus Esculentus* and Oil Extraction Technology. Masters's Thesis, Northeast Normal University, Changchun, China, 2008; pp. 1–5.
31. Liu, Z.; Jin, Z.; Zhou, X.; Bian, Y.; Bai, W. Extraction and determination of protein, fat and sugar in Cyperus. *J. Inn. Mong. Norm. Univ. Natl. (Nat. Sci. Ed.)* **1988**, *13*, 62–65.
32. Hu, B.; Zhou, K.; Liu, Y.T. Optimization of microwave-assisted extraction of oil from tiger nut (*Cyperus esculentus* L.) and its quality evaluation. *Ind. Crops Prod.* **2018**, *115*, 290–297. [CrossRef]
33. Linssen, J.P.H.; Kielman, G.M.; Cozijnsen, J.L.; Pilnik, W. Comparison of chufa and olive oils. *Food Chem.* **1988**, *28*, 279–285. [CrossRef]
34. Muhammad, N.O.; Bamishaiye, E.I.; Bamishaiye, O.M.; Usman, L.A.; Salawu, M.O.; Nafiu, M.O.; Oloyede, O.B. Physicochemical properties and fatty acid composition of *Cyperus esculentus* (Tiger Nut) tuber oil. *Biorearch Bull.* **2011**, *5*, 51–54
35. Ezeh, O.; Gordon, M.H.; Niranjan, K. Tiger nut oil (*Cyperus esculentus* L.): A review of its composition and physic-chemical properties. *Eur. J. Lipid Sci. Technol.* **2014**, *116*, 783–794. [CrossRef]
36. Shi, X.; Tian, L.; Xue, L.; Wang, X.; Zhu, J. Analysis and comparison of main fatty acids among Cyperus oil and other vegetable oils. *Food Ind.* **2016**, *37*, 52–54.
37. Duan, L.; Han, M., Wei, Z.; Wu, X. Microwave assisted extraction and fatty acid analysis of Cyperus oil. *Cereals Oils* **2021**, *34*, 59–62.
38. Olabiyi, A.A.; Oboh, G.; Adefegha, S.A. Effect of dietary supplementation of tiger nut (*Cyperus esculentus* L.) and walnut (*Tetracarpidium conophorum* müll. Arg.) on sexual behavior, hormonal level, and antioxidant status in male rats. *J. Food Biochem.* **2017**, *41*, 66–75. [CrossRef]
39. Chao, X.; Li, Y.; Jing, S.; Zhai, Y.; Liu, G.; Zhang, J.; Qi, K.; Zhu, X.; Buwizogher, A. Analysis of Nutrient Composition and Bioactivity of *Cyperus esculentus* (*C. esculentus* L.) before and after Germination. *Sci. Technol. Food Ind.* **2021**, *42*, 327–333. (In Chinese)
40. Jebasingh, D.; Venkataraman, S.; Jackson, D.D. Physiochemical and toxicological studies of the medicinal plant *Cyperus rotundus* L. (Cyperaceae). *Int. J. Appl. Res. Nat. Prod.* **2012**, *5*, 1–8.
41. Chen, Z.; Hu, X.; Liu, P.; Bai, C.; Liu, G. Research Progresses on Cultivation and Utilization of *Cyperus esculentus*. *Chin. J. Trop. Agric.* **2017**, *37*, 56–60.
42. Liu, H.; Wang, L.; Liu, Y. Development status of Cyperus products in China. *Farm Prod. Process.* **2021**, *7*, 97–99.
43. Hu, G. Study on Synthesis of Anthraquinones and Their Cytotoxic Activity against Tumor Cells. Master's Thesis, Lanzhou University, Lanzhou, China, 2014; pp. 5–11.
44. Achoribo, E.S.; Ong, M.T. Tiger nut (*Cyperus esculentus*): Source of natural anticancer drug? Brief review of existing literature. *Euromediterranean Biomed. J.* **2017**, *12*, 91–94.
45. Wiebe, J.C.; Lópezríos, L.; Vegamorales, T.; Gericke, N. Neurophysiological effects of a special extract of *Cyperus esculentus* L. (Cyperol). *J. Herb. Med. Res.* **2019**, *4*, 34. [CrossRef]
46. Jing, S.Q.; Wang, S.S.; Zhong, R.M.; Zhang, J.Y.; Wu, J.Z.; Tu, Y.X.; Pu, Y.; Yan, L.J. Neuroprotection of *Cyperus esculentus* L. orientin against cerebral ischemia/reperfusion induced brain injury. *Neural Regen. Res.* **2020**, *15*, 548–556. [CrossRef] [PubMed]
47. Elshamy, A.I.; Farrag, A.R.H.; Ayoub, I.M. UPLC-qTOF-MS phytochemical profile and antiulcer potential of Cyperus conglomeratus rottb. alcoholic extract. *Molecules* **2020**, *18*, 4234. [CrossRef]
48. Imo, C.; Uzochukwuw, C.D.; Mamma, E.; Mayel, M.H.; Kukoyi, A.J.; David Apaj, A. Effects of Ethanolic Extracts of Phoenix dactylifera Fruit, Cyperus esculentus Nut and Cocos nucifera Nut on Selected Indices of Kidney Function in Male Albino Rats. *J. Appl. Sci.* **2018**, *18*, 116–121. [CrossRef]
49. Nto, N.; Hassan, L.; Anyanwu, G.; Obikili, E.; Finbarrs-Bello, E.; Abireh, L. Curative effect of aqueous extract of *Cyperus esculentus* on flutamide-induced testicular dysfunction in male Wistar rats. *J. Exp. Clin. Cancer Res.* **2018**, *17*, 13. [CrossRef]
50. Jagpal, C. The in vitro use of natural antioxidant oils (*Cyperus esculentus*; *Nigella sativa*) to reduce cell sickling in sickle cell anaemia. *Biochem. Pharmacol.* **2017**, *139*, 134. [CrossRef]

51. Moral-Anter, D.; Campo-Sabariz, J.; Ferrer, R.; Martín-Venegas, R. *Cyperus esculentus* L. Tubers (Tiger Nuts) Protect Epithelial Barrier Function in Caco-2 Cells Infected by Salmonella Enteritidis and Promote Lactobacillus plantarum Growth. *Nutrients* **2020**, *13*, 71. [CrossRef]
52. Huang, M.; Wang, X.; Pang, Z. Research Status and Prospect of *Cyperus esculentus* L. *Crop Res.* **2013**, *27*, 293–301.
53. Gao, Y.; Sun, Y.; Gao, H.; Chen, Y.; Wang, X.; Xue, J.; Jia, X.; Li, R. Correction to: Ectopic overexpression of a type-II DGAT (CeDGAT2-2) derived from oil-rich tuber of *Cyperus esculentus* enhances accumulation of oil and oleic acid in tobacco leaves. *Biotechnol. Biofuels* **2021**, *14*, 51–56. [CrossRef] [PubMed]
54. Bamishaiye, E.I.; Muhammad, N.O.; Bamishaiye, O.M. Histological changes and serum lipid profile of selected rat tissues fed on *Cyperus esculentus* (tiger nut) tuber oil meal-based diet. *Der Pharma Chem.* **2010**, *2*, 90–96.
55. Moon, M.K.; Ahn, J.; Lee, H.; Ha, T.J. Anti-obesity and hypo-lipidemic effects of chufa (*Cyperus esculentus* L.) in mice fed a high-fat diet. *Food Sci. Biotechnol.* **2012**, *21*, 317–322. [CrossRef]
56. Salem, M.L.; Zommara, M.; Imaizumi, K. Dietary supplementation with *Cyperus esculentus* L. (tiger nut) tubers attenuated atherosclerotic lesion in apolipoprotein E knockout mouse associated with inhibition of inflammatory cell responses. *Am. J. Immunol.* **2005**, *1*, 60–67. [CrossRef]
57. Oluwajuyitan, T.D.; Ijarotimi, O.S. Nutritional, antioxidant, glycaemic index and Antihyperglycaemic properties of improved traditional plantain-based (Musa AAB) dough meal enriched with tiger nut (*Cyperus esculentus*) and defatted soybean(*Glycine max*)flour for diabetic patients. *Heliyon* **2019**, *5*, 22–29. [CrossRef] [PubMed]
58. Saheed, S.; Oladipipo, A.E.; Abraham, B.F. Four weeksdaily dose oral administration assessment of *Cyperus esculentus* L. aqueous extract on key metabolic markers of Wistar rats. *Pharmacologia* **2016**, *7*, 125–133. [CrossRef]
59. Onuoha, N.; Ogbusua, N.; Okorie, A.; Ejike, C.E. Tiger nut (*Cyperus esculentus* L.) 'milk' as a potent 'nutri-drink' for the prevention of acetaminophen-induced hepatotoxicity in a murine model. *J. Intercult. Ethnopharmacol.* **2017**, *6*, 290–295. [CrossRef]
60. Hassanein, H.M.; Nazif, N.M.; Aboutabl, E.A.; Hammouda, F.M. In-vitro hepatoprotection study, free radical scavenging activity and gc/ms analysis of *Cyperus esculentus* essential oils from hydrodistillation and mae techniques. *Planta Med.* **2013**, *79*, 312–320. [CrossRef]
61. Prakash, N.; Ragavan, B. Phytochemical observation and antibacterial activity of *Cyperus esculentus* L. *Anc. Sci. Life* **2009**, *28*, 16–20.
62. Abdel Karim, M.; Fath Elrahman, A. GC-MS analysis and antimicrobial activity of sudanese *Cyperus esculentus* (Cyperaceae) fixed oil. *Int. J. Adv. Res.* **1930**, *2016*, 256–263.
63. Su, L.; Wang, S.; Jing, S. Preliminary study on microcirculation and anticoagulant activity of alcohol extract of Cyperus (*Cyperus esculentus* L.). *Food Ind.* **2017**, *38*, 200–203.
64. Jing, S.; Ouyang, W.; Ren, Z.; Ma, Z. In vitro and in vivo antioxidant properties of *Cyperus esculentus* oil from Xinjiang, China. *J. Sci. Food Agric.* **2013**, *93*, 505–509. [CrossRef] [PubMed]
65. Willis, S.; Jackson, C.; Verghese, M. Effects of processing on antioxidant capacity and metabolizing enzyme inhibition of tiger nut tubers. *Food Nutr. Sci.* **2019**, *10*, 1132–1141. [CrossRef]
66. Badejo, A.A.; Olawoyin, B.; Salawu, S.O.; Fasuhanmi, O.S.; Boligon, A.A.; Enujiugha, V.N. Antioxidative potentials and chromatographic analysis of beverages from blends of glutenfree acha (*Digitaria exilis*) and tiger nut (*Cyperus esculentus*) extracts. *J. Food Meas. Charact.* **2017**, *11*, 2094–2101. [CrossRef]
67. Ademosun, A.O.; Oboh, G. Inhibition of carbohydrate hydrolyzing enzymes associated with type 2 diabetes and antioxidative properties of some edible seeds in vitro. *Int. J. Diabetes Dev. Ctries.* **2015**, *35*, 516–521. [CrossRef]
68. Adebayo, S.F.; Arinola, S.O. Effect of germination on the nutrient and antioxidant properties of tiger nut (*Cyperus esculentus*). *Biol. Agric. Healthc.* **2017**, *7*, 88–94.
69. Jing, S.; Wang, S.; Li, Q.; Zheng, L.; Yue, L.; Fan, S.; Tao, G. Dynamic high pressure microfluidization-assisted extraction and bioactivities of *Cyperus esculentus* (*C. esculentus* L.) leaves flavonoids. *Food Chem.* **2016**, *192*, 319–327. [CrossRef]
70. Papadi, A.; Oluwale, B.; Ismail, K. Allelopathic Effect of Some Weeds on the Germination of Seeds of Selected Crops Grown in Akwa Ibom State, Nigeria. *World J. Agric. Res.* **2013**, *1*, 59–64.
71. Drost, D.C.; Doll, J.D. The Allelopathic Effect of Yellow Nutsedge (*Cyperus esculentus*) on Corn (*Zea mays*) and Soybeans (*Glycine max*). *Weed Sci.* **1980**, *28*, 229–233. [CrossRef]
72. Reinhardt, C.F.; Bezuidenhout, S.R. Growth Stage of *Cyperus esculentus* Influences Its Allelopathic Effect on Ectomycorrhizal and Higher Plant Species. *J. Crop Prod.* **2001**, *4*, 323–333. [CrossRef]
73. Oamen, H.P.; Idu, M. Allelopathic effects of aqueous extract of Cyperus esculentus L. on germination and seedling growth of some bean seeds. *Asian J. Microbiol. Biotechnol. Environ. Sci.* **2009**, *11*, 705–710.
74. Singh, C.M.; Angiras, N.N.; Singh, S.D. Allelopathic Effects of Some Weeds on Wheat Seed Germination. *Indian J. Weed Sci.* **1988**, *20*, 63–66.
75. Sani, J.J.; Rahimi, A.; Hosseini, S.M.; Mirik, A.A.M. Allelopathic Effect of Nutsedge Weeds (*Cyperus esculentus* L.) and Redroot Pigweed (*Amaranthus retroflexus* L.) Plants Extracts on Germination and Seedling Growth of Rapeseed (*Brassica napus*). *Weed Sci.* **2012**, *6*, 27–36.

76. Patterson, M.G.; Buchanan, G.A.; Street, J.E.; Crowley, R.H. Yellow Nutsedge (*Cyperus esculentus*) Competition with Cotton (*Gossypium hirsutum*). *Weed Sci.* **1980**, *28*, 327–329. [CrossRef]
77. Avicor, S.W.; Wajidi, M.F.F.; Achoribo, E.S.; Ong, M.T.; Hamzah, S.N. Tiger nut (*Cyperus esculentus*) as a potential phytoinsecticide: Larvicidal activity of crude extracts on Aedes aegypti and Culex quinquefasciatus (Diptera: Culicidae). *Trop. Biomed.* **2021**, *38*, 186–191.

plants

MDPI

Review

Allelopathy and Allelochemicals of *Solidago canadensis* L. and *S. altissima* L. for Their Naturalization

Hisashi Kato-Noguchi * and Midori Kato

Department of Applied Biological Science, Faculty of Agriculture, Kagawa University, Miki, Kagawa 761-0795, Japan
* Correspondence: kato.hisashi@kagawa-u.ac.jp

Abstract: *Solidago canadensis* L. and *Solidago altissima* L. are native to North America and have naturalized many other continents including Europa and Asia. Their species is an aggressive colonizer and forms thick monospecific stands. The evidence of the allelopathy for *S. canadensis* and *S. altissima* has accumulated in the literature since the late 20th century. The root exudates, extracts, essential oil and rhizosphere soil of *S. canadensis* suppressed the germination, growth and the arbuscular mycorrhizal colonization of several plants, including native plant species. Allelochemicals such as fatty acids, terpenes, flavonoids, polyphenols and their related compounds were identified in the extracts and essential oil of *S. canadensis*. The concentrations of total phenolics, total flavonoids and total saponins in the rhizosphere soil of *S. canadensis* obtained from the invasive ranges were greater than those from the native ranges. Allelochemicals such as terpenes, flavonoids, polyacetylene and phenols were also identified in the extracts, essential oil and the rhizosphere soil in *S. altissima*. Among the identified allelochemicals of *S. altissima*, the *cis*-dehydromatricaria ester may be involved in the allelopathy considering its growth inhibitory activity and its concentration in the rhizosphere soil. Therefore, the allelopathy of *S. canadensis* and *S. altissima* may support their invasiveness, naturalization and formation of thick monospecific stands. This is the first review article focusing on the allelopathy of both of *S. canadensis* and *S. altissima*.

Keywords: allelochemical; invasive species; monospecific stand; mycorrhizal colonization; *Solidago*; phytotoxicity; rhizosphere soil

Citation: Kato-Noguchi, H.; Kato, M. Allelopathy and Allelochemicals of *Solidago canadensis* L. and *S. altissima* L. for Their Naturalization. *Plants* **2022**, *11*, 3235. https://doi.org/10.3390/plants11233235

Academic Editors: Margot Schulz and Vincenzo Tabaglio

Received: 22 October 2022
Accepted: 23 November 2022
Published: 25 November 2022

1. Introduction

Solidago canadensis sensu lato (s.l.), belonging Asteraceae, is an erect rhizomatous perennial plant, native to North America. *S. canadensis* s.l. was introduced to Europe as an ornamental plant in the 17th to 18th centuries. The species spread from the gardens to the natural environments, and has extended its habitats in Central and Eastern Europe. It expanded its habitat at a rate of 741 km^2 per year in Europe [1]. The species has also been introduced and naturalized in many other countries such as Australia, Brazil, China, India, New Zealand and Japan [2–6].

The species expands its habitat though seed distribution and rhizome expansion. The rhizomes arise near the base of the shoots in autumn and produce aerial stems from their apex in the following spring. The stems are not branched, and bear triple-nerved, lanceolate, alternate leaves which are found along the stems and roots at the base of the shoots. The rhizome systems contribute to expanding the species' community and to form thick monospecific stands [7,8]. Shoot density in the established stands of the species was reported to be 309 shoots per m^2 [3]. In addition, oil-filled cavities, which contain terpenes and/or lipids, were randomly distributed in the rhizomes [9]. These compounds may have some biological functions such as allelopathy. The species is a prolific seed producer. Its inflorescence forms broad pyramidal panicles, which contain numerous florets (Figure 1). A single plant produces 1000–20,000 light-winged achenes which contain seeds. The achenes disperse easily by wind, water and human activities. The germination rate is 30–75%, depending on the conditions [7,8,10,11]. The seed distribution may contribute to establishing the populations of *S. canadensis* s.l. in new habitats.

Figure 1. *Solidago canadensis* s.l.

This species has adapted to a wide range of soil fertility and water potential [8,11,12], as well as been colonized into disturbed areas such as abandoned fields, roadsides, riverbanks and forest edges [11,13,14]. The species established its population on an agriculture field in two years after the abandonment [15,16]. Once established, the population remained dominant over 30 years [8,17–20]. It was also reported that the species showed great impact on the native plant diversity in the introduced ranges [21–23]. Owing to the potential of the species for the rapid expansion and the formation of thick monospecific stands in the introduced ranges, as well as its impact on the environments, *S. canadensis* s.l. has been designated as a harmful invasive plant species [4,24].

The *S. canadensis* complex is a highly variable species. *S. canadensis* s.l. contains *S. canadensis* L. as *S. canadensis* subsp. *canadensis* (L.), and *S. altissima* L. as *S. canadensis* subsp. *altissima* (L.) O.Bolòs et Vigo [2,25,26]. *S. canadensis* and *S. altissima* are very similar taxa. The field experiments also showed that the competitive abilities of *S. altissima* and *S. canadensis* against other plant species was similar [27]. However, they could be distinguished by their morphological traits such as shoot length and flowering time [2,26,28,29]. The chromosome number between *S. canadensis* (diploid; $2n = 18$) and *S. altissima* (hexaploid; $2n = 54$) is also different [7,29,30]. The native ranges of both species in North America are not exactly the same [8,26]. In addition, the experimental crossing of *S. canadensis* and *S. altissima* could not bear viable seeds [30], which indicates a genetic barrier between both taxa. However, owing to the lack of consistency in the identification of the species, the separation of both species has been considered to be very problematic [2,28,29]. For example, most European populations of both species were described to be *S. altissima* [3,25], although macro-morphological analyses indicate that *S. canadensis* is a common species in Europa [29,31]. *S. altissima* was also mentioned as a synonym of *S. canadensis* [32,33]. There may have been a misidentification of the species.

It was reported that invasive plants are often allelopathic and inhibit the germination and growth of the native plant species in the invasive ranges through specific secondary metabolites defined as allelochemicals [34–43]. *S. canadensis* showed an allelopathic activity on the sugar maple seedlings in the field, greenhouse and laboratory conditions [44]. The *cis*-Dehydromatricaria ester was identified in the extracts of *S. altissima* as its allelochemical [45]. The evidence of the allelopathy of *S. canadensis* and *S. altissima* has been accumulated since the late 20th century, and their allelopathy was often implicated in the potential of their invasiveness and naturalization. However, there has been no review paper focusing on the allelopathy of both *S. canadensis* and *S. altissima*. This review provides an overview of the allelopathy and allelochemicals in *S. canadensis* and *S. altissima*, as well as a discussion on the involvement of allelopathy in the invasiveness and naturalization of the species. Despite the potential misidentification of *S. canadensis* and *S. altissima* described earlier, due to the fact that both species are very similar and the identification of the species is problematic [2,28,29], this paper followed the identification of the species in the publications because it is impossible to confirm their identification.

2. Allelopathy of *S. canadensis*

Allelopathy is the chemical interaction between donor plants and recipient plants through allelochemicals. Allelochemicals are produced in some plant parts and released into the vicinity of the donor plants, including their rhizosphere soil either by the root exudation, rainfall leachates, volatilization from the plant parts or decomposition processes of plant residues [46–49]. Several investigations in field conditions showed that *S. canadensis* reduced the number and biodiversity of the native plant community in its invaded ranges [50]. The invasion level of *S. canadensis* correlated negatively with the taxonomic diversity of the native plant community, and positively with the invasibility of the community [51]. Those observations may imply the involvement of allelopathy in the interaction between *S. canadensis* and native plant species to some extent. Many researchers have evaluated the allelopathic activity of the root exudates, rhizosphere soil, residues and plant extracts of *S. canadensis* (Table 1).

2.1. Allelopathy of Root Exudate and Plant Residue

Root exudates of *S. canadensis*, which were obtained from its aeroponic culture, significantly suppressed the growth of two Asian original plant species; *Gnaphalium affine* D.Don and *Xanthium sibiricum* Patrin ex Widder., two of America origin; *Conyza canadensis* (L.) Cronquist and *Celosia argentea* L., two of tropical origin; *Aster subulatus* Michx. and *Sesbania cannabina* (Renz.) Poir. and a cosmopolitan species; *Eclipta prostrata* (L.) L. The suppression rate was similar in all plant species [52]. Root exudates of *S. canadensis* also showed the growth inhibition of *Arabidopsis thaliana* (L.) Heynh. [53]. When the seeds of seven European native plant species were sown into the *S. canadensis* cultivated soils with or without activated carbon, the germination of five species such as *Dactylis glomerata* L., *Lythrum salicaria* L., *Stachys officinalis* (L.) Trevis. and *Trifolium pratense* L. were significantly suppressed in activated carbon-free plots than in activated carbon plots. Although the germination rate was not significantly different between both plots, the biomass of *Arrhenatherum elatius* (L.) P.Beauv. ex J. et C.Presl in the activated carbon plots after three months of sowing was two times greater than that in the activated carbon-free plots [53]. Activated carbon is a widely used material to investigate allelopathy because it adsorbs allelochemicals in the plant rhizosphere soil [34,54]. In addition, aqueous extracts of the rhizosphere soil of *S. canadensis* inhibited the germination and growth of *Digitaria sanguinalis* (L.) Scop. and *Amaranthus retroflexus* L., and the inhibitory activity was greater in the extracts of the soil obtained from the invasive ranges of *S. canadensis* (China) than that from its native ranges (USA) [55]. These observations suggest that certain allelochemicals, which may cause growth inhibition, would be released into the rhizosphere soil as root exudates of *S. canadensis*, and the released allelochemicals in the soil may be greater in the invasive ranges than those in the native ranges.

Crushed stems, leaves and rhizomes of *S. canadensis* were mixed with soil and water and kept at 20/15 °C (12/12 h light/dark condition), and the mixture was filtered after 45 days. The obtained filtrate suppressed the germination and growth of *Raphanus sativus* L. and *Triticum aestivum* L. [56]. This observation also suggests that certain allelochemicals may be released into the rhizosphere soil during the decomposition process of plant residues of *S. canadensis*.

2.2. Allelopathy of Plant Extract

Some plant tissues may contain allelochemicals, since allelochemicals are synthesized and stored in certain plant tissues until their release into the environment [46–49]. Many investigations on the allelopathic activity of the extracts from different plant parts of *S. canadensis* have been conducted. Aqueous extracts of the leaves of *S. canadensis* inhibited the germination and root growth of *Raphanus sativus* L. and *Lactuca sativa* L. [57], as well as those of *Triticum aestivum* L. and *Setaria viridis* (L.)P.Beauv. [58]. The extracts also suppressed the germination, growth and chlorophyll content of *Trifolium pratense* L. and *Raphanus sativus* L., and increased their electrolyte leakage from the cell membrane of the seedlings [59,60].

The fresh leaves and stems of *S. canadensis* were soaked in water for 48 h, and the obtained soaking water showed the inhibitory activity on the germination and growth of *Raphanus sativus* L. and *Triticum aestivum* L. [54]. Aqueous extracts of the above-ground parts of *S. canadensis* suppressed the germination and growth of *Lactuca sativa* L. [61], as well as those of *Digitaria sanguinalis* (L.) Scop. and *Amaranthus retroflexus* L. [55]. The inhibitory activity was greater in the plant extracts obtained from the heavily invaded stands than in those obtained from the lightly invaded stands [62], and in the plant extracts obtained from the invasive ranges than those from the native ranges [55].

Aqueous extracts of the above-ground parts and roots of *S. canadensis* inhibited the germination and growth of *Zoysia japonica* Steud, and the extracts of the above-ground parts significantly stimulated malondialdehyde and peroxidase activity [62]. The extracts of the stems, roots, blossoms and seeds of *S. canadensis* suppressed the germination and growth of *Brassica napus* L. and *Lolium perenne* L. [63], and the extracts of the roots and rhizomes of *S. canadensis* also inhibited the root growth of *Raphanus sativus* L. and *Lactuca sativa* L. [57].

Aqueous ethanol extracts of the roots and rhizomes of *S. canadensis* inhibited the germination and growth of *Trifolium repens* L., *Trifolium pratense* L., *Medicago lupulina* L., *Suaeda glauca* (Brunge) Brunge, *Plantago virginica* L., *Kummerowia stipulacea* (Maxium.) Makino, *Festuca arundinacea* Schreb., *Ageratum conyzoides* L., *Portulaca oleracea* L. and *Amaranthus spinosus* L. [64]. Aqueous ethanol extracts of the above- and below-ground parts of *S. canadensis* suppressed the germination of *Kummerowia striata* (Thunb.) Schindl., and the inhibitory activity was greater in the plant extracts collected from the invasive ranges of *S. canadensis* than those from its native ranges [65]. Aqueous and ethanol extracts of the leaves, stems and rhizomes of *S. canadensis* inhibited the germination and growth of *Morus alba* L., *Pharbitis nil* (L.) Roth, *Triticum aestivum* L. and *Brassica campestris* L., and the inhibition was grater in the ethanol extracts than in the aqueous extracts [66].

Investigations on the aqueous and ethanol extracts of every part of *S. canadensis* showed the allelopathic activity on the germination, growth, chlorophyll content, electrolyte leakage and/or some enzyme activities of several plant species, including the native plant species. The inhibitory activity was greater in the plant extracts obtained from the invasive ranges of *S. canadensis* than in those from its native ranges, and in the extracts collected from the heavily invaded stands than in those collected from the lightly invaded stands. These observations suggest that whole parts of *S. canadensis* may contain water and ethanol extractable allelochemicals, which may cause the inhibition. In addition, the plants grown in the invasive ranges and heavily invaded stands may contain more allelochemicals than the plants in the native ranges and lightly invaded stands.

2.3. Effects of the Extract on Arbuscular Mycorrhizal Fungi

The rhizomes of *S. canadensis* were soaked in water for 24 h, and the obtained soaking water caused the suppression of the arbuscular mycorrhizal colonization of *Echinochloa crus-galli* (L.) P.Beauv., *Kummerowia striata* (Thnb.) Schindl. and *Ageratum conyzoides* L. [67]. The field and greenhouse investigations also showed that *S. canadensis* altered the composition of the arbuscular mycorrhizal fungal population in its rhizosphere soil through the inhibition of some dominant species and the stimulation of other species. The established arbuscular mycorrhizal community increased the competitive ability and the biomass of *S. canadensis* [67–71]. This altered arbuscular mycorrhizal community also increased the mycorrhizal-mediated ^{15}N uptake in *S. canadensis*, as well as decreased the ^{15}N uptake in the native species *Kummerowia statrica* (Thunb.) Schindl. [72]. In addition, the aqueous ethanol extract of the roots and rhizomes of *S. canadensis* also suppressed the population of the soilborne pathogens, namely *Pythium ultimum* Trow and *Rhizoctonia solani* J.G. Kühn [73]. These observations indicate that the aqueous extracts *of S. canadensis* may alter the arbuscular mycorrhizal population and suppress the colonization of the native plant species. The established arbuscular mycorrhizal community enhanced the competitive ability of *S. canadensis*. Certain compounds in the extracts may be involved in the alteration of the arbuscular mycorrhizal community.

Table 1. Allelopathic activities of exudates, rhizosphere soil, residues and plant extracts of *S. canadensis*.

Source	Inhibition				Target Plant Species	Reference
	Germination	Growth	Chlorophyll	Mycorrhizal colonization		
Root exudate		✓			*Gnaphalium affine, Xanthium sibiricum, Conyza canadensis, Celosia argentea, Aster subulatus, Sesbania cannabina, Eclipta prostrata*	[52]
		✓			*Arabidopsis thaliana,*	[53]
Rhizosphere soil	✓	✓			*Dactylis glomerata, Lythrum salicaria, Stachys officinalis, Trifolium pratense*	[53]
Soil extract	✓	✓			*Digitaria sanguinalis, Amaranthus retroflexus*	[55]
Residue	✓	✓			*Raphanus sativus, Triticum aestivum*	[56]
Plant extract						
Whole part	✓	✓			*Kummerowia striata*	[65]
Leaf	✓	✓			*Raphanus sativus, Lactuca sativa*	[57]
	✓	✓			*Triticum aestivum, Setaria viridi*	[58]
			✓		*Raphanus sativus*	[59]
			✓		*Trifolium pratense*	[60]
Leaf and stem	✓	✓			*Raphanus sativus, Triticum aestivum*	[56]
Above-ground part	✓	✓			*Lactuca sativa*	[61]
	✓	✓			*Digitaria sanguinalis (L.) Scop. and Amaranthus retroflexus*	[55]
Above-ground part, root	✓	✓			*Zoysia japonica*	[62]
Stem, root, blossom, seed	✓	✓			*Brassica napus, Lolium perenne*	[63]

Table 1. *Cont.*

Source	Inhibition			Target Plant Species	Reference
Leaf, stem, rhizome	✓	✓		*Morus alba, Pharbitis nil, Triticum aestivum, Brassica campestris*	[66]
Root, rhizome		✓		*Raphanus sativus, Lactuca sativa*	[57]
	✓	✓		*Trifolium repens, Trifolium pratense, Medicago lupulina, Suaeda glauca, Plantago virginica, Kummerowia stipulacea, Festuca arundinacea, Ageratum conyzoides, Portulaca oleracea, Amaranthus spinosus*	[69]
Rhizome			✓	*Echinochloa crus-galli, Kummerowia striata, Ageratum conyzoides*	[67]

2.4. Allelochemicals

As described above, the inhibitory activity of the extracts of the plants and rhizosphere soil of *S. canadensis* obtained from the invasive ranges was greater than that obtained from the native ranges [55,65]. The concentrations of total phenolics, total flavones and total saponins in *S. canadensis* and its rhizosphere soil obtained from the invasive ranges were also greater than those from the native ranges [55,65]. These concentrations in the soil obtained from *S. canadensis*-infested stands were also greater than those in the soil obtained from *S. canadensis*-free stands [74].

Major compounds identified in the aqueous methanol extracts of the leaves and inflorescences of *S. canadensis* were chlorogenic acid, quercitrin and rutin (quercetin-3-*O*-β-rutinoside) (Figure 2) [75]. A fatty acid, *n*-hexadecanonic acid, was isolated from the aqueous ethanol extract of the stems and leaves of *S. canadensis* as an allelopathic agent. *n*-Hexadecanonic acid significantly inhibited the growth of *Triticum aestivum* L. [56]. A flavonoid, kaempferol-3-*O*-D-glucoside, was isolated from the aqueous ethanol extract of the *S. canadensis* straw, and the compound inhibited the growth of *Echinochloa colona* (L.) Link [75]. In addition, the concentration of rutin in the leaves of *S. canadensis* was greater than that of other *Solidago* species [76,77]. Some flavonoids were also identified in the aerial parts of *S. canadensis* [78].

The essential oil of many plant species was reported to have several biological activities, and to comprise volatile compounds such as terpenoids [79]. These volatile compounds often exhibited several biological activities including allelopathic activity [80,81]. Therefore, the involvement of the essential oil of *S. canadensis* in allelopathy was also investigated. The production of the essential oil was greater in *S. canadensis* obtained from the heavily invaded stands than that from mildly invaded stands [82]. The main components of the essential oil in the leaves and inflorescences of *S. canadensis* were α-pinene, trans-verbenol, limonene, bornyl acetate and β-cubebene [83], and those in the aerial parts of *S. canadensis* were α-pinene, β-pinene, germacrene D, limonene, thymol, (+)epi-bicyclosesquiphellandrene, β-cadinene, γ-cadinene, δ-cadinene, α-muurolene, γ-muurolene, α-cubebene and β-elemene [84,85]. The essential oil in the inflorescences of *S. canadensis* showed antimicrobial activity, antioxidant activity and free-radical scavenging activity, and its main components were α-pinene, germacrene D and limonene [86]. α-Pinene and β-pinene were reported to disturb the cell division through the interference of DNA synthesis, and cause membrane peroxidation [87–89]. Several monoterpenes were reported to alter soil microflora [90]. It was also reported that the essential oil of the aerial parts of *S. canadensis* inhibited the germination and radical growth of *Raphanus sativus* L. and *Lepidium sativum* L. [84].

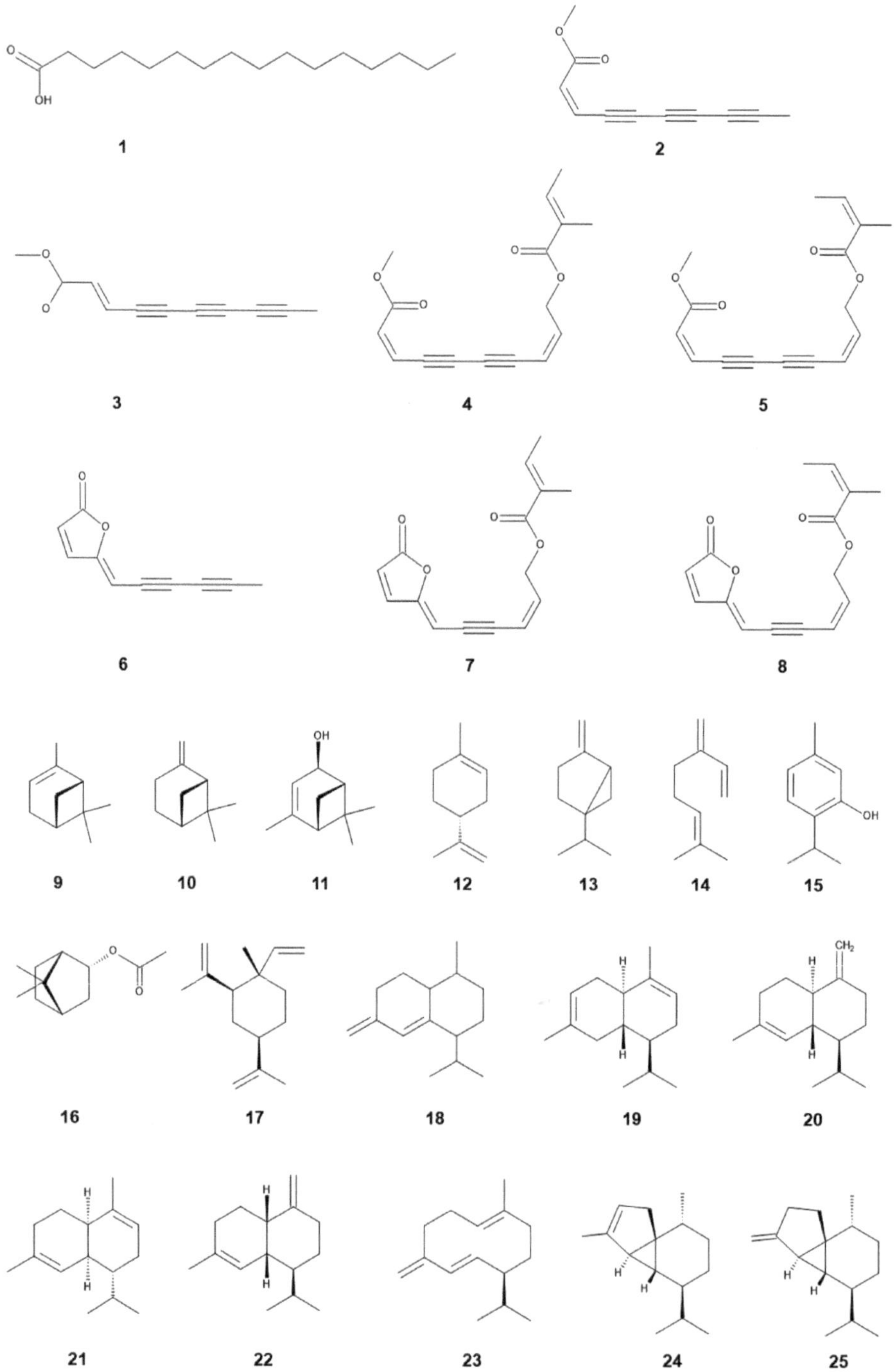

Figure 2. *Cont.*

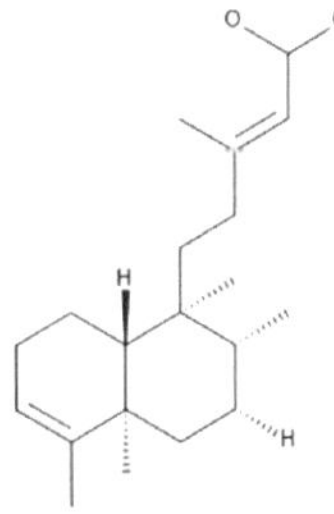
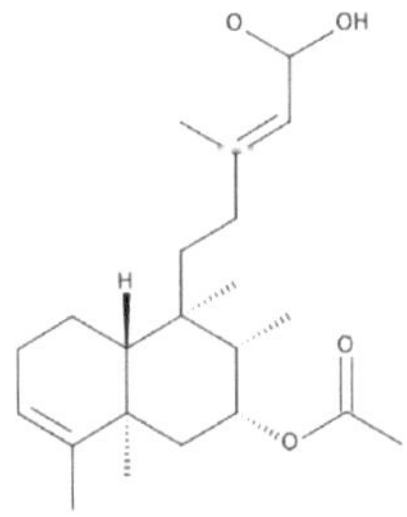
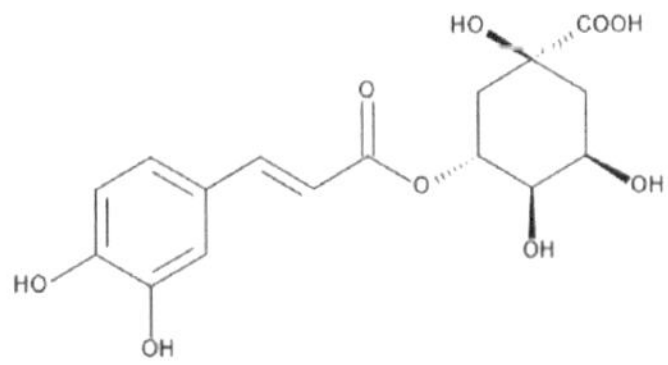

26 27 28

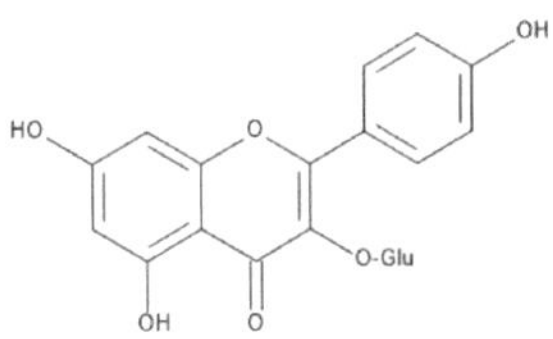
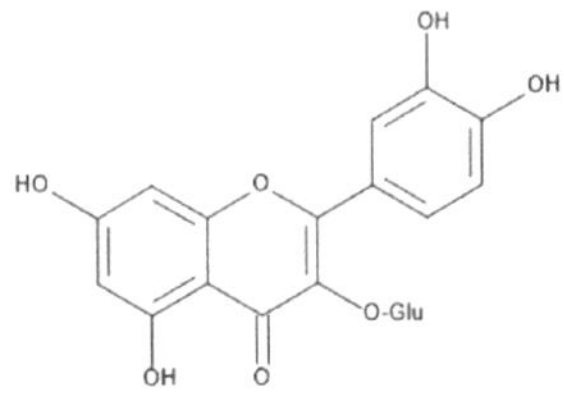
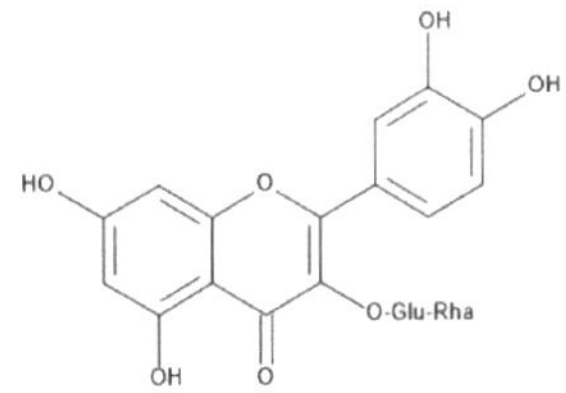

29 30 31

Figure 2. Allelochemicals identified in *S. canadensis* and *S. altissima*. Compound number and name were listed in Table 2.

2.5. Contribution of Allelopathy of S. canadensis to Its Invasiveness

All parts of the extracts of *S. canadensis*, including its rhizosphere soils, the root exudates, residues and essential oil, inhibited the germination and growth of several plant species (Table 1). These observations suggest that *S. canadensis* produces and accumulates allelochemicals in its plant tissues, and that it releases those allelochemicals into the surrounding environments of *S. canadensis*, including its rhizosphere soils through the root exudation, decomposition process of plant residues and volatilization from its essential oil. Several allelochemicals such as fatty acids, terpenes, flavonoids and polyphenols (Table 2) were identified in the extracts and essential oil of *S. canadensis*. Some of these compounds were reported to inhibit the germination and growth of other plants species, including the native plant species.

The extracts of *S. canadensis* and its rhizosphere soil obtained from its invasive ranges had greater inhibitory activity compared to the extracts and rhizosphere soil obtained from its native ranges [55,65]. The total phenolics, flavones and saponins in *S. canadensis* in the invasive ranges were also greater than those in the native ranges [55,65,74]. These observations suggest that *S. canadensis* in the invasive ranges may enhance the competitive ability through the increased production of some of these compounds. The novel weapons hypothesis suggests that the competitive ability of the invasive plants is superior to that of the native plant species due to its allelochemicals. Those allelochemicals are new to the indigenous plant species, and those indigenous plants are susceptible to the compounds [34,91]. However, there has been no available information on the specific compounds of which the production is increased in *S. canadensis* in the invasive ranges, as well as the reasons for which those productions are increased in the invasive ranges.

S. canadensis showed the alteration of the composition of the arbuscular mycorrhizal population and suppressed the mycorrhizal colonization of the native plant species in the field and greenhouse experiments. The extracts of *S. canadensis* also suppressed the mycorrhizal colonization of the native plant species. These investigations suggest that certain compounds in the extracts may be involved in the suppression. The authors also indicated the involvement of allelochemicals in the suppression. Arbuscular mycorrhizal fungi are important mycorrhiza for the most territorial plants, and their colonization enhances the ability of the host plants to absorb mineral nutrients and water, as well as increase the defense functions against pathogen attacks and stress conditions [92–94]. The suppression of the colonization reduces the potential of the native plants for nutrient and water absorption, as well as the defense function, which may cause the reduction in growth and vigor of the native plant species.

All the literature described above indicates that *S. canadensis* may suppress the regeneration process of the native plant species directly by the suppression of their germination and growth, and indirectly by the degradation of the mycorrhizal fungal mutualism of the native plant species through its allelopathy. Therefore, the allelopathy of *S. canadensis* may contribute to its increasing competitive ability and make the plant invasive. However, it is necessary to identify the allelochemicals involved in the suppression of the fungal mutualism of the native plant species. Investigations into the specific inhibitory activity of the identified allelochemicals on the native plant species, as well as their concentrations in the rhizosphere soil and/or surrounding environments are necessary to evaluate the contribution of these allelochemicals to the invasiveness of *S. canadensis*.

3. Allelopathy of *S. altissima*

There has been limited information on the allelopathic activity of the root exudates, extracts and residues of *S. altissima*, unlike *S. canadensis*, as well as its allelopathy in the field conditions. The available information is the inhibitory effect of the *n*-hexane extracts of the underground parts of *S. altissima* on the growth of *Lactuca sativa* L. [95]. However, phytochemical investigations suggest that *S. altissima* contains polyacetylenes [96,97], monoterpene and sesquiterpenes [98], diterpenes [95,98–102], triterpenes [103,104], flavonoid glycosides [105,106] and polyphenols [107]. Some of them may act as allelopathic agents of the species.

3.1. Terpene

13*E*-7α-acetoxyl kolavenic acid (solidagonic acid) was isolated from the aqueous acetone extract of *S. altissima* [95,108], and the compound showed the growth inhibitory activity on the seedlings of *Lactuca sativa* L. and *Lolium multiflorum* Lam. [95,108]. 13*E*-kolavenic acid was also isolated from its *n*-hexane extracts [95]. Although 13*E*-kolavenic acid did not show any growth inhibitory activity, it exhibited antifungal activity and insect antifeedant activity [109,110]. These activities may help the species to protect itself from fungal pathogens and insect damage, as well as to establish its stands. It was reported that young rhizomes contained many more terpenes than old rhizomes, and α-pinene, β-pinene, limonene and germacrene D were identified in the rhizomes by GC-MS analysis [98]. α-Pinene, β-pinene, sabinene, myrcene, limonene, bornyl acetate and germacrene D were also found in the essential oil of *S. altissima*. However, the species was identified as a synonym of *S. canadensis* [33].

Table 2. Allelochemicals identified in *S. canadensis* and *S. canadensis* and their sources.

Chemical Class	Compound	*Solidado canadensis*		Reference	*Solidado altissima*			Reference
		Extract	Essential oil		Extract	Soil	Essential oil	
Fatty acid	**1**: *n*-Hexadecanonic acid	✓		[56]				
Polyacetylene	**2**: *cis*-Dehydromatricaria ester				✓	✓		[45,96,98,104]
	3: *trans*-Dehydromatricaria ester					✓		[45]
	4: (2Z,8Z)-10-Tigloyloxy matricaria ester				✓			[95]
	5: (2Z,8Z)-10-Angeloyloxy matricaria ester				✓			[95,96]
	6: Dehydromatricaria lactone				✓			[95,104,111]
	7: (4Z,8Z)-10-Trigloyloxy matricaria lactone				✓			[95,111]
	8: (4Z,8Z)-10-Angeloyloxy matricaria lactone				✓			[95]
Monoterpene	**9**: α-Pinene		✓	[83–86]	✓		✓	[33,98]
	10: β-Pinene		✓	[84]	✓		✓	[33,98]
	11: *trans*-Verbenol		✓	[83]				
	12: Limonene		✓	[83–86]	✓		✓	[33,98]
	13: Sabinene						✓	[33]
	14: Myrcene						✓	[33]
	15: Thymol		✓	[84]				
	16: Bornyl acetate		✓	[83]			✓	[33]
Sesquiterpene	**16**: β-Elemene		✓	[84]				
	17: (+)epi-Bicyclosesquiphellandrene		✓	[84]				
	18: β-Cadinene		✓	[84]				
	19: γ-Cadinene		✓	[84]				
	20: *δ-Cadinene*		✓	[84]				
	21: α-muurolene		✓	[84]				
	22: γ-muurolene		✓	[84]				
	23: Germacrene D		✓	[84,86]	✓		✓	[33,98]
	24: α-cubebene		✓	[84]				
	25: β-Cubebene		✓	[83,84]				
Diterpene	**26**: 13*E*-kolavenic acid				✓			[95]
	27: 13*E*-7α-acetoxyl kolavenic acid				✓			[95,108]
Polyphenol	**28**: Chlorogenic acid	✓		[76]				
Flavonoid	**29**: Kaempferol-3-*O*-D-glucoside	✓		[75]				
	30: Quercitrin	✓		[76,77]				
	31: Rutin	✓		[76]				

3.2. Polyacetylene

Several polyacetylenes, such as *cis*-dehydromatricaria ester (*cis*-DME), dehydromatricaria lactone, (2Z,8Z)-10-tigloyloxy matricaria ester, (4Z,8Z)-10-trigloyloxy matricaria lactone, (2Z,8Z)-10-angeloyloxy matricaria ester (methyl10-[(Z)-2-methyl-2-butenoyloxy]-(2Z,8Z)-2,8-decadiene-4,6-diynoate) and (4Z,8Z)-10-angeloyloxy matricaria lactone were isolated from the stems, roots and/or rhizomes of *S. altissima* [45,95,96,98,104,111]. Polyacetylenes in Asteraceae plants were reported to be synthesized from crepenynic acid, which is formed oleic acid with linoleic acid, introducing the first acetylene bond [95,112].

Among the polyacetylenes found in *S. altissima*, *cis*-DME, dehydromatricaria lactone and the (2Z,8Z)-10-angeloyloxy matricaria ester were reported to inhibit the coleoptile growth of *Panicum crus-galli* L. var. *frumentaceum* Trin. up to 77.4–93.5%, 81.7% and 80.0% that of the control, respectively, at the concentration of 1 ppm [96,111]. The (2Z,8Z)-10-Tigloyloxy matricaria ester also inhibited the growth of *Lolium multiflorum* Lam. [95]. However, *cis*-DME among those polyacetylenes was the most studied.

The bioactive concentration of *cis*-DME caused the growth inhibition, which was of 1–20 ppm in laboratory conditions [45,95,96,98,108,111,113]. *cis*-DME suppressed the germination of *Asclepias syriaca* L., *Ambrosia artemisiifolia* L. and *Miscanthus sinensis* Anderson; the germination and growth of *Poa pratensis* L. and *Oryza sativa* L.; and the growth of *Lactuca sativa* L. and *Panicum crus-galli* L. var. *frumentaceum* Trin. Their germination was inhibited up to 20–95% that of the control by 32–60 ppm *cis*-DME, and their growth was inhibited up to 10–56% that of the control by 48–100 ppm *cis*-DME (Table 3). *Asclepias syriaca* and *Miscanthus sinensis*, respectively, are competitive species with *S. altissima* in the native range of the USA and in the invasive range of Japan. *Ambrosia artemisiifoli* and *Poa pratensis* are the invasive plant species from North America, Europa and North Asia [45,98,114,115]. However, the effectiveness of *cis*-DME on those plant species was not evidently different.

Table 3. Growth inhibitory activity of *cis*-dehydromatricaria ester.

cis-DME Concentration (ppm)	Target Plant	Germination	Growth	Reference
		(% of Control)		
32	*Asclepias syriaca*	5		[98]
48	*Poa pratensis*	20	50	[113]
50	*Oryza sativa*	88	10–15	[45]
50	*Ambrosia artemisiifolia*	25		[45]
50	*Miscanthus sinensis*	22		[45]
100	*Lactuca sativa*		30–56	[108]
100	*Panicum crus-galli*		18	[111]

The concentration of *cis*-DME was reported to be of 250–400 ppm in the roots of *S. altissima*, of 6.3 ppm in the soil under *S. altissima* [45] and of 0–17.3 ppm in the soil [116]. The concentrations in the soil varied depending on the soil properties and microbial activities [117–119]. The existence of *cis*-DME in the soil under *S. altissima* suggests that *cis*-DME may be released from *S. altissima* into the soils through the root exudation, rainfall leachates and/or the decomposition processes of plant residues. In addition, considering the bioactive concentration (1–20 ppm) described above and the soil concentration (0–17.3 ppm) of *cis*-DME, which is the same in laboratory conditions, the *cis*-DME in the soil may have some ecological function.

trans-DME was also in the soil under *S. altissima*, and its inhibitory activity was comparable of that of *cis*-DME [45]. When the ethanol solution (1%) of *cis*-DME was kept at 29 °C under 2000 lux for 2 days, 50% of *cis*-DME was isomerized into *trans*-DME, since the *cis*-form of the unsaturated ester is easily isomerized by light, pH or other conditions into the more stable *trans*-form [120]. In addition, *trans*-DME was not found in the roots and rhizomes of *S. altissima* [45]. Therefore, *cis*-DME may be isomerized by light and other environmental factors into *trans*-DME after the exudation from the plants into the soil. When we consider the involvement of DME in the allelopathy of *S. altissima*, the concentration of both *cis*-DME and *trans*-DME in the rhizosphere soil should be counted.

It was also reported that the aqueous extracts of the below-ground parts of *S. altissima* killed a pine wilt nematode (*Bursaphelenchus lignicolous* Mamiya and Kiyohara), and an active compound involved in it was determined as *cis*-DME. *cis*-DME at 10–11 ppm increased 50% of the molarity of the pine wilt nematode and a root-knot nematode (*Meloidogyne incognita* Kofild and White) [121]. The methanol extracts of the *S. altissima* rhizomes suppressed the hatching of fruit flies (*Drosophila melanogaster* Meigen), and an active compound was also determined as *cis*-DME [97]. The defense capacity of the invasive plants with natural enemies, such as pathogens, parasites and herbivores, contributes their establishment and naturalization into introduced ranges [37,122,123]. Therefore, the nematicidal and insecticidal activities of *cis*-DME may contribute to the establishment and naturalization of *S. altissima* in the introduced ranges.

3.3. Contribution of Allelopathy of S. altissima to Its Invasiveness

Unlike *S. canadensis*, the information on the allelopathic activity of *S. altissima* extracts, root exudates and residues is limited. However, polyacetylenes, monoterpene and sesquiterpenes, diterpenes, triterpenes and flavonoid glycosides were identified in its extracts and essential oil as allelopathic agents. Most potential compounds among them involved in the allelopathy may be *cis*-DME. *cis*-DME inhibited the germination and growth of several plant species at the concentration of 1–20 ppm [45,95,98,108,111,113]. Its concentration in the roots of *S. altissima* and in its rhizosphere soil was of 250–400 ppm and 0–17.3 ppm, respectively [45,116]. These observations suggest that *cis*-DME may be released into the rhizosphere soil by root exudation, rainfall leachates and/or decomposition processes of plant residues. *trans*-DME was also formed by the isomerization of *cis*-DME in the soil, and its inhibitory activity was the same as *cis*-DME. Although there has been no information of the concentration of *trans*-DME in the rhizosphere soil of *S. altissima*, the concentration of only *cis*-DME in several soils was over the concentration which was able to cause the growth inhibition [45,116]. In addition, *cis*-DME possesses nematicidal and insecticidal activities [97,113,121,124] Some flavonoids were also identified in the aerial parts of *S. canadensis* [78]. 13*E*-kolavenic acid also showed antifungal activity and insect antifeedant activity [109,110]. Therefore, these compounds including *trans*-DME may be involved in allelopathy and/or defense function against natural enemies, such as fungal pathogens, parasites and herbivores, and may contribute to their establishment and naturalization into the introduced ranges. However, it is necessary to determine the concentration of *trans*-DME in the rhizosphere soil of *S. altissima*, as well as to investigate the allelopathy of *S. altissima* in greenhouse and field conditions. It is also worth investigating the effect of *S. altissima* on the arbuscular mycorrhizal colonization of the native plant species due to the fact that *S. altissima* and *S. canadensis* are very close species.

4. Conclusions

S. canadensis and *S. altissima* are harmful invasive species naturalized in many countries, and form thick monospecific stands. Based on the literature, both species are allelopathic. The root exudate, residues, extracts, essential oil and rhizosphere soil of *S. canadensis* showed allelopathic activity on several plant species, including native plants, and suppressed the arbuscular mycorrhizal mutualism of the native plant species. Several allelochemicals were also identified in its extracts and essential oil. The concentrations of total phenolics, total flavones and total saponins in the rhizosphere soil of *S. canadensis* obtained from the invasive ranges were greater than those obtained from the native ranges. Therefore, the allelopathy of *S. canadensis* may contribute the interruption of the regeneration process of native plant species directly through the suppression of their germination and growth, and indirectly through the suppression of the mycorrhizal fungal mutualism of the native plant species. Several allelochemicals of *S. altissima* were also identified in its extracts and essential oil. Among them, *cis*-DME inhibited the germination and growth of several plant species at a concentration of 1–20 ppm, and its extractable concentration in the rhizosphere soil of *S. altissima* was of 0–17.3 ppm. *cis*-DME also exhibited nematicidal and insecticidal activities. *trans*-DME was formed by the isomerization of *cis*-DME in the soil, and its inhibitory activity was similar to that of *cis*-DME. Therefore, *cis*-DME, including *trans*-DME, may also cause the suppression of the regeneration process of the native plant species through the inhibition of their germination and growth. The allelopathy of *S. canadensis* and *S. altissima* may provide the species with competitive advantage against native plant species, and contribute to their invasiveness and naturalization.

Funding: This research received no external funding.

Institutional Review Statement: Not applicable.

Informed Consent Statement: Not applicable.

Conflicts of Interest: The authors declare no conflict of interest.

References

1. Weber, E. The dynamics of plant invasions: A case study of three exotic goldenrod species (*Solidago* L.) in Europe. *J. Biogeogr.* **1998**, *25*, 147–154. [CrossRef]
2. Weber, E. Morphological variation of the introduced perennial *Solidago canadensis* L. *sensu lato* (Asteraceae) in Europe. *Bot. J. Linn. Soc.* **1997**, *123*, 197–210. [CrossRef]
3. Weber, E. Biological flora of Central Europe: *Solidago altissima* L. *Flora* **2000**, *195*, 123–134. [CrossRef]
4. Invasive Species Compendium, Solidago canadensis. Available online: https://www.cabi.org/isc/datasheet/50599 (accessed on 12 September 2022).
5. Royal Botanical Gardens, Kew, Solidago canadensis. *Solidago altissima* L. Available online: https://powo.science.kew.org/taxon/urn:lsid:ipni.org:names:249454-1 (accessed on 12 September 2022).
6. Royal Botanical Gardens, Kew, *Solidago altissima* L. Available online: https://powo.science.kew.org/taxon/urn:lsid:ipni.org:names:285347-2 (accessed on 12 September 2022).
7. Croat, T. *Solidago canadensis* complex of the great plains. *Brittonia* **1972**, *24*, 317–326. [CrossRef]

8. Werner, P.A.; Bradbury, I.K.; Gross, R.S. The biology of Canadian weeds. 45. *Solidago canadensis* L. *Can. J. Plant Sci.* **1980**, *60*, 1393–1409. [CrossRef]
9. Curtis, J.D.; Lersten, N.R. Oil reservorirs in stem, rhizomes, and root of *Solidago canadensis* (Asteraceas, tribe Astereae). *Nor. J. Bot.* **1990**, *10*, 443–449. [CrossRef]
10. Meyer, A.H.; Schmid, B. Seed dynamics and seedling establishment in the invading perennial *Solidago altissima* under different experimental treatments. *J. Ecol.* **1999**, *87*, 28–41. [CrossRef]
11. Cornelius, R. The strategies of *Solidago canadensis* L. in relation to urban habitats. I. Resource requirements. *Acta Oecol.* **1990**, *11*, 19–34.
12. Huang, H.; Guo, S.; Chen, G. Reproductive biology in an invasive plant *Solidago canadensis*. *Front. Biol. China* **2007**, *2*, 196–204. [CrossRef]
13. Follak, S.; Eberius, M.; Essl, F.; Fürdös, A.; Sedlacek, N.; Trognitz, F. Invasive alien plants along roadsides in Europe. *EPPO Bull.* **2018**, *48*, 256–265. [CrossRef]
14. Park, J.S.; Choi, D.; Kim, Y. Potential Distribution of Goldenrod (*Solidago altissima* L.) during Climate Change in South Korea. *Sustainability* **2020**, *12*, 6710. [CrossRef]
15. Kotowska, D.; Pärt, T.; Żmihorski, M. Evaluating google street view for tracking invasive alien plants along roads. *Ecol. Indi.* **2021**, *121*, 107020. [CrossRef]
16. Shimoda, M.; Nakamoto, M. Vegetation and threatened plant dynamics of wet abandoned rice fields in Nakaikemi, Fukui, Prefecture, Japan. *Jpn. J. Ecol.* **2003**, *53*, 197–217.
17. Shimoda, S.; Wagai, R. Ecosystem dynamics after abandonment of rice paddy fields; Does aaien plant invasion enhance carbon storage? *Ecosystems* **2020**, *23*, 617–629. [CrossRef]
18. Newell, S.J.; Tramer, E.J. Reproductive strategies in herbaceous plant communities during succession. *Ecology* **1978**, *59*, 228–234. [CrossRef]
19. Bakelaar, R.G.; Odum, E.P. Community and population level response to fertilization in an old-field ecosystem. *Ecology* **1978**, *59*, 660–665. [CrossRef]
20. Maddox, G.D.; Cook, R.E.; Wimberger, P.H.; Gardescu, S. Clone structure in four *Solidago altissima* (Asteraceae) population: Rhizome connections within genotypes. *Am. J. Bot.* **1989**, *76*, 318–326. [CrossRef]
21. Dong, L.J.; Yu, H.W.; He, W.M. What determines positive, neutral and negative impacts of *Solidago canadensis* invasion on native plant species richness? *Sci. Rep.* **2015**, *5*, 16804. [CrossRef]
22. Fenesi, A.; Vágási, C.I.; Beldean, M.; Földesi, R.; Kolcsár, L.P.; Shapiro, J.T.; Török, E.; Kovács-Hostyánszki, A. *Solidago canadensis* impacts on native plant and pollinator communities in different-aged old fields. *Basic Appl. Ecol.* **2015**, *16*, 335–346. [CrossRef]
23. Zagurskaya, Y.V. Study issues of invasive species of the genus *Solidago*. *Ecosyst. Transform.* **2022**, *5*, 42–54. [CrossRef]
24. EPPO. PQR database. Paris, France: European and Mediterranean Plant Protection Organization. Available online: https://gd.eppo.int/taxon/SOOCA (accessed on 12 September 2022).
25. Weber, E.; Schmid, B. Latitudinal population differentiation in two species of *Solidago* (Asteraceae) introduced into Europe. *Am. J. Bot.* **1998**, *85*, 1110–1121. [CrossRef] [PubMed]
26. iNatuarist, Identifying Solidago altissima & Solidago canadensis. Available online: https://www.inaturalist.org/posts/19288-identifying-solidago-altissima-solidagocanadensis (accessed on 12 September 2022).
27. Szymura, M.; Szymura, T.H. Interactions between alien goldenrods (*Solidago* and *Euthamia* species) and comparison with native species in Central Europe. *Flora* **2016**, *218*, 51–61. [CrossRef]
28. Semple, J.C.; Rahman, H.; Bzovsky, S.; Sorour, M.K.; Kornobis, K.; Laphitz, R.L.; Tong, L. A multivariate morphometric study of the *Solidago altissima* complex and *S. canadensis* (Asteraceae: Astereae). *Phytoneuron* **2015**, *10*, 1–31.
29. Verloove, F.; Zonneveld, B.J.M.; Semple, J.C.S. First evidence for the presence of invasive *Solidago altissima* (Asteraceae) in Europe. *Willdenowia* **2017**, *47*, 69–75. [CrossRef]
30. Melville, M.R.; Morton, J.K. A biosystematics study of the *Solidago canadensis* (Compositae) complex. I. The Ontario populations. *Can. J. Bot.* **1982**, *60*, 976–997. [CrossRef]
31. Szymura, M.; Szymura, T.H.; Kreitschitz, A. Morphological and cytological diversity of goldenrods (*Solidago* L. and *Euthamia* Nutt.) from south-western Poland. *Biodiv. Res. Conserv.* **2015**, *38*, 41–49. [CrossRef]
32. Pisula, N.; Meiners, S.J. Allelopatyhic effcts of goldenrod species on turnover in successional communities. *Am. Midl. Nat.* **2010**, *163*, 161–172. [CrossRef]
33. Lawson, S.K.; Sharp, L.G.; Powers, C.N.; McFeeters, R.L.; Satyal, P.; Setzer, W.N. Volatile Compositions and Antifungal Activities of Native American Medicinal Plants: Focus on the Asteraceae. *Plants* **2020**, *9*, 126. [CrossRef]
34. Callaway, R.M.; Aschehoug, E.T. Invasive plants versus their new and old neighbors: A mechanism for exotic invasion. *Science* **2000**, *290*, 521–523. [CrossRef]
35. Callaway, R.M.; Ridenour, W.M. Novel weapons: Invasive success and the evolution of increased competitive ability. *Front. Ecol. Environ.* **2004**, *2*, 419–426. [CrossRef]
36. Cappuccino, N.; Arnason, J.T. Novel chemistry of invasive exotic plants. *Biol. Lett.* **2006**, *2*, 189–193. [CrossRef] [PubMed]
37. Chengxu, W.; Mingxing, Z.; Xuhui, C.; Bo, Q. Review on allelopathy of exotic invasive plants. *Procedia. Engin.* **2011**, *18*, 240–246. [CrossRef]

38. Meiners, S.J.; Kong, C.H.; Ladwig, L.M.; Pisula, N.L.; Lang, K.A. Developing an ecological context for allelopathy. *Plant. Ecol.* **2012**, *213*, 1861–1867. [CrossRef]
39. Kato-Noguchi, H. Involvement of allelopathy in the invasive potential of *Tithonia diversifolia*. *Plants* **2020**, *9*, 766. [CrossRef]
40. Kato-Noguchi, H.; Kurniadie, D. Allelopathy of *Lantana camara* as an invasive plant. *Plants* **2021**, *10*, 1028. [CrossRef]
41. Kato-Noguchi, H. Allelopathy of knotweeds as invasive plants. *Plants* **2022**, *11*, 3. [CrossRef]
42. Kato-Noguchi, H.; Kurniadie, D. Allelopathy and allelochemicals of *Leucaena leucocephala* as an invasive plant species. *Plants* **2022**, *11*, 1672. [CrossRef]
43. Kato-Noguchi, H. Allelopathy and Allelochemicals of *Imperata cylindrica* as an invasive plant species. *Plants* **2022**, *11*, 2551. [CrossRef]
44. Fisher, R.F.; Woods, R.A.; Glavicic, M.R. Allelopathic effects of golderrod and aster on young sugar maple. *Can. J. Forest Res.* **1978**, *8*, 1–9. [CrossRef]
45. Kobayashi, A.; Morimoto, S.; Shibata, Y.; Yamashita, K.; Numata, M. C_{10}-Polyacetylenes as allelopathic substances in dominants in early stages of secondary succession. *J. Chem. Ecol.* **1980**, *6*, 119–131. [CrossRef]
46. Rice, E.L. *Allelopathy*, 2nd ed.; Academic Press: Orlando, FL, USA, 1984; pp. 1–422.
47. Bais, H.P.; Weir, T.L.; Perry, L.G.; Gilroy, S.; Vivanco, J.M. The role of root exudates in rhizosphere interactions with plants and other organisms. *Annu. Rev. Plant Biol.* **2006**, *57*, 233–266. [CrossRef] [PubMed]
48. Bonanomi, G.; Sicurezza, M.G.; Caporaso, S.; Esposito, A.; Mazzoleni, S. Phytotoxicity dynamics of decaying plant materials. *New Phytol.* **2006**, *169*, 571–578. [CrossRef] [PubMed]
49. Belz, R.G. Allelopathy in crop/weed interactions—An update. *Pest. Manag. Sci.* **2007**, *63*, 308–326. [CrossRef] [PubMed]
50. Bielecka, A.; Borkowska, L.; Królak, E. Environmental changes caused by the clonal invasive plant *Solidago canadensis*. *Ann. Bot. Fennici.* **2019**, *57*, 33–48. [CrossRef]
51. Wang, C.; Cheng, H.; Wanga, S.; Weia, M.; Du, D. Plant community and the influence of plant taxonomic diversity on community stability and invasibility: A case study based on *Solidago canadensis* L. *Sci. Total Environ.* **2021**, *768*, 144518. [CrossRef]
52. Yang, B.; Li, J. Phytotoxicity of toot exudates of invasive *Solidago canadensis* on co-occurring native and invasive plant species. *Pak J. Bot.* **2022**, *54*, 1019–1024. [CrossRef]
53. Abhilasha, D.; Quintana, N.; Vivanco, J.; Joshi, J. Do allelopathic compounds in invasive *Solidago canadensis* s.l. restrain the native European flora? *J. Ecol.* **2008**, *96*, 993–1001. [CrossRef]
54. Weißhuhn, K.; Prati, D. ; Activated carbon may have undesired side effects for testing allelopathy in invasive plants. *Basic Appl. Ecol.* **2009**, *10*, 500–507. [CrossRef]
55. Yang, X.; Cheng, J.; Yao, B.; Lu, H.; Zhang, Y.; Xu, J.; Song, X.; Qiang, S. Polyploidy-promoted phenolic metabolism confers the increased competitive ability of *Solidago canadensis*. *Oikos* **2021**, *130*, 1014–1025. [CrossRef]
56. Zhang, D.; Ye, Y.; Li, L.; Domg, L. Allelopathic pathways, isolation and identification of an allelopathic substance from *Solidago canadensis* L. *Allelopath. J.* **2014**, *33*, 201–212.
57. Butcko, V.M.; Jensen, R.J. Evidence of tissue-specific allelopathic activity in *Euthamia graminifolia* and *Solidago canadensis* (Asteraceae). *Am. Midl. Nat.* **2002**, *148*, 253–262. [CrossRef]
58. Li, S.L.; Li, Z.H.; Wang, Y.F.; Xiao, R.; Pan, C.D.; Wang, Q. Preliminary study for the allelopathic effect of water extracts from *Solidago canadensis* leaves. *Adv. Mater. Res.* **2013**, *699*, 340–348. [CrossRef]
59. Możdżeń, K.; Barabasz-Krasny, B.; Zandi, P.; Kliszcz, A.; Puła, J. Effect of aqueous extracts from *Solidago canadensis* L. Leaves on germination and early growth stages of three cultivars of *Raphanus Sativus* L. var. *Radicula* Pers. *Plants* **2020**, *9*, 1549. [CrossRef] [PubMed]
60. Zandi, P.; Barabasz-Krasny, B.; Stachurska-Swakoń, A.; Puła, J.; Możdżeń, K. Allelopathic effect of invasive Canadian goldenrod (*Solidago canadensis* L.) on early growth of red clover (*Trifolium pratense* L.). *Not. Bot. Horti. Agrobo.* **2020**, *48*, 2060–2071. [CrossRef]
61. Yu, Y.; Cheng, H.; Xu, Z.; Zhong, S.; Wang, C.; Guo, E. Invasion intensity modulates the allelopathic impact of *Solidago canadensis* L. leaves and roots against *Lactuca sativa* L. during germination and early seedling stage. *Int. J. Environ. Res.* **2022**, *16*, 48. [CrossRef]
62. Sun, J.F.; Liang, Q.J.; Wu, N.; Javed, Q.; Huang, P.; Du, D.L. Allelopathic effects of aqueous extracts from different plant parts of Canada goldenrod (*Solidago canadensis* L.) on seed germination and seedling growth of Korean lawngrass (*Zoysia japonica* Steud.). *Appl. Ecol. Environ. Res.* **2022**, *20*, 1009–1022. [CrossRef]
63. Baležentien, L. Secondary metabolite accumulation and phytotoxicity of invasive species *Solidago canadensis* L. during the growth period. *Allelopath. J.* **2015**, *35*, 217–226.
64. Yang, R.Y.; Mei, L.X.; Tang, J.J.; Chen, X. Allelopathic effects of invasive *Solidago canadensis* L. on germination and growth of native Chinese plant species. *Allelopath. J.* **2007**, *19*, 241–248.
65. Yuan, Y.; Wang, B.; Zhang, S.; Tang, J.; Tu, C.; Hu, S.; Yong, J.W.H.; Chen, X. Enhanced allelopathy and competitive ability of invasive plant *Solidago canadensis* in its introduced range. *J. Plant Ecol.* **2013**, *6*, 253–263. [CrossRef]
66. Sun, B.Y.; Tan, J.Z.; Wan, Z.G.; Gu, F.G.; Zhu, M.D. Allelopathic effects of extracts from *Solidago canadensis* L. against seed germination and seedling growth of some plants. *J. Environ. Sci.* **2006**, *18*, 304–309.
67. Zhang, Q.; Yao, L.J.; Yang, R.Y.; Yang, X.Y.; Tang, J.J.; Chin, X. Potential allelopathic effects of an invasive species *Solidago canadensis* on the mycorrhizae of native plant species. *Allelopath. J.* **2007**, *20*, 71–78.
68. Sun, Z.K.; He, W.M. Evidence for enhanced mutualism hypothesis: *Solidago canadensis* plants from regular soils perform better. *PLoS ONE* **2010**, *5*, e15418. [CrossRef]

69. Betekhtina, A.A.; Mukhacheva, T.A.; Kovaleva, S.Y.; Gusevb, A.P.; Veselkin, D.V. Abundance and diversity of arbuscular mycorrhizal fungi in invasive *Solidago canadensis* and indigenous *S. virgaurea*. *Russ. J. Ecol.* **2016**, *47*, 575–579. [CrossRef]
70. Dong, L.J.; Ma, L.N.; He, W.M. Arbuscular mycorrhizal fungi help explain invasion success of *Solidago canadensis*. *Appl. Soil. Ecol.* **2021**, *167*, 103763. [CrossRef]
71. Řezáčová, V.; Řezáč, M.; Gryndler, M.; Hršelová, H.; Gryndlerová, H.; Michalová, T. Plant invasion alters community structure and decreases diversity of arbuscular mycorrhizal fungal communities. *Appl. Soil. Ecol.* **2021**, *167*, 104039. [CrossRef]
72. Yang, R.; Zhou, G.; Zan, S.; Guo, F.; Su, N.; Li, J. Arbuscular mycorrhizal fungi facilitate the invasion of *Solidago canadensis* L. in southeastern China. *Acta Oecol.* **2014**, *61*, 74–77. [CrossRef]
73. Zhang, S.; Jin, Y.; Tang, J.; Chen, X. The invasive plant *Solidago canadensis* L. Suppresses local soil pathogens through allelopathy. *Appl. Soil. Ecol.* **2009**, *41*, 215–222. [CrossRef]
74. Zhang, S.; Zhu, W.; Wang, B.; Tang, J.; Chen, X. Secondary metabolites from the invasive *Solidago canadensis* L. Accumulation in soil and contribution to inhibition of soil pathogen *Pythium ultimum*. *Appl. Soil. Ecol.* **2011**, *48*, 280–286. [CrossRef]
75. Jun, L.; Yonghao, Y.; Hongwu, H.; Liyao, D. Kaempferol-3-*O*-β-D-glucoside, a potential allelochemical isolated from *Solidago canadensis*. *Allelopath. J.* **2011**, *28*, 259–266.
76. Radusiene, J.; Marska, M.; Ivanauskas, L.; Jakstas, J.; Karpaviciene, B. Assessment of phenolic compound accumulation in two widespread goldenrods. *Ind. Crops Prod.* **2015**, *63*, 158–166. [CrossRef]
77. Likhanov, A.; Oliinyk, M.; Pashkevych, N.; Churilov, A.; Kozyr, M. The role of flavonoids in invasion strategy of *Solidago canadensis* L. *Plants* **2021**, *10*, 1748. [CrossRef]
78. Shelepova, O.; Vinogradova, Y.; Vergun, O.; Grygorieva, O.; Brindza, J. Assessment of flavonoids and phenolic compound accumulation in invasive *Solidago canadensis* L. in Slovakia. *Slovak J. Food Sci.* **2020**, *14*, 587–594.
79. Sharifi-Rad, J.; Sureda, A.; Tenore, G.C.; Daglia, M.; Sharifi-Rad, M.; Valussi, M.; Tundis, R.; Sharifi-Rad, M.; Loizzo, M.R.; Ademiluyi, A.O.; et al. Biological activities of essential oils: From plant chemoecology to traditional healing systems. *Molecules* **2017**, *22*, 70. [CrossRef] [PubMed]
80. Jassbi, A.R.; Zare, S.; Firuzi, O.; Xiao, J. Bioactive phytochemicals from shoots and roots of *Salvia* species. *Phytochem Rev.* **2016**, *15*, 829–867. [CrossRef]
81. Macías, F.A.; Mejías, F.J.R.; Molinillo, J.M.G. Recent advances in allelopathy for weed control: From knowledge to applications. *Pest Manag. Sci.* **2019**, *75*, 2413–2436. [CrossRef] [PubMed]
82. Baranová, B.; Troščáková-Kerpčárová, E.; Grul'ová, D. Survey of the *Solidago canadensis* L. morphological traits and essential oil production: Aboveground biomass growth and abundance of the invasive goldenrod appears to be reciprocally enhanced within the invaded dtands. *Plants* **2022**, *11*, 535. [CrossRef]
83. Radušienė, J.; Karpavičienė, B.; Marksa, M.; Ivanauskas, L.; Raudonė, L. Distribution patterns of essential oil terpenes in native and invasive *Solidago* species and their comparative assessment. *Plants* **2022**, *11*, 1159. [CrossRef]
84. Grul'ová, D.; Baranová, B.; Ivanova, V.; De Martino, L.; Mancini, E.; De Feo, V. Composition and bio activity of essential oils of *Solidago* spp. and their Impact on radish and garden cress. *Allelopath. J.* **2016**, *39*, 129–141.
85. Shelepova, O.; Vinogradova, Y.; Zaitchik, B.; Ruzhitsky, A.; Grygorieva, O.; Brindza, J. Constituents of the essential oil in *Solidago canadensis* L. from Eurasia. *Slovak J. Food Sci.* **2018**, *12*, 20–25. [CrossRef]
86. Marinas, I.C.; Oprea, E.; Buleandra, M.; Bleots, C.; Badea, I.R.; Anastasiu, P.; Lazar, V.; Gardus, I.D.; Chifiric, M.C. Chemical, antimicrobial, antioxidant and anti-proliferative features of the essential oil extracted from the invasive plant *Solidago canadensis* L. *Rev. Chim.* **2020**, *71*, 255–264. [CrossRef]
87. Nishida, N.; Tamotsu, S.; Nagata, N.; Saito, C.; Sakai, A. Allelopathic effects of volatile monoterpenoids produced by *Salvia leucophylla*: Inhibition of cell proliferation and DNA synthesis in the root apical meristem of *Brassica campestris* seedlings. *J. Chem. Ecol.* **2005**, *31*, 1187–1203. [CrossRef]
88. Chowhan, N.; Singh, H.P.; Batish, D.R.; Kohli, R.H. Phytotoxic effects of β-pinene on early growth and associated biochemical changes in rice. *Acta Physiol. Plant.* **2011**, *33*, 2369–2376. [CrossRef]
89. Chowhan, N.; Singh, H.P.; Batish, D.R.; Kaur, S.; Ahuja, N.; Kohli, R.K. β-Pinene inhibited germination and early growth involves membrane peroxidation. *Protoplasma* **2013**, *250*, 691–700. [CrossRef] [PubMed]
90. Yun, K.W.; Kil, B.S.; Han, D.M. Phytotoxic and antimicrobial activity of volatile constituents of *Artemisia princeps* var. *orientalis*. *J. Chem. Ecol.* **1993**, *19*, 2757–2766. [CrossRef] [PubMed]
91. Bais, H.P.; Vepachedu, R.; Gilroy, S.; Callaway, R.M.; Vivanco, J.M. Allelopathy and exotic plant invasion: From molecules and genes to species interaction. *Science* **2003**, *301*, 1377–1380. [CrossRef]
92. Smith, S.E.; Read, D.J. *Mycorrhizal Symbiosis*, 3rd ed.; Academic Press: London, UK, 2008; pp. 1–815.
93. Diagne, N.; Ngom, M.; Djighaly, P.I.; Fall, D.; Hocher, V.; Svistoonoff, S. Roles of Arbuscular Mycorrhizal Fungi on Plant Growth and Performance: Importance in Biotic and Abiotic Stressed Regulation. *Diversity* **2020**, *12*, 370. [CrossRef]
94. Tang, H.; Hassan, M.U.; Feng, L.; Nawaz, M.; Shah, A.N.; Qari, S.H.; Liu, Y.; Miao, J. The critical role of arbuscular mycorrhizal fungi to improve drought tolerance and nitrogen use efficiency in crops. *Front. Plant Sci.* **2022**, *13*, 919166. [CrossRef] [PubMed]
95. Nishidono, Y.; Tanaka, K. Comprehensive characterization of polyacetylenes and diterpenes from the underground parts of *Solidago altissima* L. and their contribution to the overall allelopathic activity. *Phytochemistry* **2022**, *193*, 112986. [CrossRef]
96. Ichihara, K.; Kawai, T.; Kaji, M.; Noda, M. A mew polyacetylene from *Solidago altissima* L. *Agric. Biol. Chem.* **1976**, *40*, 353–358. [CrossRef]

97. Kawazu, K.; Ariwa, M.; Kii, Y. An ovicidal substance, *cis*-dehydromatricaria ester from *Solidago altissima*. *Agric. Biol. Chem.* **1977**, *41*, 223–224.
98. Johnson, R.H.; Halitschke, R.; Kessler, A. Simultaneous analysis of tissue- and genotype-specific variation in *Solidago altissima* (Asteraceae) rhizome terpenoids, and the polyacetylene dehydromatricaria ester. *Chemoecology* **2010**, *20*, 255–264. [CrossRef]
99. Nishino, C.; Manabe, S.; Kazui, M.; Matsuzaki, T. Piscicidal *cis*-clerodane diterpenes from *Solidago altissima*. L.: Absolute configurations of 5α, 10α-*cis*-clerodanes. *Tetrahedron Lett.* **1984**, *25*, 2809–2812. [CrossRef]
100. Bohlmann, F.; Singh, P.; Singh, R.K.; Joshi, K.C.; Jakupovic, J. A diterpene with a new carbon skeleton from *Solidago altissima*. *Phytochemistry* **1985**, *24*, 1114–1115. [CrossRef]
101. Tori, M.; Katto, A.; Sono, M. Nine new clerodane diterpenoids from rhizomes of *Solidago altissima*. *Phytochemistry* **1999**, *52*, 487–493. [CrossRef]
102. Nishidono, Y.; Tanaka, K. New clerodane diterpenoids from *Solidago altissima* and stereochemical elucidation via ^{13}C NMR chemical shift analysis. *Tetrahedron* **2022**, *110*, 132691. [CrossRef]
103. Okano, A.; Nomura, Y.; Tezuka, T. Identification of bauerenol in *Solidago altissima*. *J. Nat. Prod.* **1983**, *46*, 750–751. [CrossRef]
104. Sawabe, A.; Minemoto, K.; Minematsu, T.; Ouchi, S.; Okamoto, T.; Morita, M.; Ouchi, S.; Okamoto, T. Characterization of the *Z* and *E* isomers of dehydromatricaria lactones. *J. Jpn. Oil Chem. Soc.* **1997**, *46*, 413–417. [CrossRef]
105. Jin, H.; Tanaka, T.; Kouno, I.; Ishimaru, K. A new kaempferol trioside from *Solidago altissima* L. *J. Nat. Med.* **2007**, *61*, 351–354. [CrossRef]
106. Wu, B.; Takahashi, T.; Kashiwagi, T.; Tebayashi, S.; Kim, C.S. New flavonoid glycosides from the leaves of *Solidago altissima*. *Chem. Pharm. Bull.* **2007**, *55*, 815–816. [CrossRef]
107. Jin, H.; Ogino, K.; Fujioka, T.; Yoshida, M.; Ishimaru, K. A new acylphloroglucinol glycoside from *Solidago altissima* L. *J. Nat. Med.* **2008**, *62*, 199–201. [CrossRef]
108. Sawabe, A.; Minemoto, K.; Ouchi, S.; Okamoto, T. Effects of acetylenes and terpenoids from *Solidago altissima* L. on seed germination. *J. Jpn. Oil Chem. Soc.* **1999**, *48*, 139–142. [CrossRef]
109. Salah, M.A.; Bedir, E.; Toyang, N.J.; Khan, I.A.; Harries, M.D.; Wedge, D.E. Antifungal clerodane diterpenes from *Macaranga monandra* (L) Muell. et Arg. (Euphorbiaceae). *J. Agric. Food Chem.* **2003**, *51*, 7607–7610. [CrossRef] [PubMed]
110. Morimoto, M. Chemical defense against insects in *Heterotheca subaxillaris* and three Orobanchaceae species using exudates from trichomes. *Pest Manag. Sci.* **2019**, *75*, 2474–2481. [CrossRef] [PubMed]
111. Ichihara, K.; Kawai, T.; Noda, M. Polyacetylenes of *Solidago altissima* L. *Agric. Biol. Chem.* **1978**, *42*, 427–431. [CrossRef]
112. Minto, R.E.; Blacklock, B.J. Biosynthesis and function of polyacetylenes and allied natural products. *Prog. Lipid Res.* **2008**, *47*, 233–306. [PubMed]
113. Uesugi, A.; Kessler, K. Herbivore exclusion drives the evolution of plant competitiveness via increased allelopathy. *New Phytol.* **2013**, *198*, 916–924. [CrossRef]
114. Carson, W.; Root, R. Herbivory and plant species coexistence: Community regulation by an outbreaking phytophagous insect. *Ecol. Monogr.* **2000**, *70*, 73–99. [CrossRef]
115. Yamada, T. Miscanthus. In *Industrial Crops. Handbook of Plant Breeding*; Cruz, V.M.V., Dierig, D.A., Eds.; Springer: New York, NY, USA, 2015; Volume 9, pp. 43–66.
116. Uesugi, A.; Johnson, R.; Kessler, K. Context-dependent induction of allelopathy in plants under competition. *Oikos* **2019**, *128*, 1492–1502. [CrossRef]
117. Ito, I.; Kobayashi, K.; Yoneyama, T. Fate of dehydromatricaria ester added to soil and its implications for the allelopathic effect of *Solidago altissima* L. *Ann. Bot.* **1998**, *82*, 625–630. [CrossRef]
118. Kaur, H.; Kaur, R.; Kaur, S.; Baldwin, I.T.; Inderjit. Taking ecological function seriously: Soil microbial communities can obviate allelopathic effects of released metabolites. *PLoS ONE* **2004**, *4*, e4700. [CrossRef]
119. Kobayashi, K.; Koyama, H.; Shim, I. Relationship between behaviour of dehydromatricaria ester in soil and the allelopathic activity of *Solidago altissima* L. in the laboratory. *Plant Soil* **2004**, *259*, 97–102. [CrossRef]
120. Tsao, R.; Eto, M. Light-activated plant growth inhibitory activity of *cis*-dehydromatricaria ester, rose bengal and fluoren-9-one on lettuce (*Lactuca sativa* L.). *Chemosphere* **1996**, *32*, 1307–1317. [CrossRef]
121. Sakai, H.; Yoneda, K. Possible dual roles of an allelopathic compound, *cis*-dehydromatricaria ester. *J. Chem. Ecol.* **1981**, *8*, 185–193. [CrossRef] [PubMed]
122. Mack, R.M. Predicting the identity and fate of plant invaders: Emergent and emerging approaches. *Biol. Conserv.* **1996**, *78*, 107–121. [CrossRef]
123. Keane, R.M.; Crawley, M.J. Exotic plant invsions and the enemy release hypothesis. *Trends Ecol. Evol.* **2002**, *17*, 164–169. [CrossRef]
124. Ismail, M.; Kowsar, A.; Javed, S.; Choudhary, M.I.; Khan, S.W.; Abbas, Q.; Tang, Y.; Wang, W. The antibacterial, insecticidal and nematocidal activities and toxicity studies of *Tanacetum falconeri* Hook.f. Turk. *J. Pharm. Sci.* **2021**, *18*, 744–751.

plants

MDPI

Review

Abutilon theophrasti's Resilience against Allelochemical-Based Weed Management in Sustainable Agriculture – Due to Collection of Highly Advantageous Microorganisms?

Vincenzo Tabaglio [1], Andrea Fiorini [1], Tracy M. Sterling [2] and Margot Schulz [3,*]

[1] Department of Sustainable Crop Production DI.PRO.VE.S., Section Agronomy and Plant Biotechnologies, Università Cattolica del Sacro Cuore, Via Emilia Parmense 84, 29122 Piacenza, Italy

[2] Department of Land Resources & Environmental Sciences, Montana State University, Bozeman, MT 59717, USA

[3] IMBIO Institute of Molecular Physiology and Biotechnology of Plants, University of Bonn, Karlrobert-Kreiten Str. 13, 53115 Bonn, Germany

* Correspondence: ulp509@uni-bonn.de

Abstract: *Abutilon theophrasti* Medik. (velvetleaf) is a problematic annual weed in field crops which has invaded many temperate parts of the world. Since the loss of crop yields can be extensive, approaches to manage the weed include not only conventional methods, but also biological methods, for instance by microorganisms releasing phytotoxins and plant-derived allelochemicals. Additionally, benzoxazinoid-rich rye mulches effective in managing common weeds like *Amaranthus retroflexus* L. have been tested for this purpose. However, recent methods for biological control are still unreliable in terms of intensity and duration. Rye mulches were also ineffective in managing velvetleaf. In this review, we present the attempts to reduce velvetleaf infestation by biological methods and discuss possible reasons for the failure. The resilience of velvetleaf may be due to the extraordinary capacity of the plant to collect, for its own survival, the most suitable microorganisms from a given farming site, genetic and epigenetic adaptations, and a high stress memory. Such properties may have developed together with other advantageous abilities during selection by humans when the plant was used as a crop. Rewilding could be responsible for improving the microbiomes of *A. theophrasti*.

Keywords: *Abutilon theophrasti*; biological control; allelopathy; fungi; bacteria; secondary metabolites; benzoxazinoids; rye mulch

Citation: Tabaglio, V.; Fiorini, A.; Sterling, T.M.; Schulz, M. *Abutilon theophrasti*'s Resilience against Allelochemical-Based Weed Management in Sustainable Agriculture – Due to Collection of Highly Advantageous Microorganisms? *Plants* **2023**, *12*, 700. https://doi.org/10.3390/plants12040700

Academic Editor: Mónica Fernández-Aparicio

Received: 20 December 2022
Revised: 30 January 2023
Accepted: 1 February 2023
Published: 4 February 2023

1. Introduction

While its center of origin is uncertain, *Abutilon theophrasti* Medik. (velvetleaf) is thought to be native to China and India [1]. It was introduced across the Northern hemisphere as a fiber crop or protein food source and may have spread as a seed contaminant [1–4]. It is now a troublesome weed in major agricultural crops, such as in maize, soybeans, and cotton, and also grows in crops such as citrus, peach, and asparagus [5]. It has expanded its range to Africa, Australia, and New Zealand [6–9], and was recently identified as naturalized in Libya [10]. The species has regained interest as a fiber crop [7,11], taking up one of its old uses in Asia and for a short time in the USA, aside from its use as a medicinal plant [1,12]. In China and Kashmir, the seed is a valuable food because of its high protein content [2,8,9,13]. With the first archeological find of 7000-year-old velvetleaf seeds in a storage jar in southern Hungary, there is now evidence that velvetleaf was harvested, prepared, and stored by the Tisza culture, further supporting velvetleaf's use as food, fiber, and medicine [14].

However, velvetleaf remains a very competitive weed. Depending on water availability, velvetleaf density, and herbicide efficacy, maize yield loss can range up to 80% [5,15,16], up to 50% in cotton [17], and up to 70% in soybean [15,18,19], causing large economic losses [2,6,20]. Competitive maize hybrids have been evaluated as a strategy to manage velvetleaf [21–23]; nevertheless, velvetleaf must be controlled within maize's V3 to V8 growth stages to avoid negative effects on crop yield. Therefore, mixtures of synthetic herbicides are instead commonly recommended as pre- and post-treatments to manage velvetleaf [10,20,22]. However, herbicides can be expensive, and may persist in the environment or create selection pressures which result in the evolution of resistant biotypes. Four velvetleaf biotypes resistant to PSII herbicides have been independently reported to date [24].

Several *A. theophrasti* characteristics contribute to its competitive abilities and include its relatively large, hard-coated seeds, fast germination and emergence of the seedlings throughout the growing season, vigorous growth and large leaves, and high seed production, with long viability leading to persistent soil seed banks [2,25]. One plant can produce 8,000 seeds, with a significant percentage of them still viable for more than 50 years [2,13,25]. Moreover, velvetleaf is better colonized by arbuscular mycorrhizal fungi than other weeds, allowing velvetleaf greater access to nitrogen [26,27]. When in competition with maize, the genes of hexaploid velvetleaf involved in carbon utilization, photosynthesis, red light signaling, cell division, secondary metabolism, and phosphate availability are preferentially and more highly expressed [28]. In turn, numerous genes are downregulated in maize when grown near velvetleaf, such as those involved in photosynthesis, nutrient utilization, or signal transduction [29]. In addition, velvetleaf releases allelochemicals which are inhibitory to alfalfa, maize, and other crops (see below). In the 1980s, seed-associated endophytic microorganisms were suspected of enhancing velvetleaf´s fitness [30].

Therefore, in this article emphasizing sustainable agriculture, we shortly review results that underpin the hypothesis that *Abutilon* may organize its variable seed-coat and root-associated microbiomes with high flexibility. Additionally, this dependency on microbial diversity in site-specific soils allows for velvetleaf survival across diverse environments, for instance in those rich in benzoxazinoids and related allelochemicals.

2. Biological Control of *Abutilon theophrasti* by Microorganisms

Approaches to biological control of *A. theophrasti* include the use of microorganisms, plant-derived allelochemicals, mycoherbicides, or other microbial secondary metabolites. Fungi like *Verticillium dahliae, Alternaria abutilonis, Cercospora abutilonis, Cladosporium herbarum, Colletotrichum malvarum, Fusarium lateritium*, and others have been studied for this purpose [2,30–36], (Table 1). However, many of these fungi are not host specific and also infest crops (for instance, *Cercospora* leaf spot disease of sugar beet). In other cases, any biological activity damaging the weed is dependent on environmental conditions, and therefore, an unreliable tool for weed management. For instance, *F. lateritium*, a pathogenic agent of velvetleaf, reduced its growth in greenhouse experiments and in the field, but velvetleaf suppression varied from year to year, ranging from 46 to only 27% [35].

Similarly, an *Abutilon*-specific pathovar of *Colletotrichum coccodes* was genetically engineered to overexpress the oxaloacetate acetylhydrolase gene, creating a hypervirulent strain which was active at lower humidity and lethal to velvetleaf, but only through its three-leaf stage just after emergence [37]. Oxalate inhibits callose synthase by chelating the co-factor calcium, leading to less callose which weakens plant defenses [37]. Infection by other phytopathogens was also facilitated by higher oxalate levels and by exogenously applied oxalate [38–42]. Moreover, high oxalate amounts change the physicochemical properties of the soil, which may influence adjacent crops [43].

Another example in which pathogen efficacy for biological control of weeds was enhanced includes work which introduced the necrosis-inducing extracellular protein Nep1 from *Fusarium oxysporum*. This protein induces necrosis via ethylene synthesis and has been described as a successful biological control agent against weeds, including *A. theophrasti* [44,45]. However, genetically upgraded phytopathogens are generally not allowed anywhere in organic agriculture systems, nor in conventional agriculture in most EU countries.

Microbial secondary metabolites with phytotoxic properties against *Abutilon* produced by *Streptomyces* species [46] are able to inhibit velvetleaf germination and seedling growth by 100%. Even so, the velvetleaf seeds used in this study were surface-sterilized, disallowing any degradation or inactivation of bioherbicides by seed-surface-colonizing microorganisms, most likely reduced, if not eliminated under sterile conditions. Thus, the inhibition of velvetleaf growth found in laboratory experiments may not occur under field conditions. In other studies, hydrogen cyanide-producing *Pseudomonas putida* ATH-1RI/9 and *Acidovorax delafieldii* ATH2-2RS/1 suppress the growth of velvetleaf [47,48], though it is not known whether the bacterial-produced hydrogen cyanide was in concentrations lethal to soil organisms. It is important to note that many microorganisms can degrade hydrogen cyanide [49,50], which would remove any phytotoxic effects in the soil environment.

Overall, biological control of velvetleaf is not commonly used. Accordingly, velvetleaf is not mentioned as a target for biological control in relevant recent reviews [51,52]. As implied by Chee-Sanford et al. [53], a detailed knowledge of the entire ecosystem is a prerequisite when microorganisms will be used for the weed management. In general, it can be said that these biological control methods are still unreliable in terms of intensity and consistency, especially against weeds with gradual emergence in the field, like *A. theophrasti*.

Table 1. Microorganisms with biocidal effects on velvetleaf.

Microorganisms	Mode of Action	Note	Reference
Fungi			
Alternaria abutilonis	leaf spot		[2]
Cercospora abutilonis *Cerospora* spec. complex?	Leaf spot	species needs additional identification	[2]
Cercospora althaeina Sac.	Leaf rot		[2]
Cladosporium herbarum	secondary leaf spot		[2]
Colletotrichum coccodes	foliar pathogen, plant death; suppression of plant defenses	enhancement of virulence by mannose and oxalic acid, NEP1 for hypervirulent transgenic strain	[2,39,40,45]
Colletotrichum malvarum	leaf spot		[2]
Fusarium lateritium	growth suppression		[2,35]
Fusarium oxysporum	24-kDa protein Nep1 induces necrosis		[44]
Macrophomina phaseoli	stem rot		[2]
Phyllosticta althaeina Sacc.	Leaf rot		[2]
Phymatotrichum omnivorum	root rot		[2]
Puccinia heterospora	rust	possible mycoherbicide	[2]
Verticillium dahliae, V. nigrescens	wilt	suggested as a biocontrol agent, unspecific	[2]
Bacteria			
Acidovorax delafieldii ATH2-2RS/1	inhibits growth		[48]
Pseudomonas putida ATH-1RI/9	inhibits growth		[48]
Pseudomonas spec isolates Pp001 and Pf239	Reduction of seedlings emergence weaken seedlings		[32]
Streptomyces isolates	inhibits germination, growth		[46]

Natural Compounds—Monoterpenoids, Benzyl-Isothiocyanate, Phenolics, and Sorgoleone in the Management of Abutilon theophrasti

Monoterpenes can be phytotoxic when applied in high dosage [54,55]. Vaughn and Spencer tested several volatile monoterpenes from the essential oils of aromatic plants for velvetleaf control in comparison to crops like maize, wheat, alfalfa, cucumber, and soybean [56]. While seed germination of Italian ryegrass (*Lolium multiflorum* Lam.), large crabgrass, (*Digitaria sanguinalis* (L.) Scop.), and redroot pigweed (*Amaranthus retroflexus* L.) was completely inhibited by many of the essential oils, velvetleaf exhibited a tolerance similar to that found with maize and cucumber. Only fenchone, citronellol, and citral totally suppressed velvetleaf germination, but fenchone was non-selective, also inhibiting maize, wheat, and alfalfa germination. The monoterpenes also inhibited growth of maize seedlings less than cucumber and alfalfa, but greater than soybean. Consistently, most volatile monoterpenes were not specific enough to only inhibit velvetleaf germination or seedling growth.

Wolf et al. described inhibitory effects of benzyl isothiocyanate from *Carica papaya* L. seeds on velvetleaf germination and seedling growth [57]. The *C. papaya* seeds contain not only the glucosinolate glucotropaeolin but also cyanogenic glycosides; in this study, the velvetleaf seeds were surface-sterilized. The inhibition of velvetleaf by benzyl isothiocyanate was also reported by Tollsten and Bergstrom [58]. In a study of Shettle and Balke [59], effects of salicylic acid, p-hydroxybenzoic acid, caffeine, hydroquinone, and umbelliferone on shoot dry-weight accumulation of several weeds and crops were investigated in a greenhouse trial. Except for hydroquinine, velvetleaf growth was only inhibited by the compounds when exposed to high amounts (56.0 kg ha^{-1}). Exudates from roots of competitive soybean cultivars have been studied for velvetleaf suppression as well; just 20 of 280 tested cultivars were able to reduce the dry weight of velvetleaf plants and did so up to 15% [60]. Dayan et al. demonstrated negative effects of sorgoleone, an allelochemical of *Sorghum bicolor* (L.) Moench on velvetleaf seedling emergence and rates of photosynthetic electron transfer in leaf discs [61]; however, the seedlings were again grown from surface-sterilized seeds. Apparently, there are presently no plant-derived compounds or extracts available which have sufficient allelopathic potential to manage velvetleaf in the field, including the benzoxazinoid-derived compounds (see below) and others [62].

Many secondary metabolites can modulate, even impoverish, species abundancies and the biodiversity of the original soil microbiome. Therefore, the use of plant and microbial allelochemicals as potential bioherbicides can lead to unexpected negative effects on valuable soil microorganisms. This consequence can be a problem when other crops relying on microbial species are subsequently cropped. Plant secondary metabolites such as glucosinolate breakdown products, p-coumaric acid, vanillic acid, salicylic acid, the flavonoids catechin and quercetin, and benzoxazinoids, gramine, and others affect microbe growth in laboratory and field studies [63–72]. Recovery of a rich biodiversity of microbial communities, for instance after fallow periods, is of utmost importance, not only for soil quality and fertility [73], but also to catalyze the complete metabolization of important plant and microbial allelochemicals such as catechol and polyphenols, including catechin and dihydroquercetin (see below) by specialized microorganisms and under special soil conditions [74].

3. Resilience of *Abutilon theophrasti* against Benzoxazinoid Allelochemicals from Rye (*Secale cereale* L.) Mulches

Benzoxazinoids are characteristic compounds found in maize and other gramineous species, including rye or wheat. Benzoxazinoids and phenoxazinone spontaneously produced from 2-aminophenol, a microbial degradation of benzoxazolinone, are phytotoxic [75,76]. These allelochemicals have the potential to help manage defined weed species [77–79]. Furthermore, rye cultivation as cover crop is increasingly gaining ground, both in conventional and, even more, in organic farming systems, due to agro-environmental benefits [80]. For instance, benzoxazinoids contained in rye mulch can manage numerous weeds such as *Chenopodium album* (L.), *Amaranthus retroflexus* (L.), *Portulaca oleracea* (L.), and others. At the same time, benzoxazinoid content in rye biomass depends on cultivar and growth stage [81–83]. In this respect, the termination timing of rye cover crop may influence the allelochemical concentrations in rye mulches which decrease with plant age [83], but, if delayed, also allow more biomass production and a concomitant physical control of weed seedlings emergence [84].

Therefore, several approaches have been undertaken to manage *A. theophrasti* using rye mulch. Mulches were prepared from different rye cultivars differing in their benzoxazinoid content, ranging from about 200 to 550 μg g^{-1}, and were used for pot experiments [83,85] (Figure 1). The mulches effectively reduced seedling growth of three weeds (*C. album*, *A. retroflexus*, *P. oleracea*); however, none of the mulches inhibited the growth of velvetleaf seedlings (Figure 1). Interestingly, the number of emerged velvetleaf seedlings in the mulched pots was 35% higher than in the control pots on average, though not statistically different. These results corroborate previous outcomes in other pot and field studies [82,85]. The failure to manage velvetleaf with rye mulches and velvetleaf's subsequent success in maize fields point to efficient strategies of the plant to cope with benzoxazinoid allelochemicals.

Percentage of velvetleaf seedlings emerged 30 days after sowing, in presence of rye mulches or not control

Cultivar	Born	Fasto	For	Mat	Nik	Prim	Pro	Tri	C
% em.	118	106	125	142	120	123	120	106	100
Cultivar	**Duk.**	**Fasto**	**Fore**	**Hell**	**Hym**	**Prim.**	**Pro**	**Sito**	**C**
% em.	70	65	63	69	61	69	63	68	49

Figure 1. (**A**,**B**): Preparation of rye mulches at the Università Cattolica del Sacro Cuore research station in Piacenza, Italy. (**C**): Rye mulch covered and control planter boxes for velvetleaf germination experiments in the greenhouse. (**D**,**E**): Emergence of velvetleaf seedlings (white circles) in pots with rye mulch (**E**), only seedlings overtopping the rye mulch cover are circled) in comparison to the control (**D**). (**F**): Flowering velvetleaf plant in a maize field. Table—Abbreviations: % em: percentage of emerged velvetleaf seedlings; cultivars: For, Forestier; Mat, Matador; Nik, Nikita, Prim, Primizia; Pro, Protector; Tri, Treviso; Duk, Dukato; Fore, Forestal; Hell, Hellvus; Hym, Hymonta; Sito, Sito70; C: control.

3.1. *Microorganisms Associated with Abutilon theophrasti and Microbial Role in Benzoxazinoid Tolerance*

A number of fungal species are known to colonize the seed coat of *A. theophrasti*, including *Cordyceps sinensis*, *Cephaliophora tropica*, *Mandurella mycetomatis*, *Actinomucor elegans*, *Aspergillus flavus*, *Trichoderma viridae*, *Penicillium* spp., *Fusarium* spp., *Alternaria alternata*, and *Verticillium* spp. [30,31,53], while *Cladosporium cladosporioides* and *Epicoccum purpurascens* are seed-borne fungi (Table 2). Bacteria isolated from the seed surface were identified as species belonging to the genera *Bacillus*, *Flavobacterium*, and *Pseudomonas*. More than half of the seeds contained seed-borne bacteria belonging to the genera *Pseudomonas*, *Alcaligenes*, and *Flavobacteria* [30]. There is also evidence for site-specific composition of seed-colonizing microorganisms. For example, velvetleaf seeds collected from three sites (Meleti, Pralboino, and Gambara) in Northern Italy were tested to verify the presence of endophytic fungi [86]. Seeds were surface-sterilized and then incubated at 25 °C for germination. *Alternaria alternata* was identified in 27% of the radicles and *Fusarium* spp. was identified in 36% of the radicles from seeds collected at Meleti site. In 42% of radicles from seeds collected at the Pralboino site, *Fusarium* species were present. Seeds from the Gambara site led to seedlings without these fungal endophytes. The isolation of fungi from 6-day-old radicles suggests migration from the seed to the root system. *Alternaria alternata* and *Fusarium* species, frequently associated with *A. theophrasti* seeds, are often endophytic [87,88].

Table 2. Microorganisms known to be associated with velvetleaf. Locations: 1 Missouri Agronomy Research Center, 2 Osage County, MO, USA; 3 Meleti, 4 Gambara, 5 Pralboino, Italy; 6 Gradiste, Serbia; 7 ASSET agricultural fields IL, USA; 8 USSEU undisturbed site IL, USA.

Microorganisms	Seed Surface	Within Seeds	Location	Activity	Reference
Fungi					
Actinomucor elegans, strain AF157119	x x		6 7	protects *Abutilon* seedlings	[53,89–91]
Alternaria alternata	x	x	1, 2, 3		[30,86]
Aspergillus flavus	x		7		[53]
Cephaliophora tropica strains	x		8		[53]
Cladosporium cladosporioides	x		1, 2		[30]
Cordyceps sinensis	x		7, 8		[53]
Epicoccumpurpurascens	x		1, 2	inhibits seed-borne bacteria	[30]
Fusarium spp.	x	x	1, 2, 3, 4, 5	moderate antagonism toward test fungi	[30,86]
Madurella mycetomatis	x		8		[53]
Bacteria					
Alcaligenes spec. isolates			1, 2	highly antagonistic to seed fungi	[30]
Bacillus spec. isolates			1, 2		[30]
Bacillus subtilis			1, 2		[30]
Flavobacterium spec.			1, 2		[30]
Pseudomonas spec. isolates			1, 2		[30]

Actinomucor elegans was found on velvetleaf seeds purchased from Herbiseed (Twyford, UK), which were harvested from the Herbiseed farm Gradiste, Serbia, in 2011 and 2012. *Actinomucor elegans* was not found on seeds harvested in 2013 from another Herbiseed farm site. These findings support the assumption that the habitat of the parent plant is crucial for the variable microbial species composition associated with *A. theophrasti* seeds. It is known that root microbiomes contain microorganisms attracted from the soil microbial community [89]. *Actinomucor elegans*, re-isolated from velvetleaf roots, was identified as a "consortium", harboring numerous bacteria, for instance, *Pantoea ananatis* and *Stenotrophomonas maltophilia* and the yeast *Papiliotrema baii*. This consortium has a larger metabolic potential than its isolated members [90].

Incubations of the *A. elegans* consortium with benzoxazolinone (BOA), the first degradation product of benzoxazinone, did not inhibit fungal growth or lead to any degradation or detoxification products within one week. Additionally, phenoxazinone, a phytotoxin derived from 2-aminophenol, a microbial degradation product of BOA, did not impair fungal growth. Even under nutrient deficiencies, the consortium did not use these compounds as a C or N source [91,92]. Nevertheless, the consortium responded to high BOA concentrations by enhancing the glucosyl ceramide content of membranes isolated from the consortium [93]. The consortium released tryptophan as the most prominent compound into the medium. Tryptophan release increased in the presence of BOA during the first week of incubation but dropped below the control level with increasing incubation times over four weeks [92]. It is doubtful that the released tryptophan is used for auxin production by the Trp-dependent IAA biosynthetic pathway when the consortium colonizes velvetleaf roots, since its root exudates have the potential to reduce IAA production significantly. This possibility is important since high IAA amounts are deleterious to velvetleaf growth [94]. The consortium had no inhibitory effect of germination or seedling growth (Figure 2). In contrast, *A. elegans* hyphae seems to protect the seedlings ([90,95], and see below).

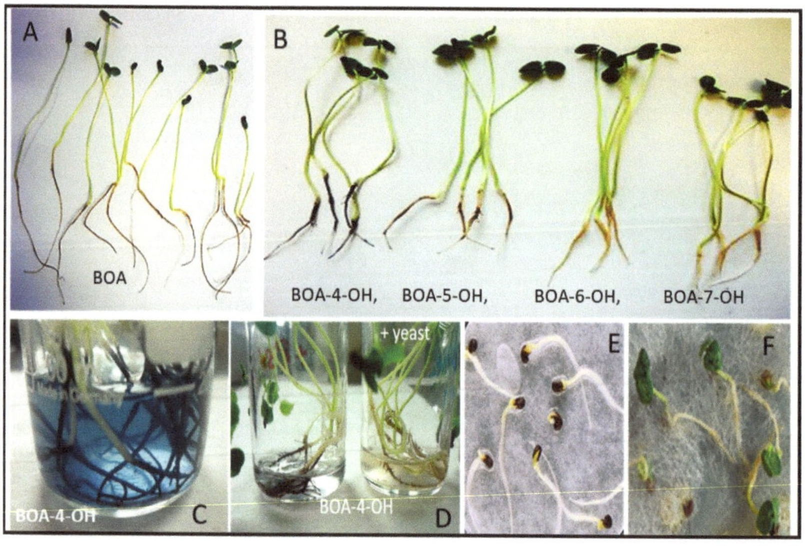

Figure 2. (**A**): Seedlings after 24 h BOA incubation, leaves are not impaired. (**B**): Incubations of seedlings with the four BOA-OH isomers. Seedlings are shown after the incubation: Only BOA-4-OH led to roots covered with a bluish-black polymer. (**C**): Roots exposed to BOA-4-OH during incubation. (**D**): Addition of the yeast *Papiliotrema baii*, a member of the *Actinomucor elegans* consortium, prevents polymer covers on the root surfaces. (**E**): *A. elegans* consortium has no damaging effects on germinating velvetleaf. (**F**): Healthy velvetleaf seedlings are cocooned by *A. elegans* hyphae.

Isolated *P. ananatis* and the entire consortium, but not the isolated yeast, are able to nitrate the plant-produced detoxification intermediate BOA-6-OH, resulting in 6-hydroxy-5-nitrobenzo[*d*]oxazol-2(3*H*)-one (NO2-BOA-6-OH). In contrast, *P. baii* formed only a dimer of BOA-6-OH [90,96]. NO2-BOA-6-OH was produced by velvetleaf root surface proteins and was also identified in the incubation media of velvetleaf when supplemented with BOA-6-OH. This nitroaromatic compound in concentrations of 8 to 86 μM enhanced *A. theophrasti* root growth at pH 5 but had no effect on shoot growth. The I_{50} value was 300 μM for root growth at pH 5 and 700 μM for shoot growth at pH range of 4 to 6. Velvetleaf root surface proteins that included microbial sources were able to reduce NO2-BOA-6-OH up to 90% when assayed in vitro with the compound. In contrast, *Lepidium sativum* (L.) root and shoot growth was inhibited by 50% and 90%, respectively, with 86 μM NO2-BOA-6-OH [96].

NO2-BOA-6-OH? I_{50} values were 20 μM at pH 6.0 for *L. sativum* root growth and 10 μM for its shoot growth [96]. Thus, NO2-BOA-6-OH can be deleterious for other plants, but apparently not for velvetleaf grown from non-sterilized seeds.

3.2. Defense Strategies against Benzoxazolinone (BOA)—Uptake, Detoxification, Membrane Repair, and Polymerizations at the Root Surface

Velvetleaf detoxifies benzoxazolinone (BOA) differently than other important summer weeds, such as *Chenopodium album*, *Amaranthus retroflexus*, and *Portulaca oleracea*, but also many others. While BOA uptake by velvetleaf roots was similar to uptake by *C. album*, *A. retroflexus* and *P. oleracea*, velvetleaf roots produced remarkably fewer detoxification products when 20 to 40 μmol BOA was applied. Additionally, less BOA was transported to the shoots where almost no detoxification products were found in velvetleaf shoots, in contrast to the above-mentioned weed species [97].

Velvetleaf seedlings not colonized by the *A. elegans* consortium exhibited a different detoxification behavior by accumulating higher amounts of BOA-6-O-glucoside in roots and shoots during incubation with 500 μM BOA [90,97]. When colonized by the consortium, more BOA-5-O-glucoside than BOA-6-O-glucoside accumulated in velvetleaf roots, suggesting that the consortium influences the position of BOA hydroxylation. Without the fungus, velvetleaf seedling length was affected by BOA-6-OH and BOA-5-OH in comparison to the control, while BOA-6-OH led to a higher inhibition than BOA-5-OH. Velvetleaf seedlings grown from seeds not associated with *A. elegans* from the Meleti field also had more BOA-6-O-glucoside than BOA-5-O-glucoside.

The responsible glucosyltransferase activity could be extracted from plant cell wall material whereas the cytosolic fraction was inactive. The plant-produced intermediates BOA-6-OH and BOA-5-OH, as well as the synthetic isomers BOA-4- and -7-OH, led to brownish and black roots but otherwise had no effect on the phenotype, although all these compounds also induced H_2O_2 production, indicated by high catalase activity at the root surfaces [90] (Figure 2). High catalase activity was found in the bacterial isolates and a yeast belonging to the *A. elegans* consortium (see above). On root surfaces, peroxidase, laccase, and phenol oxidase activities could also be measured, so at least a portion of these activities are from root-colonizing microorganisms [90]. Acting in concert, the activities are responsible for BOA-OH polymer formation at the root surface (Figure 2). It is unknown whether these polymers permanently cover roots under natural conditions, because microorganisms, for instance the yeast *P. baii*, associated with the consortium, remove BOA-4-OH over time, also preventing its polymerization.

Incubation of velvetleaf seedlings with BOA-6-OH, which has a higher toxicity than BOA, damages membrane lipids and fatty acids in the root tips. Decreased content of α-linoleic acid (18:2), phosphatidylinositol, and phosphatidylcholine were measured in root tips within 30 min of incubation. However, membrane repair was as rapid as found with maize root tips and control levels were reestablished within one hour after exposure to the compound [98]. Thus, even when BOA-6-OH would accumulate in velvetleaf root tips, which had not yet been observed, the plant has powerful mechanisms for rapid membrane repair.

Like many other natural compounds, benzoxazinoids are considered templates for the design of new herbicides. Substitutions such as hydroxylation, especially in position 6, activate BOA, allowing further substitutions, for instance, in the case of 6-hydroxylation, in which nitration in position 5 reduces the phytotoxicity for *Abutilon*.

4. Secondary Metabolites of Velvetleaf

The allelopathic activity of aqueous and methanolic extracts from all parts of *A. theophrasti* has often been studied. In particular, exudates of the glandular trichomes of *Abutilon* are phytotoxic [18,99]. Inhibitory effects of aqueous extracts from velvetleaf plants on maize, tomato, sugar beet, oilseed rape, oats, cucumber, radish, soybean, and sunflower have also been described [100–102]. Paszkowski and Kremer [103] identified the flavonoids delphinidin, cyanidin, quercetin, myricetin, (+), (–)- catechin, and -epicatechin in extracts from velvetleaf seed coats. Kaempferol 3-O-β-(6″-p-coumaroyl) glucopyranoside was identified in methanolic extracts of above-ground velvetleaf tissue [104]. Tian et al. [105] reported gallic acid, protocatechuic acid, tannin, catechin, vanillic acid, caffeic acid, ferulic acid, rutin, and quercetin in extracts from the root, stem, leaves, seeds, and exocarp. Hibiscuslide C isolated from velvetleaf above-ground tissue had antifungal activity in *Candida albicans* [106]. In addition, Mamadalieva et al. [107] isolated (6S,9R)-roseoside and (6S,9S)-roseoside from the aerial parts of the plant (Figure 3).

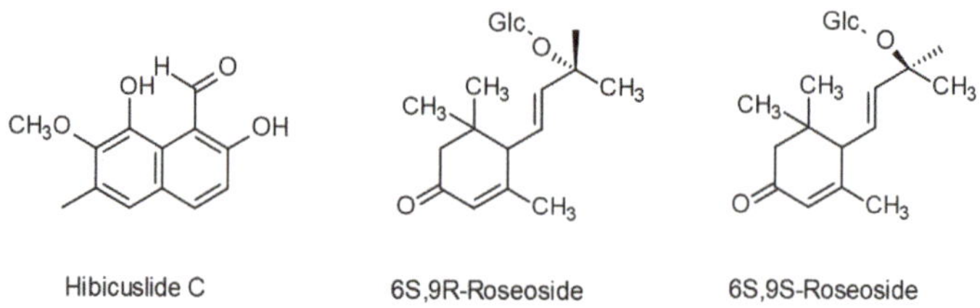

Figure 3. Less known secondary metabolites found in extracts of *A. theophrasti:* hibiscuslide C and roseosides, identified by Hwang et al. and Mamadalieva et al. [106,107].

Aqueous extracts of velvetleaf seed coats can inhibit multiple soil fungi. Single flavonoids from velvetleaf leaves and stems had negative effects on fungi with seed-decomposing properties [104]. It is possible that not all bioactive secondary metabolites of velvetleaf are yet identified, but almost all of those known are widespread across plants and can act as allelochemicals. Additionally, the more recently identified roseosides are not specific to velvetleaf. They have been found in *Adina* and a *Polygonum* species and in *Centella erecta* [108–110]. Except for the roseosides and hibiscuslide C, all the identified phenolic compounds are known for complete degradation by bacteria (Figure 4).

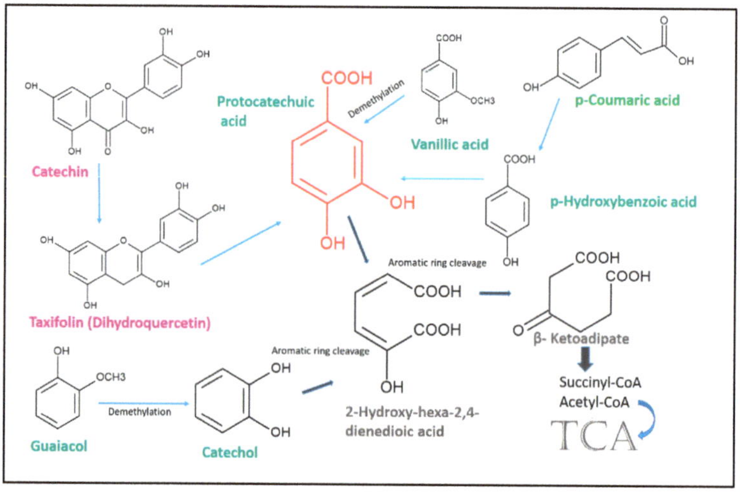

Figure 4. Some steps of the microbial degradation of catechin and taxifolin as members of the flavonoids (magenta), p-coumaric acid, a cinnamic acid derivative (green), and several simple phenolic compounds (petrol). Many phenolic compounds are degraded via the central intermediate protocatechuic acid, and others via catechol. Catechol and protocatechuic acid (red structure) undergo either ortho- or meta- cleavage of the aromatic ring, resulting in 2-hydroxy-hexa-2,4-dienedioic acid and β-ketoadipate. These compounds are converted to succinyl-CoA and acetyl-CoA for entering the citrate cycle (TCA).

Consistently, these compounds in their monomeric form do not accumulate appreciably in the soil, but perhaps do so within the rhizosphere in amounts that are sufficient for allelopathic inhibitions or other functions. More likely, the compounds in their specific combination and amount may have defined chemotactic properties, attracting microorganisms valuable for velvetleaf fitness at a given growing site. Additionally, degradation intermediates, when sufficiently stable, may have defined functions, as was shown for protocatechuic acid derived from catechin [69]. Protocatechuic acid is phytotoxic, but the compound also has antioxidant activity [111]. Microbial conversions and degradation products of the parental phenolic secondary metabolites increase the number of bioactive derivatives dramatically, at least temporarily, but their roles, particularly short-term ones, in the plant–microbe interaction are understudied. Is the production of common and unusual microbial conversion products one reason for the deleterious impacts on other plants, although these plants synthesize the same compounds and exude them? Presently, this question cannot be answered. Since many phenolic compounds have antioxidant properties, they may also protect beneficial, root-colonizing microorganisms against oxidative stress when attacked by pathogens or when faced with antibiotics produced by competitive species used for allelopathic activity. When highly concentrated, the protection provided by phenolic compounds with antioxidant properties will reverse and become toxic.

5. Considerations and Further Perspective

This short review demonstrates the extraordinary capability of *A. theophrasti* to withstand attempts for its biological control, for instance by allelochemicals. Velvetleaf seems equipped with a combination of highly efficient, unique strategies, in addition to those mentioned in the introduction, for high tolerance against allelochemicals such as benzoxazinoids. One of these strategies is certainly featured by attracting the most suitable soil microorganisms, thus avoiding competitive conflicts among the attracted species and the already established species. This strategy allows enormous flexibility and an optimal adaptation to different growing sites with specific pedo-climatic conditions and interspecific interactions between micro- and macro-biota. The mode of attraction is unclear, but the exuded secondary metabolites, perhaps specific mixtures of them, are certainly involved. The microbe-*Abutilon* interactions may be of crucial importance for the success of the plant.

Another strategy encompasses the translocation of allelochemical detoxification and elimination into the cell wall compartment and to the root surface, where microorganisms take over the work. This strategy seems to be highly developed in velvetleaf and presents a distinctive difference to many other weeds and crops. The assumption is in line with the finding that bentazon metabolites were not detected in velvetleaf cells prepared from calli produced in axenic culture, in the velvetleaf hypocotyl sections, or in the media used for incubations. In contrast, soybean cells from suspension cultures detoxified bentazon to 6-*O*-ß-glucose-bentazon and 8-*O*-ß-glucose-bentazon [112]. It can also be hypothesized that the plant has evolved mechanisms for suppressing the release of phytotoxins produced by microorganisms that intrude the rhizosphere. The inhibition of bacterial IAA production points in this direction.

Other reasons for *Abutilon's* fitness may be related to its old use as a crop, where plants with desired new properties were selected by humans for further cultivation [1], while escapes of the cultured plant to the wild might allow velvetleaf more opportunity to select suitable microorganisms from its soil. During rewilding, domesticated *Abutilon* could reestablish or even improve its microbiomes for re-attaining biodiversity levels previously lost or diminished during domestication. Diminished microbiome diversity as a result of cultivation is a known phenomenon for some crops [113,114], and rewilding to improve microbiomes is being explored to strengthen modern crops [115], if such restoration is still possible given the loss of microbial biodiversity due to human activities [116]. When experiencing stress conditions during rewilding, *Abutilon* might have also evolved a strong transgenerational stress memory that includes epigenetics, overaccumulation of proteins involved in managing stress, in dynamic proteome modulation and rapid protein interactions to adapt quickly to stress conditions [117], while concurrently maintaining high seed production. Presently speculative, *Abutilon* may collect preferentially microorganisms with a well-developed stress memory [118], although the molecular mechanisms of selective capture are enigmatic.

Ghanizadeh and James studied the establishment of velvetleaf in different regions of New Zealand and found that the invasion dynamics depend on climate niches with different local temperatures and daily solar radiation [9]. For this reason, velvetleaf is problematic only in one region in the North Island where temperatures are higher, suggesting that there are limiting factors for velvetleaf and its ability to adapt [9], although the influence of soil microbiomes was not included in the study. It is certainly worthwhile to study the genetic, biochemical, and physiological traits of this weed more intensively to disclose vulnerable features and strong attributes important for survival. In this sense, *A. theophrasti* could be a role model for the development of new strategies and environmentally acceptable technologies that strengthen crops when in competition with weeds.

Author Contributions: Conceptualization, M.S. and V.T.; data curation, V.T., M.S. A.F. and T.M.S.; writing, M.S., V.T. and T.M.S. All authors have read and agreed to the published version of the manuscript.

Funding: This research received no external funding.

Data Availability Statement: Not applicable.

Conflicts of Interest: The authors declare no conflict of interest.

References

1. Cardina, J. *Lives of Weeds: Opportunism, Resistance, Folly*; Cornell University Press: Ithaca, NY, USA, 2021; Chapter 3; pp. 73–100, ISBN 9781501758980.
2. Warwick, S.I.; Black, L.D. The biology of Canadian Weeds. 90. *Abutilon theophrasti. Can. J. Plant Sci.* **1988**, *68*, 1069–1085. [CrossRef]
3. Stegink, S.J.; Spencer, N.R. Using Protein Electrophoresis to Investigate the Phylogeny of Velvetleaf (*Abutilon theophrasti*). *Weed Sci.* **1988**, *36*, 172–175. Available online: https://www.jstor.org/stable/4044866 (accessed on 31 January 2023). [CrossRef]
4. Rojas-Sandoval, J. Abutilon theophrasti (Velvet Leaf). Available online: https://cabidigitallibrary.org/terms-and-conditions (accessed on 3 March 2022).
5. Cortés, J.A.; Mendiola, M.A.; Castejón, M. Competition of velvetleaf (*Abutilon theophrasti*) weed with cotton (*Gossypium hirsutum* L.). Economic damage threshold. *Span. J. Agric. Res.* **2010**, *8*, 391–399. [CrossRef]
6. Rain, A.S.; Cârciu, G. *Abutilon theophrasti* L. (velvetleaf): A problem-weed for grain maize. *J. Hortic. Sci. Biotechnol.* **2018**, *22*, 124–127. Available online: https://journal-hfb.usab-tm.ro/2018/Volum%2022(3)%20PDF/21Rain%20A.%202.pdf (accessed on 31 January 2023).
7. Follak, S.; Aldrian, U.; Schwarz, M. Spread dynamics of *Abutilon theophrasti* in Central Europe. *Plant Protect. Sci.* **2014**, *50*, 157–163. [CrossRef]
8. Tanji, A.; Taleb, A. New weed species recently introduced into Morocco. *Weed Res.* **1997**, *37*, 27–31. [CrossRef]
9. Ghanizadeh, H.; James, T.K. Behaviour of *Abutilon theophrasti* in Different Climatic Niches: A New Zealand Case Study. *Front. Plant Sci.* **2022**, *13*, 885779. [CrossRef]
10. Mahklouf, M.H.; El-Ahamir, S.M. The First Record of *Abutilon theophrasti* (Malvaceae) in the Flora of Libya. *Phytotaxa* **2019**, *402*, 259–264. [CrossRef]
11. Scheliga, M.; Brand, U.; Türk, O.; Gruber, S.; Petersen, J. Field Trials Indicating the Potential of *Abutilon theophrasti* (Medic.) as a New Fibre Crop in Southwest Germany. *J. Adv. Agric.* **2019**, *10*, 1726–1739. [CrossRef]
12. Bensky, D.; Gamble, A.; Kaptchuk, T.J. *Chinese Herbal Medicine: Materia Medica*; Eastland Press: Seattle, WA, USA, 1993; pp. 142–143. ISBN 0939616157.
13. Spencer, N.R. Velvetleaf, *Abutilon theophrasti* (Malvaceae). History and Economic Impact in the United States. *Econ. Bot.* **1984**, *38*, 407–416. Available online: https://www.jstor.org/stable/4254678 (accessed on 31 January 2023). [CrossRef]
14. Medovic, A.; Horvath, F. Content of a Storage Jar from the Late Neolithic Site of Hódmezővásárhely-Gorzsa, South Hungary: A Thousand Carbonized Seeds of *Abutilon theophrasti* Medic. *Veg. Hist. Archaeobot.* **2012**, *21*, 215–220. [CrossRef]
15. Vaughn, L.G.; Bernards, M.L.; Arkebauer, T.J.; Lindquist, J.L. Corn and Velvetleaf (*Abutilon theophrasti*) Growth and Transpiration Efficiency under Varying Water Supply. *Weed Sci.* **2016**, *64*, 596–604. Available online: https://www.jstor.org/stable/26420728 (accessed on 31 January 2023). [CrossRef]
16. Lindquist, J.L.; Mortensen, D.A. Tolerance and Velvetleaf (*Abutilon theophrasti*) Suppressive Ability of Two Old and Two Modern Corn (*Zea mays*) Hybrids. *Weed Sci.* **1998**, *46*, 569–574. [CrossRef]
17. Ma, X.; Yang, J.; Wu, H.; Jiang, W.; Ma, Y.; Ma, Y. Growth Analysis of Cotton in Competition with Velvetleaf (*Abutilon theophrasti*). *Weed Technol.* **2016**, *30*, 123–136. [CrossRef]

18. Sterling, T.M.; Putnam, A.R. Possible role of glandular trichome exudates in interference by velvetleaf (*Abutilon theophrasti*). *Weed Sci.* **1987**, *35*, 308–314. Available online: https://www.jstor.onarg/stable/4044589 (accessed on 31 January 2023). [CrossRef]
19. De Sanctis, J.; Jhala, A. Interaction of dicamba, fluthiacet-methyl, and glyphosate for control of velvetleaf (*Abutilon theophrasti*) in dicamba/glyphosate–resistant soybean. *Weed Technol.* **2021**, *35*, 761–767. [CrossRef]
20. Striegel, A.; Eskridge, K.M.; Lawrence, N.C.; Knezevic, S.Z.; Kruger, G.R.; Proctor, C.A.; Hein, G.L.; Jhala, A.J. Economics of herbicide programs for weed control in conventional, glufosinate, and dicamba/glyphosate-resistant soybean across Nebraska. *Agron. J.* **2020**, *112*, 5158–5179. [CrossRef]
21. Travlos, I.; Kanatas, P.; Economou, G.; Kotoulas, V.; Chachalis, D.; Tsioros, S. Evaluation of velvetleaf interference with maize hybrids as influenced by relative time of emergence. *Exp. Agric.* **2012**, *48*, 127–137. [CrossRef]
22. Tataridas, A.; Kanatas, P.; Chatzigeorgiou, A.; Zannopoulos, S.; Travlos, I. Sustainable Crop and Weed Management in the Era of the EU Green Deal: A Survival Guide. *Agronomy* **2022**, *12*, 589. [CrossRef]
23. Barnes, E.R.; Lawrence, N.C.; Knezevic, S.Z.; Rodriguez, O.; Irmak, S.; Jhala, A.J. Weed control and response of yellow and white popcorn hybrids to herbicides. *Agron. J.* **2020**, *112*, 458–469. [CrossRef]
24. Heap, I. The International Survey of Herbicide Resistant Weeds. Available online: www.weedscience.org (accessed on 12 December 2022).
25. Cardina, J.; Norquay, H.M. Seed Production and Seedbank Dynamics in Subthreshold Velvetleaf (*Abutilon theophrasti*) Populations. *Weed Sci.* **1997**, *45*, 85–90. Available online: https://www.jstor.org/stable/4045717 (accessed on 31 January 2023). [CrossRef]
26. Vatovec, C.; Jordan, N.; Huerd, S. Responsiveness of certain agronomic weed species to arbuscular mycorrhizal fungi. *Renew. Agric. Food Syst.* **2005**, *20*, 181–189. [CrossRef]
27. O'Neil, C.M.; Nash, J.; Tiemann, L.K.; Miesel, J.R. Mycorrhizal Symbioses Enhance Competitive Weed Growth in Biochar and Nutrient-Amended Soils. *Front. Agron.* **2021**, *3*, 731184. [CrossRef]
28. Horvath, D.P.; Llewellyn, D.; Clay, S.A. Heterologous Hybridization of Cotton Microarrays with Velvetleaf (*Abutilon theophrasti*) Reveals Physiological Responses due to Corn Competition. *Weed Sci.* **2007**, *55*, 546–557. Available online: https://www.jstor.org/stable/4539616 (accessed on 31 January 2023). [CrossRef]
29. Horvath, D.; Gulden, R.; Clay, S. Microarray analysis of late-season velvetleaf (*Abutilon theophrasti*) effect on corn. *Weed Sci.* **2006**, *54*, 983–994. [CrossRef]
30. Kremer, R.J.; Hughes, L.B., Jr.; Aldrich, R.J. Examination of Microorganisms and Deterioration Resistance Mechanisms Associated with Velvetleaf Seed. *Agron. J.* **1984**, *76*, 745–749. [CrossRef]
31. Kremer, R.J. Antimicrobial Activity of Velvetleaf (*Abutilon theophrasti*) Seeds. *Weed Sci.* **1986**, *34*, 617–622. [CrossRef]
32. Kremer, R.J.; Kennedy, A.C. Rhizobacteria as biocontrol agents of weeds. *Weed Technol.* **1996**, *10*, 601–609. [CrossRef]
33. Boyetchko, S. Impact of soil microorganisms on weed biology and ecology. *Phytoprotection* **1996**, *77*, 41–56. [CrossRef]
34. Kremer, R.J.; Spencer, N.R. Impact of a Seed-Feeding Insect and Microorganisms of Velvetleaf (*Abutilon theophrasti*) Seed Viability. *Weed Sci.* **1989**, *37*, 211–216. Available online: https://www.jstor.org/stable/4044846 (accessed on 31 January 2023). [CrossRef]
35. Boyette, C., Walker, H. Evaluation of *Fusarium lateritium* as a Biological Herbicide for Controlling Velvetleaf (*Abutilon theophrasti*) and Prickly Sida (*Sida spinosa*). *Weed Sci.* **1986**, *34*, 106–109. [CrossRef]
36. Chee-Sanford, J.C.; Williams, M.M.; Davis, A.S.; Sims, G.K. Do microorganisms influence seed-bank dynamics? *Weed Sci.* **2006**, *54*, 575–587. [CrossRef]
37. Meir, S.; Larroche, C.; Al-Ahmad, H.; Gressel, J. Fungal transformation of *Colletotrichum coccodes* with bacterial oahA to suppress defenses of *Abutilon theophrasti*. *Crop. Prot.* **2009**, *28*, 749–755. [CrossRef]
38. Gressel, J.; Michaeli, D.; Kampel, V.; Amsellem, Z.; Warshawsky, A. Ultralow calcium requirements of fungi facilitate use of calcium regulating agents to suppress host calcium-dependent defenses, synergizing infection by a mycoherbicide. *J. Agric. Food Chem.* **2002**, *50*, 6353–6360. [CrossRef]
39. Ahn, B.; Paulitz, T.; Jabaji-Hare, S.; Watson, A. Enhancement of *Colletotrichum coccodes* virulence by inhibitors of plant defense mechanisms. *Biocontrol Sci. Technol.* **2005**, *15*, 299–308. [CrossRef]
40. Dutton, M.V.; Evans, C.S. Oxalate production by fungi: Its role in pathogenicity and ecology in the soil environment. *Can. J. Microbiol.* **1996**, *42*, 881–895. [CrossRef]
41. Briere, S.C.; Watson, A.K.; Hallett, S.G. Oxalic acid production and mycelia biomass yield of *Sclerotinia minor* for the formulation enhancement of a granular turf bioherbicide. *Biocontrol Sci. Technol.* **2000**, *10*, 281–289. [CrossRef]
42. Watson, A.K.; Ahn, E.S. *Better Control of Weeds with Enhanced Bioherbicides. Enhancing Biocontrol Agents and Handling Risks*; Vurro, M., Gressel, J., Butt, T., Harman, G., Pilgeram, A., Leger, R.S., Nuss, D., Eds.; IOS Press: Amsterdam, The Netherlands, 2001; pp. 106–113.
43. Syed, S.; Buddolla, V.; Lian, B. Oxalate Carbonate Pathway—Conversion and Fixation of Soil Carbon—A Potential Scenario for Sustainability. *Front. Plant Sci.* **2020**, *11*, 591297. [CrossRef]
44. Jennings, J.C.; Apel-Birkhold, P.C.; Bailey, B.A.; Anderson, J.D. Induction of Ethylene Biosynthesis and Necrosis in Weed Leaves by a *Fusarium oxysporum* Protein. *Weed Sci.* **2000**, *48*, 7–14. Available online: http://www.jstor.org/stable/4046151 (accessed on 31 January 2023). [CrossRef]
45. Amsellem, Z.; Cohen, B.C.; Gressel, J. Engineering hypervirulence in a mycoherbicidal fungus for efficient weed control. *Nat. Biotechnol.* **2002**, *10*, 1035–1039. [CrossRef]

46. Šantrić, L.; Radivojević, L.; Gajić Umiljendić, J.; Sarić-Krsmanović, M. Allelopathic effects of actinobacterial isolates on seed germination and early seedling growth of velvetleaf (*Abutilon theophrasti* Medik.). *Pestic. Phytomed.* **2022**, *37*, 41–48. [CrossRef]
47. Kremer, R.J.; Begonia, M.F.T.; Stanley, L.; Lanham, E.T. Characterization of rhizobacteria associated with weed seedlings. *Appl. Environ. Microbiol.* **1990**, *56*, 1649–1655. [CrossRef] [PubMed]
48. Owen, A.; Zdor, R. Effect of cyanogenic rhizobacteria on the growth of velvetleaf (*Abutilon theophrasti*) and corn (*Zea mays* L.) in autoclaved soil and the influence of supplemental glycine. *Soil Biol. Biochem.* **2001**, *33*, 801–809. [CrossRef]
49. Siegien, I.; Bogatek, R. Cyanide action in plants-from toxic to regulatory. *Acta Physiol. Plant.* **2006**, *28*, 483–497. [CrossRef]
50. Silva-Avalos, J.; Richmond, M.G.; Nagappan, O.; Kunz, D.A. Degradation of the metal-cyano complex tetracyanonickelate(II) by cyanide-utilizing bacterial isolates. *Appl. Environ. Microbiol.* **1990**, *56*, 3664–3670. [CrossRef]
51. Harding, D.P.; Raizada, M.N. Controlling weeds with fungi, bacteria and viruses: A review. *Front. Plant Sci.* **2015**, *6*, 659. [CrossRef] [PubMed]
52. Cheng, L.; DiTommaso, A.; Kao-Kniffin, J. Opportunities for Microbiome Suppression of Weeds Using Regenerative Agricultural Technologies. *Front. Soil Sci.* **2022**, *2*, 838595. [CrossRef]
53. Chee-Sanford, J.C. Weed Seeds as Nutritional Resources for Soil Ascomycota and Characterization of Specific Associations Between Plant and Fungal Species. *Biol. Fertil. Soils* **2008**, *44*, 763–771. [CrossRef]
54. Verdeguer, M.; Sánchez-Moreiras, A.M.; Araniti, F. Phytotoxic Effects and Mechanism of Action of Essential Oils and Terpenoids. *Plants* **2020**, *9*, 1571. [CrossRef]
55. Preusche, M.; Vahl, M.; Riediger, J.; Ulbrich, A.; Schulz, M. Modulating Expression Levels of TCP Transcription Factors by *Mentha x piperita* Volatiles—An Allelopathic Tool to Influence Leaf Growth? *Plants* **2022**, *11*, 3078. [CrossRef]
56. Vaughn, S.; Spencer, G. Monoterpenes as parent structures for herbicides. *Weed Sci.* **1993**, *41*, 114–119. [CrossRef]
57. Wolf, R.B.; Spencer, G.F.; Kwolek, W.F. Inhibition of Velvetleaf (*Abutilon theophrasti*) Germination and Growth by Benzyl Isothiocyanate, a Natural Toxicant. *Weed Sci.* **1984**, *32*, 612–615. [CrossRef]
58. Tollsten, L.; Bergstrom, G. Headspace volatiles of whole plants and macerated plant parts of *Brassica* and *Sinapis*. *Phytochemistry* **1988**, *27*, 4013–4018. [CrossRef]
59. Shettle, N.; Balke, E.N. Plant Growth Response to Several Allelopathic Chemicals. *Weed Sci.* **1983**, *31*, 293–298. [CrossRef]
60. Rose, S.J.; Burnside, O.C.; Specht, J.E.; Swisher, B.A. Competition and Allelopathy Between Soybeans and Weeds. *Agron. J.* **1984**, *76*, 523–528. [CrossRef]
61. Dayan, F.E.; Howell, J.L.; Weidenhamer, J.D. Dynamic root exudation of sorgoleone and its in planta mechanism of action. *J. Exp. Bot.* **2009**, *60*, 2107–2117. [CrossRef]
62. Godlewska, K.; Ronga, D.; Michalak, I. Plant extracts—Importance in sustainable agriculture. *Ital. J. Agron.* **2021**, *16*, 1851. [CrossRef]
63. Lebeis, S.L.; Paredes, S.H.; Lundberg, D.S.; Breakfield, N.; Gehring, J.; McDonald, M.; Malfatti, S.; Glavina del Rio, T.; Jones, C.D.; Tringe, S.G.; et al. Salicylic acid modulates colonization of the root microbiome by specific bacterial taxa. *Science* **2015**, *349*, 860–864. [CrossRef]
64. Siebers, M.; Rohr, T.; Ventura, M.; Schütz, V.; Thies, S.; Kovacic, F.; Jaeger, K.E.; Berg, M.; Dörmann, P.; Schulz, M. Disruption of microbial community composition and identification of plant growth promoting microorganisms after exposure of soil to rapeseed-derived glucosinolates. *PLoS ONE* **2018**, *13*, e0200160. [CrossRef]
65. Hansen, J.C.; Schillinger, W.F.; Sullivan, T.S.; Paulitz, T.C. Soil microbial biomass and fungi reduced with canola introduced into long-term monoculture wheat rotations. *Front. Microbiol.* **2019**, *10*, 1488. [CrossRef]
66. Hu, P.; Hollister, E.B.; Somenahally, A.C.; Hons, F.M.; Gentry, T.J. Soil bacterial and fungal communities respond differently to various isothiocyanates added for biofumigation. *Front. Microbiol.* **2015**, *5*, 729. [CrossRef] [PubMed]
67. Zhou, X.; Wu, F. Vanillic acid changed cucumber (*Cucumis sativus* L.) seedling rhizosphere total bacterial, *Pseudomonas* and *Bacillus* spp. communities. *Sci. Rep.* **2018**, *8*, 4929. [CrossRef] [PubMed]
68. Zhou, X.; Zhang, J.; Pan, D.; Xin, G.; Jin, X.; Chen, S.; Wu, F. p-Coumaric can alter the composition of cucumber rhizosphere microbial communities and induce negative plant-microbial interactions. *Biol. Fert. Soils* **2018**, *54*, 363–372. [CrossRef]
69. Wang, C.-M.; Li, T.-C.; Jhan, Y.-L.; Weng, J.-H.; Chou, C.-H. The Impact of Microbial Biotransformation of Catechin in Enhancing the Allelopathic Effects of *Rhododendron formosanum*. *PLoS ONE* **2013**, *8*, e85162. [CrossRef] [PubMed]
70. Hu, L.; Robert, C.A.M.; Cadot, S.; Zhang, X.; Ye, M.; Li, B.; Manzo, D.; Chervet, N.; Steinger, T.; Van der Heijden, M.G.A.; et al. Root exudate metabolites drive plant-soil feedbacks on growth and defense by shaping the rhizosphere microbiota. *Nat. Comm.* **2018**, *9*, 2738. [CrossRef]
71. Kudjordjie, E.N.; Sapkota, R.; Steffensen, S.K.; Fomsgaard, I.S.; Nicolaisen, M. Maize synthesized benzoxazinoids affect the host associated microbiome. *Microbiome* **2019**, *7*, 59. [CrossRef]
72. Schütz, V.; Frindte, K.; Cui, J.; Zhang, P.; Hacquard, S.; Schulze-Lefert, P.; Knief, C.; Schulz, M.; Dörmann, P. Differential Impact of Plant Secondary Metabolites on the Soil Microbiota. *Front. Microbiol.* **2021**, *12*, 666010. [CrossRef]
73. Lehman, R.M.; Cambardella, C.A.; Stott, D.E.; Acosta-Martinez, V.; Manter, D.K.; Buyer, J.S.; Maul, J.E.; Smith, J.L.; Collins, H.P.; Halvorson, J.J.; et al. Understanding and Enhancing Soil Biological Health: The Solution for Reversing Soil Degradation. *Sustainability* **2015**, *7*, 988–1027. [CrossRef]
74. McGivern, B.B.; Tfaily, M.M.; Borton, M.A.; Kosina, S.M.; Daly, R.A.; Nicora, C.D.; Purvine, S.O.; Wong, A.R.; Lipton, M.S.; Hoyt, D.W.; et al. Decrypting bacterial polyphenol metabolism in an anoxic wetland soil. *Nat. Commun.* **2021**, *12*, 2466. [CrossRef]

75. Macías, F.A.; Marín, D.; Oliveros-Bastidas, A.; Castellano, D.; Simonet, A.M.; Molinillo, J.M. Structure-activity relationships (SAR) studies of benzoxazinones, their degradation products and analogues. Phytotoxicity on standard target species (STS). *J. Agric. Food Chem.* **2005**, *53*, 538–548. [CrossRef]
76. Krogh, S.; Mensz, S.; Nielsen, S.; Mortensen, A.; Christophersen, C.; Fomsgaard, I. Fate of Benzoxazinone Allelochemicals in Soil after Incorporation of Wheat and Rye Sprouts. *J. Agric. Food Chem.* **2006**, *54*, 1064–1074. [CrossRef] [PubMed]
77. Schulz, M.; Marocco, A.; Tabaglio, V.; Macias, F.A.; Molinillo, J.M.G. Benzoxazinoids in rye allelopathy—From discovery to application in sustainable weed control and organic farming. *J. Chem. Ecol.* **2013**, *39*, 154–174. [CrossRef] [PubMed]
78. La Hovary, C.; Baerson, S.R.; Williamson, J.D.; Danehower, D.A.; Ma, G.; Pan, Z.; Mask, M.M.; Burton, J.D. Phytotoxicity and benzoxazinone concentration in field grown cereal rye (*Secale cereale* L.). *Int. J. Agron.* **2016**, *2016*, 6463826. [CrossRef]
79. Hussain, M.I.; Araniti, F.; Schulz, M.; Baerson, S.; Vieites-Álvarez, Y.; Rempelos, L.; Bilsborrow, P.; Durán, A.G.; Chinchilla, N.; Macías, F.A.; et al. Benzoxazinoids in wheat allelopathy—From discovery to application for sustainable weed management. *Environ. Exp. Bot.* **2022**, *202*, 104997. [CrossRef]
80. Fiorini, A.; Maris, S.C.; Abalos, D.; Amaducci, S.; Tabaglio, V. Combining no-till with rye (*Secale cereale* L.) cover crop mitigates nitrous oxide emissions without decreasing yield. *Soil Tillage Res.* **2020**, *196*, 104442. [CrossRef]
81. Gavazzi, C.; Schulz, M.; Marocco, A.; Tabaglio, V. Sustainable weed control by allelochemicals from rye cover crops: From the greenhouse to field evidence. *Allelopath. J.* **2010**, *25*, 259–274.
82. Tabaglio, V.; Marocco, A.; Schulz, M. Allelopathic cover crop of rye for integrated weed control in sustainable agroecosystems. *Ital. J. Agron.* **2013**, *8*, 35–40. [CrossRef]
83. Boselli, R.; Anders, N.; Fiorini, A.; Ganimede, C.; Faccini, N.; Marocco, A.; Schulz, M.; Tabaglio, V. Improving weed control in sustainable agroecosystems: Role of cultivar and termination timing of rye cover crop. *Ital. J. Agron.* **2021**, *16*, 1807. [CrossRef]
84. Reberg-Horton, S.C.; Burton, J.D.; Danehower, D.A.; Ma, G.; Monks, D.W.; Murphy, J.P.; Ranells, N.N.; Williamson, J.D.; Creamer, N.G. Changes over time in the allelochemical content of ten cultivars of rye (*Secale cereale* L.). *J. Chem. Ecol.* **2005**, *31*, 179–193. [CrossRef]
85. Tabaglio, V.; Gavazzi, C.; Schulz, M.; Marocco, A. Alternative weed control using the allelopathic effect of natural benzoxazinoids from rye mulch. *Agron. Sustain. Dev.* **2008**, *28*, 397–401. [CrossRef]
86. Ganimede, C.; Marocco, A.; Schulz, M.; Tabaglio, V. Investigation of Endophytic Fungi Associated to *Abutilon theophrasti* Root. In Proceedings of the 7th World Congress on Allelopathy, Vigo, Spain, 28 July–1 August 2014.
87. DeMers, M. *Alternaria alternata* as endophyte and pathogen. *Microbiology* **2022**, *168*, 001153. [CrossRef]
88. Toghueo, R.M.K. Bioprospecting endophytic fungi from *Fusarium* genus as sources of bioactive metabolites. *Mycology* **2019**, *11*, 1–21. [CrossRef] [PubMed]
89. Prudence, S.; Newitt, J.; Worsley, S.F.; Macey, M.C.; Murrell, J.C.; Lehtovirta-Morley, L.E.; Hutchings, M.I. Soil, senescence and exudate utilisation: Characterisation of the Paragon var. spring bread wheat root microbiome. *Environ. Microbiome* **2021**, *16*, 12. [CrossRef]
90. Schulz, M.; Sicker, D.; Schackow, O.; Hennig, L.; Yurkov, A.; Siebers, M.; Hofmann, D.; Disko, U.; Ganimede, C.; Mondani, L.; et al. Interspecies-Cooperations of *Abutilon theophrasti* with Root Colonizing Microorganisms Disarm BOA-OH Allelochemicals. *Plant Signal. Behav.* **2017**, *12*, e1358843. [CrossRef]
91. Haghi Kia, S. Unraveling *Abutilon theophrasti's* Tolerance to the Allelochemical Benzoxazolinone. Master's Thesis, Rheinische Friedrich-Wilhelms-Universität, Bonn, Germany, 2013.
92. Kia, S.H.; Schulz, M.; Ayah, E.; Schouten, A.; Müllenborn, C.; Paetz, C.; Schneider, B.; Hofmann, D.; Disko, U.; Tabaglio, V.; et al. *Abutilon theophrasti*'s defense against the allelochemical benzoxazolin-2(3H)-one: Support by *Actinomucor elegans*. *J. Chem. Ecol.* **2014**, *40*, 1286–1298. [CrossRef] [PubMed]
93. Kia, S.H.; Wewer, V.; Hölzl, G.; Schulz, M. Assessing the Effect of Benzoxazolinone on the Plant–Associated Zygomycete *Actinomucor elegans* and Its Possible Role in a Weed's Tolerance to Allelochemicals. In Proceedings of the 10th International Mycological Congress (Imc10), Bangkok, Thailand, 3–8 August 2014; Available online: http://www.imc10.com (accessed on 31 January 2023).
94. Nadane, A. Effect of Velvetleaf Root Exudates on Indole-3-Acetic Acid Production by *Rhizobium rubi* AT3-4RS/6. Ph.D. Thesis, Andrews University, Berrien Springs, MI, USA, 2012. Available online: https://digitalcommons.andrews.edu/honors/24 (accessed on 31 January 2023).
95. Schulz, M. (University of Bonn, Bonn, Germany); Hofmann, D. (Forschungszentrum Jülich GmbH, Jülich, Germany). Personal communication, 2022.
96. Schulz, M.; Sicker, D.; Schackow, O.; Hennig, L.; Hofmann, D.; Disko, U.; Ventura, M.; Basyuk, K. 6-Hydroxy-5-nitrobenzo[d]oxazol- 2(3H)-one—A degradable derivative of natural 6-hydroxybenzoxazolin-2(3H)-one produced by *Pantoea ananatis*. *Commun. Integr. Biol.* **2017**, *10*, e1302633. [CrossRef] [PubMed]
97. Schulz, M.; Marocco, A.; Tabaglio, V. BOA detoxification of four summer weeds during germination and seedling growth. *J. Chem. Ecol.* **2012**, *38*, 933–946. [CrossRef]
98. Laschke, L.; Schütz, V.; Schackow, O.; Sicker, D.; Hennig, L.; Hofmann, D.; Dörmann, P.; Schulz, M. Survival of Plants during Short-Term BOA-OH Exposure: ROS Related Gene Expression and Detoxification Reactions Are Accompanied with Fast Membrane Lipid Repair in Root Tips. *J. Chem. Ecol.* **2022**, *48*, 219–239. [CrossRef]

99. Sterling, T.M.; Houtz, R.L.; Putnam, A.R. Phytotoxic Exudates from Velvetleaf (*Abutilon theophrasti*) Glandular Trichomes. *Am. J. Bot.* **1987**, *74*, 543–550. [CrossRef]
100. Radivojević, L.; Sarić-Krsmanović, M.; Umiljendić, J.G.; Šantrić, L. Allelopathic effects of invasive weed species *Abutilon theophrasti* Medik., *Ambrosia artemisiifolia* L., *Datura stramonium* L. and *Xanthium strumarium* L. on tomato. *Pestic. Fitomed.* **2019**, *34*, 183–191. [CrossRef]
101. Tian, M.; Li, Q.; Zhao, W.; Qiao, B.; Shi, S.; Yu, M.; Li, X.; Li, C.; Zhao, C. Potential Allelopathic Interference of *Abutilon theophrasti* Medik. Powder/Extract on Seed Germination, Seedling Growth and Root System Activity of Maize, Wheat and Soybean. *Agronomy* **2022**, *12*, 844. [CrossRef]
102. Novak, N.; Novak, M.; Barić, K.; Šćepanović, M.; Ivić, D. Allelopathic potential of segetal and ruderal invasive alien plants. *J. Cent. Eur. Agric.* **2018**, *19*, 408–422. [CrossRef]
103. Paszkowski, W.L.; Kremer, R.J. Biological activity and tentative identification of flavonoid components in velvetleaf (*Abutilon theophrasti* Medik.) seed coats. *J. Chem. Ecol.* **1988**, *14*, 1573–1582. [CrossRef] [PubMed]
104. Balah, M.A.; Nassar, M.I. Allelopathic Constituents from *Abutilon theophrasti* Aerial Parts to Other Weeds. *Res. J. Agric. Sci.* **2011**, *7*, 243–250.
105. Tian, C.; Yang, C.; Zhang, D.; Han, L.; Liu, Y.; Guo, W.; Fan, R.; Liu, M. Antibacterial and antioxidant properties of various solvent extracts of *Abutilon theophrasti* Medic. leaves. *Pak. J. Pharm. Sci.* **2017**, *3*, 767–772.
106. Hwang, J.H.; Jin, Q.; Woo, E.R.; Lee, D.G. Antifungal property of hibicuslide C and its membrane-active mechanism in *Candida albicans. Biochimie* **2013**, *95*, 1917–1922. [CrossRef]
107. Mamadalieva, N.Z.F.; Sharopov, F.; Girault, J.-P.; Wink, M.; Lafont, R. Phytochemical analysis and bioactivity of the aerial parts of *Abutilon theophrasti* (Malvaceae), a medicinal weed. *Nat. Prod. Res.* **2014**, *28*, 1777–1779. [CrossRef]
108. Itoh, A.; Fujii, K.; Tomatsu, S.; Takao, C.; Tanahashi, T.; Nagakura, N.; Chen, C.-C. Six secoiridoid glucosides from *Adina racemosa. J. Nat. Prod.* **2003**, *66*, 1212–1216. [CrossRef]
109. Murai, Y.; Kashimura, S.; Tamezawa, S.; Hashimoto, T.; Takaoka, S.; Asakawa, Y.; Kiguchi, K.; Murai, F.; Tagawa, M. Absolute Configuration of (6S,9S)-Roseoside from *Polygonum hydropiper. Planta Med.* **2001**, *67*, 480–481. [CrossRef]
110. Rumalla, C.S.; Ali, Z.; Weerasooriya, A.D.; Smillie, T.J.; Khan, I.J. A new triterpene glycoside from *Centella erecta. Fitoterapia* **2010**, *81*, 751–754. [CrossRef]
111. Zhang, S.; Gai, Z.; Gui, T.; Chen, J.; Chen, Q.; Li, Y. Antioxidant Effects of Protocatechuic Acid and Protocatechuic Aldehyde: Old Wine in a New Bottle. *Evid. Based Complement Alternat. Med.* **2021**, *8*, 6139308. [CrossRef]
112. Sterling, T.M.; Balke, N.E. Use of Soybean (*Glycine max*) and Velvetleaf (*Abutilon theophrasti*) Suspension-Cultured Cells to Study Bentazon Metabolism. *Weed Sci.* **1988**, *36*, 558–565. Available online: https://www.jstor.org/stable/4044806 (accessed on 31 January 2023). [CrossRef]
113. Reese, A.T.; Chadaideh, K.S.; Caroline, E.; Diggins, C.E.; Schell, L.D.; Beckel, M.; Callahan, P.; Ryan, R.; Thompson, E.M.; Carmody, R.N. Effects of domestication on the gut microbiota parallel those of human industrialization. *eLife* **2021**, *10*, e60197. [CrossRef]
114. Soldan, R.; Fusi, M.; Cardinale, M.; Daffonchio, D.; Preston, G.M. The effect of plant domestication on host control of the microbiota. *Commun. Biol.* **2021**, *4*, 936. [CrossRef] [PubMed]
115. Raaijmakers, J.M.; Kiers, E.T. Rewilding plant microbiomes. *Science* **2022**, *378*, 599–600. [CrossRef] [PubMed]
116. Berg, G.; Cernava, T. The plant microbiota signature of the Anthropocene as a challenge for microbiome research. *Microbiome* **2022**, *10*, 54. [CrossRef]
117. García-Campa, L.; Guerrero, S.; Lamelas, L.; Meijon, M.; Hasbún, R.; Cañal, M.J.; Valledor, L. Chloroplast proteomics reveals transgenerational cross-stress priming in *Pinus radiata. Environ. Exp. Bot.* **2022**, *202*, 105009. [CrossRef]
118. Lagage, V.; Uphoff, S. Pulses and delays, anticipation and memory: Seeing bacterial stress responses from a single-cell perspective. *FEMS Microbiol. Rev.* **2020**, *44*, 565–571. [CrossRef] [PubMed]

plants

MDPI

Review

The Potential of Allelochemicals from Microalgae for Biopesticides

Livia Marques Casanova [1,*], Andrew Macrae [2], Jacqueline Elis de Souza [1], Athayde Neves Junior [1] and Alane Beatriz Vermelho [1,*]

1 Biotechnology Center-Bioinovar, Institute of Microbiology Paulo de Goes, Universidade Federal do Rio de Janeiro (UFRJ), Rio de Janeiro 21941-902, RJ, Brazil
2 Sustainable Biotechnology and Microbial Bioinformatics Laboratory, Institute of Microbiology Paulo de Goes, Universidade Federal do Rio de Janeiro (UFRJ), Rio de Janeiro 21941-902, RJ, Brazil
* Correspondence: liviacasanova@micro.ufrj.br (L.M.C.); abvermelho@micro.ufrj.br (A.B.V.)

Abstract: Improvements in agricultural productivity are required to meet the demand of a growing world population. Phytopathogens, weeds, and insects are challenges to agricultural production. The toxicity and widespread application of persistent synthetic pesticides poses a major threat to human and ecosystem health. Therefore, sustainable strategies to control pests are essential for agricultural systems to enhance productivity within a green paradigm. Allelochemicals are a less persistent, safer, and friendly alternative to efficient pest management, as they tend to be less toxic to non-target organisms and more easily degradable. Microalgae produce a great variety of allelopathic substances whose biocontrol potential against weeds, insects, and phytopathogenic fungi and bacteria has received much attention. This review provides up-to-date information and a critical perspective on allelochemicals from microalgae and their potential as biopesticides.

Keywords: bioherbicide; biocide; biopesticide; allelochemical; microalgae

Citation: Casanova, L.M.; Macrae, A.; de Souza, J.E.; Neves Junior, A.; Vermelho, A.B. The Potential of Allelochemicals from Microalgae for Biopesticides. *Plants* **2023**, *12*, 1896. https://doi.org/10.3390/plants12091896

Academic Editors: Margot Schulz and Vincenzo Tabaglio

Received: 10 February 2023
Revised: 3 May 2023
Accepted: 4 May 2023
Published: 6 May 2023

1. Introduction

Allelopathy is an ecological phenomenon in which allelochemicals are produced under certain environmental conditions. The term 'allelopathy' was introduced by Molisch in 1937 and originated from the Greek words 'allelon' (mutual) and "pathos" (harm). The term means one organism influences others through a chemical mode [1].

Allelochemicals are toxic secondary metabolites produced by plants, microalgae, bacteria, and fungi that influence the development of other organisms. These substances can be harmful to humans, other microalgae, and organisms, including phytoplankton, animals, and submerged macrophytes. Photosynthesis and enzyme activity inhibition, damage to cell structures, and other physiological alterations such as inhibition of respiration, protein synthesis, and gene expression have been observed [2,3].

Allelopathy is much studied in plants, but allelochemicals started to be reviewed in microalgae by Inderjit and Dakshini in 1994 [2,4]. Microalgae are a widely diversified group of prokaryotic and eukaryotic photosynthetic microorganisms. More than 50,000 microalgae species are distributed in rivers, lakes, oceans, and even in terrestrial ecosystems. Cyanobacteria are prokaryotic microalgae in which membrane-bound organelles such as plastids and mitochondria are missing, while eukaryotic microalgae rely on these structures to control cellular functions [5,6]. Many common microalgae taxa are included in the latter group, such as Chlorophyta (green algae) and Bacillariophyta (diatoms) [5,6]. Pyrrophyta (dinoflagellates) are mostly microplankton, they are eukaryotic and live in aquatic environments, be it salt, brackish or fresh water. Around half of all dinoflagellates are photosynthetic, while the other half are heterotrophic; some live as mixotrophic and obligate symbionts/parasites. They are usually reported as microalgae;

indeed, the most toxic microalgae are dinoflagellates [7] The non-algal dinoflagellates are primarily predators of other microorganisms or parasites of animals or giant protists [8].

One of the fields of study for allelochemicals in microalgae began with harmful algal blooms (HABs). HABs are a consequence of an ecosystem imbalance in marine and freshwater systems worldwide [9]. HABs affect water usage, including fisheries, water resources, and recruitment. The poisoning caused by the allelochemicals affects different organisms and leads to economic losses [10]. The use of allelochemicals as biocontrol agents has been suggested as a means of improving microalgae cultivation and controlling contamination [11]. An example of a mode of action is the generation of reactive oxygen species (ROS). *Nostoc spongiaeforme* produces a pigment named nostocine A; this pigment is highly cytotoxic for several microalgae. The compound accelerates ROS formation in *Chlorella reinhardtii.* Nostocine A penetrates the target and is reduced by NADPH; when oxygen levels are high, reduced nostocine A is oxidated, generating superoxide radical anion (O^{2-}) that may cause cytotoxicity [12].

The bioactivity of allelochemicals and their toxic or beneficial effects have also been the focus of research in medicine and agriculture [13], and for their anticancer [14] and antimicrobial properties (Figure 1) [15].

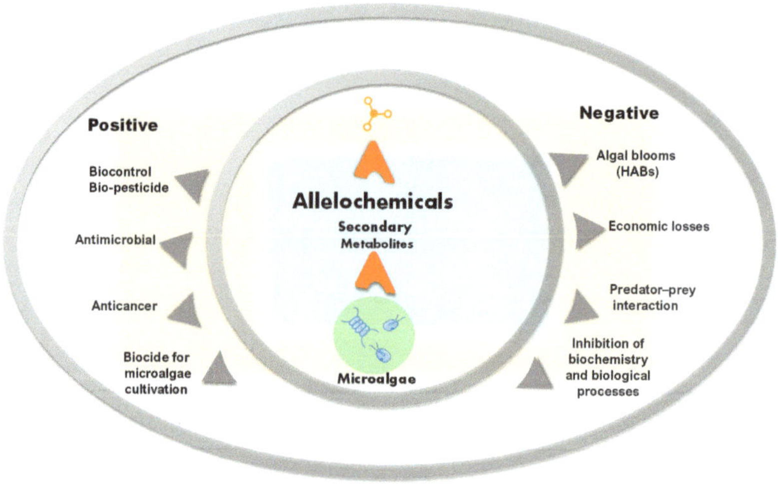

Figure 1. Schematic representation of positive and negative aspects of allelochemicals from microalgae.

Toxic compounds that are used to protect crops, and to control weeds, insects and disease, are known as pesticides. The extensive and widespread use of synthetic pesticides, however, poses several long-term health and environmental risks. Often, pesticides are also toxic to humans and animals, resulting in carcinogenicity, teratogenicity, and the disruption of hormonal systems. Pesticides that persist in the environment for years result in the contamination of water and food sources [16,17]. Another challenging problem for agriculture is the emergence of pesticide resistance in target species [18]. In a scenario of increasing pollution on the planet and of climate changes triggering environmental problems, it is urgent that sustainable, green methods for the biocontrol of pests are found. Environmentally-friendly allelochemicals from microalgae have been described as bioherbicides. These compounds arise as an alternative source of natural biocides to reduce the use of toxic synthetic biocides/herbicides [19].

This review aims to examine allelopathy in microalgae, focusing on the potential applications of allelochemicals as herbicides, insecticides, and in their action against phytopathogens.

2. Allelopathy in Microalgae

As mentioned previously, the term allelopathy was originally used to describe biomolecules produced by plants. However, it is now openly used with other organisms and has been studied in microalgae [19,20]. Microalgae organisms produce numerous secondary metabolites, including fatty acids, alkaloids, amino acids, and peptides, which are investigated and characterized as natural products by chemists and described as novel bio compounds. In the environment, these molecules are secreted for communication, defence, and adaptation purposes [21]. More knowledge is needed in order to understand the complexity of these bioactive chemicals and their interaction with and effect on ecosystems [19,22].

A recent review paper showed that *Microcystis, Nodularia, Karenia, Alexandrium (A. catenella, A. fundyense,* and others species), *Skeletonema, Chlorella* (*C. vulgaris, C. sorokiniana, C. ellipsoidea*), *Chlamydomonas,* and *Dunaliella* displayed allelopathic activity against other microalgae and cyanobacteria [23]. One strategy to study the allelopathic effects of different microalgae agents is to analyse the biochemicals present in mono- and co-cultures [24]. Wang et al. [24] observed that the dinoflagellate *Scrippsiella* acuminata had its growth significantly inhibited when in co-cultivation and/or in the presence of exudates (cell-free and sonicated-cell filtrates) from three diatoms (*Chaetoceros curvisetus, Phaeodactylum tricornutum* and *Skeletonema dohrnii*), suggesting the diatoms developed significant competitive advantage over other antagonists. Furthermore, the authors detected several types of volatile organic compounds (VOCs) in the sonicated cell-free filtrates. Polyunsaturated aldehydes (PUAs) were found in all three diatoms that showed inhibitory effects on the growth of *S. acuminata* and the diatoms themselves. Generally, the production of microalgae VOCs is influenced by environmental conditions, such as nutrient availability, temperature, and light incidence [25,26].

Studies show that allelochemicals of microalgae have either a positive or negative impact on other organisms. Those which can cause inhibitory effects are suitable for application as biopesticides, and as a new source of antimicrobials or bio-herbicide agents [19,23], especially because these biocompounds present increased target specificity [27]. The liberation of species-specific allelochemicals is desired in agriculture and is an important focus for future studies because they can be responsible for modulating the ecosystem under favourable conditions to an allelopathic microalgae. An example of this was observed in a study where, by alternating the nutrients conditions and balance of nitrogen and phosphorus sources, the allelopathic interaction between strains of *Anabaena* spp. and *Microcystis* spp. seemed to control their relative abundance and their dominance in that habitat [28].

An interesting point to remark upon here is that one of several new applications of microalgae is as a bioenergy source and an alternative to fossil fuels [29–31]. Nevertheless, one of the elements that industry must be alert to is the accumulation of allelopathic molecules released into the water by microalgae themselves. These compounds, when in high concentrations due to high biomass, along with other biomolecules can trigger positive or negative effects on other cells, leading to auto-stimulation or self-inhibition [32–34]. Lu and colleagues (2020) described studies that focused on water reuse for microalgae cultivation, and stated that several factors, including allelopathy mechanisms, cause cell damage, aggregation, and programmed cell death (PCD) (negative effects) or, on the other hand, the liberation of growth regulators and removal of growth inhibitors (positive effects). Bacterial agents can have a significant influence on allelopathy affecting microalgae biomass tanks [33,35].

Some allelopathic molecules have been characterized in the scientific literature as having a role in algal community growth regulation. Satake et al. [36] described a 19 membered ring-structured molecule called alexandrolide, which was isolated from dinoflagellate *Alexandrium catenella* and inhibited diatoms (*Skeletonema costatum* and *Chattonella antiqua*) but did not inhibit other dinoflagellates. Marennine, a blue-green polyphenolic pigment produced by diatom *Haslea ostrearia*, was described as being responsible for inhibiting the growth of other diatom species and self-stimulating the producer species in oyster-ponds [37]. By elucidating the properties of these compounds, more studies can be performed to apply these chemicals for industrial and human purposes, such as in additives in agricultural technology [38].

3. Allelochemicals from Microalgae

Microalgae are the source of a remarkable diversity of biologically active compounds, which can be involved in allelopathic interactions. The main chemical classes of microalgal allelochemicals are: alkaloids, fatty acids and derivatives, polyketides, peptides, phenolics, and terpenoids.

3.1. Alkaloids

Alkaloids are a prominent class of allelochemicals with biocidal properties. They consist of nitrogen-containing heterocyclic substances, in which the nitrogen atom is usually derived from an amino acid [39,40]. Alkaloids have been reported mainly in cyanobacteria and dinoflagellates. In the former, indole alkaloids are the most frequent. For instance, 12-*epi*-hapalindole E isonitrile (Figure 2A; **1**), from *Fischerella* sp., is an indole alkaloid with activity against green microalgae and other microorganisms [41]. Another example is the toxin calothrixin A (Figure 2A; **2**) from *Calothrix* sp [41].

Domoic acid (Figure 2A; **3**) is a well-known biotoxin from the bloom-generating diatom *Pseudo-nitzschia* sp. which has a pyrrolidine skeleton derived from glutamic acid [42–44]. Another example of an alkaloid from diatoms is saxitoxin (Figure 2A; **4**), a potent neurotoxin with a complex skeleton whose nitrogen atoms are derived from arginine. It has been reported in cyanobacterial species as well [45–48].

3.2. Fatty acids and Derivatives

Extracellular free fatty acids from microalgae have a possible allelopathic effect, controlling the growth of other microalgae and harmful organisms. Green microalgae, such as *Chlorella* sp. and *Botryococcus braunii*, are reported to secrete mixtures of common fatty acids (e.g., oleic and linoleic acids; Figure 2B) under certain conditions, which may favor their dominance [49,50]. Additionally, unsaturated fatty acids are reported to possess antimicrobial activity [51].

Substances derived from fatty acids can also act as allelochemicals. Polyunsaturated fatty acids (PUFAs), after various oxidative transformations, originate polyunsaturated aldehydes (PUAS) (e.g., 2*E*,4*E*-decadienal, 2*E*,4*E*-heptadienal, and 2*E*,4*E*-octadienal; Figure 2B), which play a role in communication and chemical defense in diatoms [11,39,52]. They exert a deleterious effect on grazers and competing microalgal species, and stimulate organic-matter-recycling bacteria [52].

3.3. Polyketides

Polyketides are a large group of structurally diverse substances biosynthesized from carboxylic acid precursors through repeated cycles of condensation, reduction and dehydration reactions catalyzed by multifunctional enzymatic complexes called polyketide synthases (PKS). These substances are found mainly in dinoflagellates and cyanobacteria and are frequently associated with harmful blooms [39,53]. Examples include maitotoxin from the dinoflagellate *Gambierdiscus* sp., the largest and most toxic known polyketide [54,55], and also brevetoxins from *Karenia brevis* (e.g., brevetoxin A; Figure 2C; **10**), a dinoflagellate associated with "red tides" [56]. Another remarkable example are trichophycins (e.g., trichophycin A; Figure 2C; **11**), linear chlorinated polyketides from cyanobacteria from the genus *Trichodesmium*, which produce frequent blooms in the Pacific Ocean [57,58].

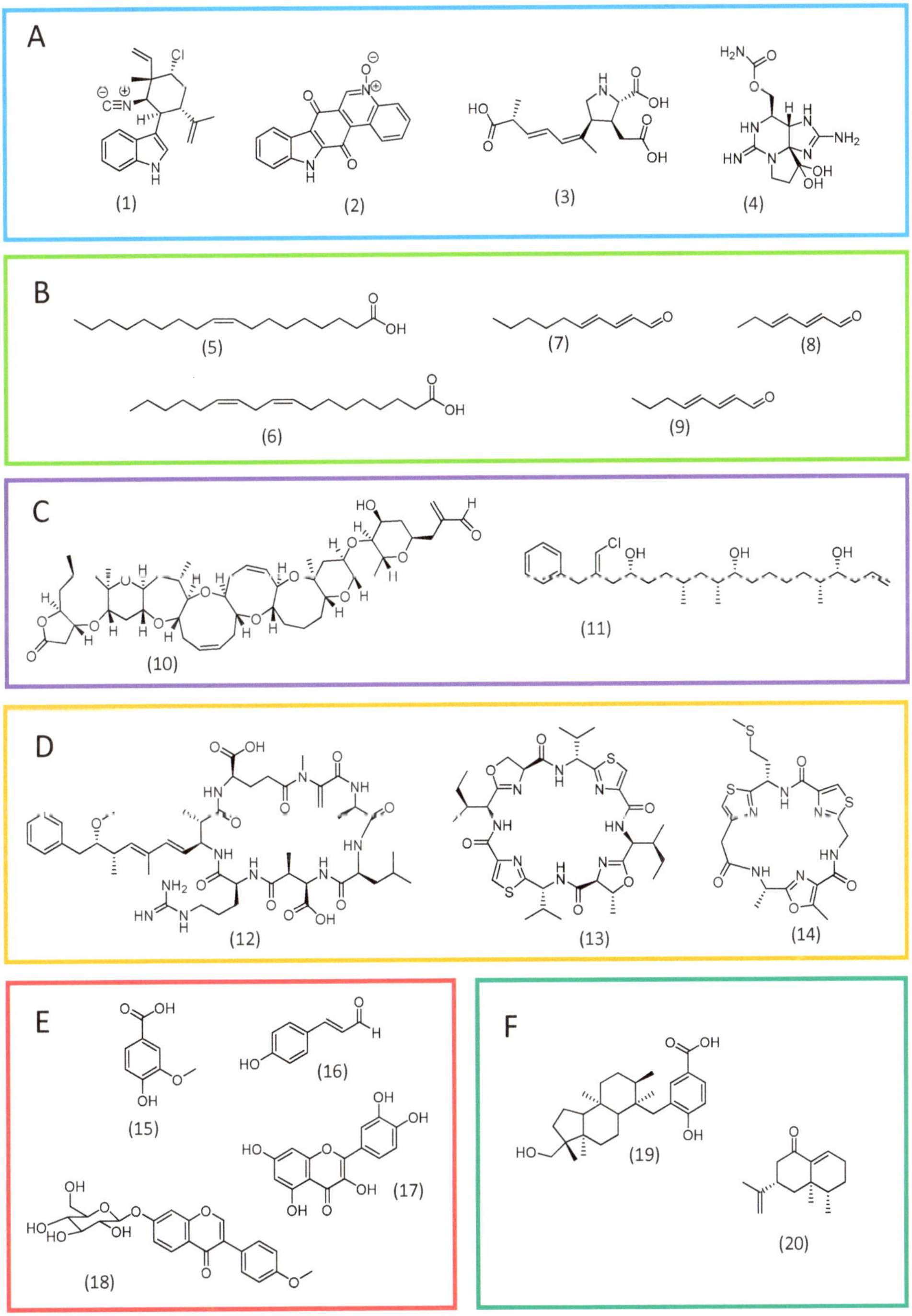

Figure 2. Examples of allelochemicals from microalgae. Alkaloids (**A**): 12-*epi*-hapalindole E isonitrile (**1**), calothrixin A (**2**), domoic acid (**3**), and saxitoxin (**4**). Fatty acids and derivatives (**B**): oleic acid (**5**), linoleic acid (**6**), 2*E*,4*E*-decadienal (**7**), 2*E*,4*E*-heptadienal (**8**), and 2*E*,4*E*-octadienal (**9**). Polyketides (**C**): brevetoxin A (**10**) and trichophycin A (**11**). Peptides (**D**): microcystin-LR (**12**), patellamide A (**13**), and tenuecyclamide C (**14**). Phenolic substances (**E**): vanillic acid (**15**), p-coumaric acid (**16**), quercetin (**17**) and ononin (**18**). Terpenoids (**F**): comnostin A (**19**) and eremophilone (**20**).

3.4. Peptides

A significant amount of cyanobacterial allelochemicals are peptides. Most of them belong to the group of non-ribosomal peptides, which are biosynthesized by multifunctional enzyme complexes, the non-ribosomal peptide synthases (NRPS). NPRS are similar to PKS and assemble sequentially common amino acids but also unusual amino acids [12,39,59]. For example, microcystins are cyclic non-ribosomal peptides produced by cyanobacteria of the genus *Anabaena*, *Microcystis*, *Planktothrix*, *Oscillatoria*, and others. They are toxic for microalgae, aquatic plants, and for mammals and birds as well [12,60]. Microcystin-LR (Figure 2D; **12**) is the most frequent of them [60].

Ribosomal peptides also occur in cyanobacteria. For instance, cyanobactins are small cyclic peptides formed solely from proteinogenic amino acids and reported in many genus, such as *Anabaena*, *Lyngbya*, *Microcystis*, *Nostoc*, and *Trichodesmium* [59,61]. They possess antimicrobial activity and probably have the function of inhibiting the growth of competing and harmful microorganisms. Patellamide A, from *Prochloron*, and tenuecyclamide C, from *Trichodesmium erythraeum*, are representatives of the group (Figure 2D; **13** and **14**) [61].

3.5. Phenolic Compounds

Phenolic compounds represent a large group of secondary metabolites, which play multiple roles in photosynthetic organisms, including in defense strategies and in communication with interacting species [62,63]. Phenolic substances are biosynthesized mainly by the phenylpropanoid pathway, having phenylalanine or tyrosine as precursors. In microalgae, knowledge about their biosynthetic pathways is still scarce, however, a recent study has shown that the enzymatic apparatus for key phenolic biosynthetic intermediates is well conserved in all major algal divisions, although the ability to biosynthesize some subclasses of phenolics seems to be restricted to some groups [62]. For instance, phenolic acids such as vanillic, ferulic, and coumaric acids are widespread. Flavonoids are frequent in green algae (e.g., quercetin and apigenin) and cyanobacteria, in which unusual structures have been reported (e.g., ononin from *Synechocystis*) (Figure 2E) [62].

3.6. Terpenoids

Terpenoids are a prominent family of organic substances biosynthesized from C_5 precursors, which are present in all living organisms [64]. Substances belonging to this group can also act as allelochemicals. Examples are comnostins, extracellular antibacterial diterpenoids produced by cyanobacteria *Nostoc* sp. [65], and the sesquiterpene eremophilone (Figure 2F), produced by cyanobacteria *Calothrix* sp. and toxic to invertebrates [66].

3.7. Volatile Organic Compounds

A specific group of allelochemicals is the volatile organic compounds (VOCs). VOCs are a broad group of compounds that encompass substances from different chemical classes, ranging from small molecules such as methane, to terpenoids and compounds derived from PUFAs including hydrocarbons, aldehydes, and ketones [25]. With a complex classification, the World Health Organization (WHO) defines VOCs as having boiling point ranges that have to fall between 50 °C and 100 °C up to 240–260 °C [67].

VOCs produced by microalgae have been widely demonstrated to have different ecological functions, including allelopathy. Environmental factors such as light, temperature, nutrition, and abiotic stresses affect VOCs emission [68]. Functions such as defense mechanisms, stress response, intra- and inter-species communication, modulating predator-prey interaction, and affecting biochemistry, metabolism, and physiology processes are described. These emissions from microalgae occur in vivo, in cells undergoing senescence or apoptosis, or when they are perishing under predator attack [69,70]. The role and the possible impact that these compounds have has been extensively studied and, in the future, they could be exploited for industrial purposes [67].

4. Allelochemicals from Microalgae for Biopesticides

Allelochemicals are considered safer and eco-friendly alternatives for efficient pest management as they have different and diversified modes of action compared with synthetic pesticides and are more easily degradable and less toxic to non-target organisms [71–74]. The potential use of allelochemicals from microalgae as biopesticides has attracted much attention lately [19,74–77]. In this section, we will present and discuss the potential use of allelochemicals from microalgae for the biocontrol of insects, weeds, and phytopathogens.

4.1. Allelochemicals against Phytopathogens

Allelochemicals from microalgae are known for their activity against a large spectrum of microorganisms, as evidenced by several reports [78–81]. Examples include hapalindole T, an antibacterial alkaloid from *Fischerella* sp. [82], nostofungicine, a lipopeptide with fungicide activity isolated from *Nostoc* sp. [83], and eicosapentaenoic acid (EPA), an antimicrobial fatty acid from *Phaeodactylum tricornutum* [84]. For this reason, extracts from microalgae have been evaluated for their potential against phytopathogens with encouraging outcomes.

Senousy et al. [85] evaluated the methanolic extracts of ten microalgal strains against phytopathogenic fungi. Among them, three stood out: the cyanobacteria *Anabaena* sp. HSSASE11 and *Oscillatoria nigroviridis* HSSASE 15, which were active against the fungi *Botryodiplodia theobromae* and *Pythium ultimum*, respectively, and the Chlorophyta *Dunaliella* sp. HSSASE13 which was active against *Fusarium solani* [85]. Similarly, Schmid et al. [86] assayed aqueous extracts from five microalgae for their activity against fungal phytopathogens. The Chlorophyta *Scenedesmus obliquus* showed the best results, inhibiting *Sclerotium rolfsii* by 32% [86]. Another Chlorophyta, *Chlorella vulgaris*, was active against the phytopathogenic bacteria *Stenotrophomonas maltophilia*, which affects potato and green vegetables. The researchers evaluated three types of extracts: aqueous, ethanolic, and methanolic. Only the latter was active against the bacteria [87].

Interestingly, microalgae can also be active against phytopathogenic nematodes. Suspended algal cultures from *Scenedesmus obliquus*, *Chlorella vulgaris*, and *Anabaena oryzae* were able to inhibit the root knot nematode (*Meloidogyne incognita*), a pathogen for banana cultures [88].

The activity of microalgae extracts against phytopathogenic organisms is frequently attributed to metabolites such as phenolics, alkaloids, and peptides [86,87,89]; however, few have conducted a thorough investigation to identify these compounds. In the study of Senousy et al. [85] mentioned above, the authors observed a positive correlation between the activity of the microalgal extracts against *Fusarium solani* and their phenolics and flavonoid content. These results indicate the probable chemical class of the bioactive substances. Phenolic substances also seem to be responsible for the antifungal activity of *Nannochloropsis* sp and *Spirulina* sp. Extracts composed of free phenolic acids from both microalgae exhibited significant inhibitory activity of *Fusarium* complex fungal pathogens. Both extracts had chlorogenic acid, a cinnamic acid derivative, (Figure 3; **21**) as their major component [90]. In another study, a phenolic substance isolated from culture filtrates of the cyanobacteria *Calothrix elenkinii* was reported as bioactive against the oomycete *Pythium debaryanum*. This substance, identified as 3-acetyl-2-hydroxy-6-methoxy-4-methyl benzoic acid (Figure 3; **22**), induced multiple morphological abnormalities in the hyphae of the microorganism [91].

Even though the antifungal mechanisms of the phenolic compounds mentioned above are not known, mechanistic studies on cinnamic and benzoic acid derivatives have been reported in the scientific literature. The antifungal activity of trans-cinnamaldehyde is related to the biosynthesis of cell wall components (glucan and chitin) [92,93]. In addition, methyl trans-cinnamate acts as an inhibitor of tyrosinase enzymes, which makes it interesting as a food anti-browning agent, but it also presented antimicrobial activity against bacteria and fungi [94]. Moreover, benzoic acid and its derivatives generally have multiple effects on fungal cells such as, for example, against *Alternaria solani* and *Eutypa lata* phytopathogens and other fungal genera. Several authors have discussed the fact that these phenolic substances generally affect cell respiration by altering cytochrome pathways, causing energy reduction by unbalancing intracellular benzoate levels, and inducing intracellular acidification, leading to protein aggregation and oxidative stress, for example. It is important to highlight that the effect of antifungal action depends on the concentration and molecular arrangement of these compounds [95–97].

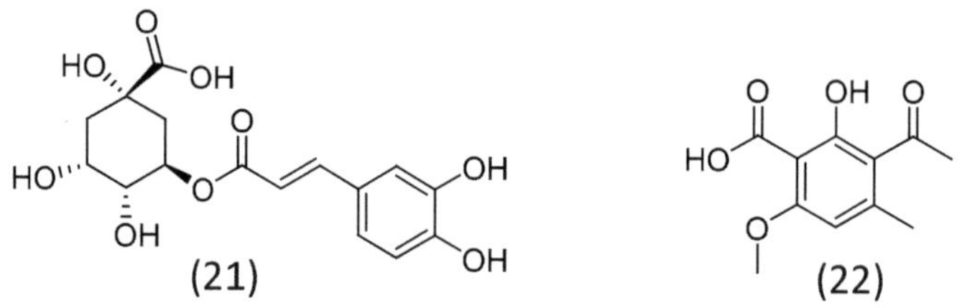

Figure 3. Phenolic substances active against phytopathogens: chlorogenic acid (**21**) and 3-acetyl-2-hydroxy-6-methoxy-4-methyl benzoic acid (**22**).

4.2. Allelochemicals as Herbicides

The focus on the herbicide potential of metabolites from microalgae has also had some promising results. Allelochemicals identified as having phytotoxic activity belong to different chemical classes, but most of them were detected in cyanobacteria. Examples are depicted in Figure 4 and will be explored in this section. This is the case for cyanobacterin (Figure 4; **23**), a phenolic substance isolated from *Scytonema hofmanni* in 1982. The substance was demonstrated to be a potent inhibitor of the photosynthetic electron transport in photosystem II. The inhibition of photosynthesis is a potential mechanism of herbicides [98,99]. The peptide nostocyclamide (Figure 4; **24**), from *Nostoc* sp., and the alkaloids fischerellins (Figure 4; **26**), from *Fischerella*, were also shown to inhibit photosystem II [99].

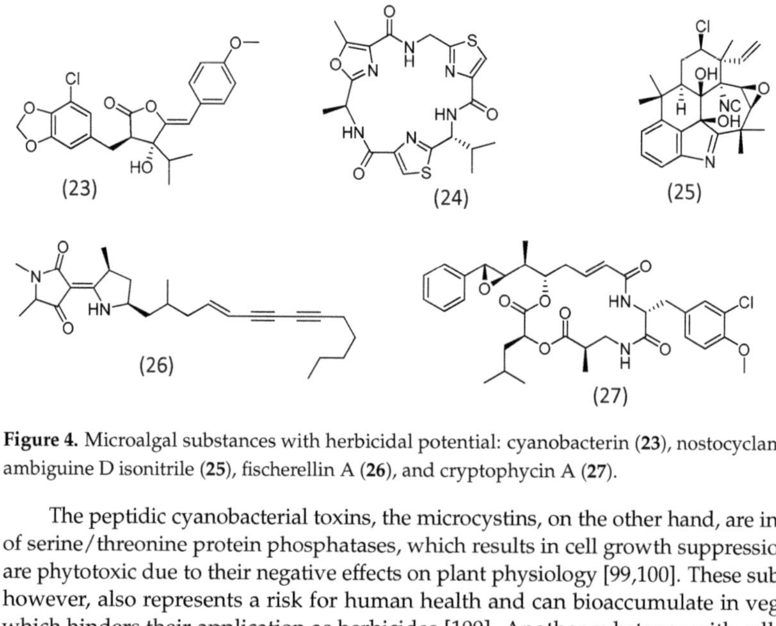

Figure 4. Microalgal substances with herbicidal potential: cyanobacterin (**23**), nostocyclamide (**24**), ambiguine D isonitrile (**25**), fischerellin A (**26**), and cryptophycin A (**27**).

The peptidic cyanobacterial toxins, the microcystins, on the other hand, are inhibitors of serine/threonine protein phosphatases, which results in cell growth suppression. They are phytotoxic due to their negative effects on plant physiology [99,100]. These substances, however, also represents a risk for human health and can bioaccumulate in vegetables, which hinders their application as herbicides [100]. Another substance with cell growth inhibitory activity is ambiguine D isonitrile (Figure 4; **25**), isolated from *Hapalosiphon* spp. This indole alkaloid was shown to suppress lettuce growth, an effect probably related to cell death induction by the overproduction of reactive oxygen species (ROS) and mitosis inhibition. Its potential as an herbicide remains to be further investigated [101].

Other substances with cell growth inhibitory activity may also exert herbicidal effects. For instance, cryptophycins (Figure 4; **27**), polyketides from *Nostoc* strains, which suppress microtubule polymerization. Methanolic extracts of a cryptophycin producer *Nostoc* strain hindered the growth of the stems and, in particular, the roots of a grass mixture [102], an effect that may be related to the cryptophycin content. Microalgae, especially cyanobacteria, are a rich source of cell division inhibitors, which have been primarily assessed for their antitumor or antimicrobial properties. These compounds, however, could also be useful as herbicides and represent an almost unexplored research field [99].

Metabolites from eukaryotic microalgae have been scarcely investigated for their herbicidal activity. However, some studies can be found, for example, VOCs from the green microalgae *Chlorella vulgaris* have evidenced a phytotoxic effect against barley. The mixture of volatiles consisted of hydrocarbons, carboxylic acids, aldehydes, alcohols, and ketones [103].

4.3. Allelochemicals as Insecticides

Allelochemicals produced by microalgae may be useful in the defense against plant predators. Thus, microalgae have been proposed as a source of allelochemicals with insecticide properties [104,105]. The good larvicidal performance of microalgae has also stimulated new technologies, such as the exploitation of the insecticide properties of *Chlamydomonas reinhardtii* extracts in the development of preparations based on nano/microparticles. The extract of the microalgae combined with microparticles of zinc oxide was able to double the mortality of mealworm beetle larvae (*Tenebrio molitor*) compared to zinc oxide alone [106].

Biofilm-forming cyanobacteria showed defense mechanisms against insects that hinder feeding, such as the formation of colonies, filaments, mucilage, spines, and production of allelochemicals. The chemical compounds produced by their biofilm is considered more effective than other defense mechanisms. *Fischerella* strain 43,239 was selected in a cyanobacterial screening for insecticidal activity against the larvae of dipteran *Chironomus riparius* (harlequin fly). The toxic active fraction from *Fischerella* 43,239 had its chemical profile assessed by mass spectrometry, which provided evidence of the presence of indole compounds, considered possible bioactive substances [107]. Later, another study with *Fischerella* 43,239 identified indole alkaloids (Figure 5; **28–31**) with insecticide activity against the larvae of *Chironomus riparius*. Among the isolated hapalindoles, 12-*epi*-hapalindole J isonitrile was able to increase the larvae mortality at 26 μM within 48 h [108].

Extracts from cyanobacteria were tested against larvae of *Aedes aegypti*. Among the samples tested, *Microcystis aeruginosa* 205 and *Anabaena circinalis* 86 were the most active. The fractionation of these extracts showed the presence of anatoxin A and microcystins, which were toxic for *Aedes aegypti* larvae [109]. Microcystins are cyclic peptides isolated from cyanobacteria that exhibit high toxicity to insects [110]. These compounds were shown to form lesions in the midgut epithelial cells of *Aedes aegypt* larvae [111]. Anatoxin-A (Figure 5; **32**) is another insecticidal secondary metabolite from cyanobacteria that acts as an acetylcholine analogue. The action mechanisms of these compounds is similar to organophosphate and carbamate insecticides; the inhibition of the acetylcholinesterase enzyme (AchE) thus disrupts the nervous system of the insects [112]. An extract of cyanobacteria rich in anatoxin- A exhibited potential insecticide activity against *Nauphoeta cinerea* cockroaches. It is speculated that the compound caused cholinergic intoxication affecting the insect locomotion [113].

The insecticide potential of *Amphora coffeaeformis* (diatom) and *Scenedesmus obliquus* (Chlorophyta) were investigated against *Culex pipiens* mosquito larvae. The acetonic extract of these microalgae showed larvicidal activity. Among them, the extracts of *Amphora coffeaeformis* was the most active (LC_{50} = 513.63 μg/mL), while the lethal concentration of *Scenedesmus obliquus* was 855.66 μg/mL. The analysis of the extracts by HPLC-DAD detected the presence of phenolic compounds such as benzoic acids (gallic acid, methyl gallate, and syringic acid), cinnamic acids (caffeic acid and cinnamic acids), and flavonoids (taxifolin, kaempferol, and naringenin) [114]. The insecticidal properties of some of the phenolic compounds identified have already been described. Gallic acid (Figure 5; **33**) affects the growth of *Spodoptera litura* (tobacco cutworm) larvae and is toxic to *Aphis craccivora* (an aphid) [115,116]. Caffeic acid (Figure 5; **34**), a cinnamic acid derivative, inhibits the digestive protease of *Helicoverpa armigera* (polyphagous pest) [117]. Moreover, kaempferol (Figure 5; **35**) was able to cause 82% mortality of *Mythimna separata* (pest of cereal crops) in the concentration of 1 mg/mL [118].

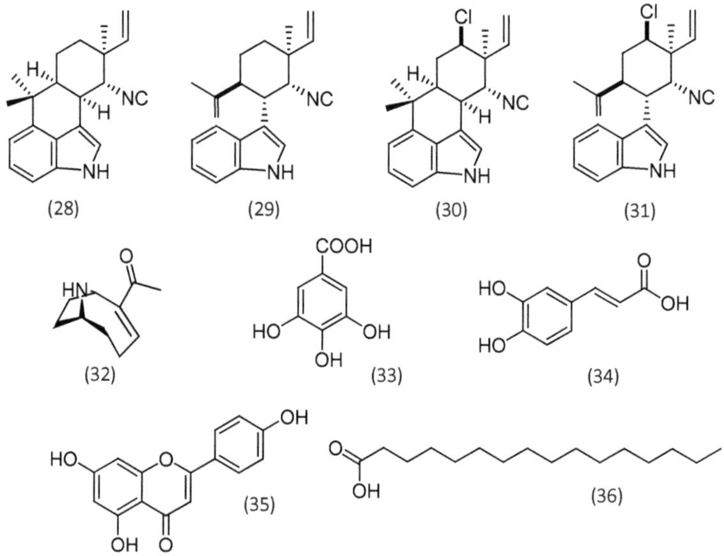

Figure 5. Microalgal substances with larvicidal or insecticidal activity. 12-*epi*-hapalindole J isonitrile (**28**), 12-*epi*-hapalindole C isonitrile (**29**), hapalindole L (**30**), 12-*epi*-hapalindole E isonitrile (**31**), anatoxin- A (**32**), gallic acid (**33**), caffeic acid (**34**), kaempferol (**35**), and *n*-hexadecenoic acid (**36**).

Unsaturated fatty acids were also detected in *Amphora coffeaeformis* and *Scenedesmus obliquus* extracts by gas chromatography coupled with mass spectrometry (CG/MS). Fatty acids can also act against insects. Hexadecanoic acid methyl ester, 7- hexadecanoic acid methyl ester, and 9-octadecenoic acid (Z)-methyl ester were detected as major compounds in *Amphora coffeaeformism*, while in *Scenedesmus obliquus* extract (Z)-9-octadecenoic acid methyl ester, hexadecanoic acid (Figure 5; **36**) methyl ester, and 7-hexadecenoic acid methyl ester were the major compounds. It is speculated that the high content of phenolics compounds and unsaturated fatty acids in these extracts are responsible for their larvicidal activity. [114].

A study described unsaturated and saturated fatty acids as being responsible for the insecticidal activity of ethanolic extracts from microalgae against *Spodoptera littoralis* (cotton leafworm) larvae [119]. The bioassay-guided fractionation of chloroform and methanol extracts from *Chlorella* sp. showed that fatty acids (*n*-hexadecanoic acid, di-*N*-octyl phthalate, 1-hexacosanol, oleic acid, and 1-docosene) are among the larvicidal compounds in the active fractions. Some fatty acids detected in these fractions had their insecticidal activity reported [120]. Additionally, another study described *n*-hexadecanoic (Figure 5; **36**) acid as a potent larvicidal against *Culex quinquefasciatus*, *Anopheles stephensi*, and *Aedes aegypti* mosquitoes [121]. Fatty acids have also been evaluated in vitro against *Melanaphis sacchari* (an aphid) Zehntner larvae. Among them, linoleic acid showed 85% of mortality rate (LC_{50} = 1181 ppm) [122].

The possible mechanism of larvicidal activity of fatty acids was investigated by Perumalsamy et al. [123]. The authors evaluated some fatty acids for two common modes of insecticidal activity: the inhibition of AchE and the interference with the octopamine signaling pathways (cAMP level enhancement). While linoleic and linolenic acids were able to both inhibit AchE and enhance cAMP levels, palmitic and oleic acid inhibited AchE and had no significant effects on cAMP levels. Elaidic and arachidic acids where ineffective as AchE inhibitors but considerably enhanced cAMP levels. These results indicate that linoleic and linolenic acids might have a dual mode of action [123].

5. Safety and Toxicity of Allelochemicals and Synthetic Pesticides

The herbicide resistance (HR) plant list continues to grow and is listed on the site: www.weedscience.org (accessed on 1 May 2023) [124]. New herbicides and pesticides are necessary to feed the world's population and even more so as pests become resistant to currently available pesticides. Pesticides are, by nature, toxic and the challenge is to increase their specificity and reduce their persistence. Pesticides can cause adverse effects on the quality of the agricultural products they are used to protect. Pesticide residues can produce long-term adverse effects on the health of humans and animals and the stability of ecosystems by poisoning water sources. These chemicals can be the cause of neurotoxicity, endocrine disruption, Parkinson's disease, cancers, and type 2 diabetes [125]. For instance, organochlorine pesticides, used in agriculture since the 1950s, are listed as a persistent organic pollutant (POP) and forbidden in most countries. Nevertheless, high concentrations of these pesticides and their transformation products (TPs) and degradation products, such as dichlorodiphenyltrichloroethane (DDT), are very persistent and have long half-lives. All of them are still detected in agricultural soils [126]. Modern formulations, called current-use pesticides (CUPs), are relatively safe to non-target species. The objective of modern formulations is to avoid persistent compounds that bioaccumulate and reduce their non-target toxic properties. However, they are capable of migrating over long distances and CUPs have been reported in air and surface water samples in the Arctic [127].

Allelochemicals as a source of biopesticides have been studied in greater detail from plants than from microalgae. However, several compounds found in plants with potential use as biopesticides are extensively present and described in microalgae, such as organic acids, esters, alkaloids, flavonoids, and terpenoids [19,76,128]. The multi-kingdom effects of some secondary allelopathic plant metabolites are known. Despite the original definition, that addresses plant–plant interactions, several compounds exhibit allelopathy and toxicity to other organisms. Currently, the term allelopathy is now used in this broader context by the International Society of Allelopathy and has been since the 1990s [129].

Plant allelochemicals typically present partial water solubility enabling easy application without surfactants and some of them are known to have similar modes of action to synthetic herbicides. Non-halogenated molecules decrease the environmental half-life while preventing soil accumulation. They have eco-environmental chemical structures with a higher nitrogen and oxygen content [130]. Modern technologies allow for their effective use and management and confer fewer environmental problems in soil due to the faster degradability of allelochemicals. However, rapid degradation of one of the chemical groups can decrease the compound's bioactivity [131].

The diversity of allelochemicals makes them promising targets for discovering novel, specific target sites in plants or other organisms. Besides this, allelochemicals are also characterized by multi-site action, which differs from the synthetic pesticides, and affords them high specificity. In the case of synthetic herbicides, before they are licensed for use, they are tested to evaluate that their phytotoxic activity is in the range between 10^{-5} and 10^{-7} M. The chemical characterization of the molecular structure, the study of the mode/s of action, and their half-life in soil need to be established. It is also necessary to evaluate their influence on microbial ecology and other organisms, their potential toxic properties on human health, and the economic viability of their production on commercial scales [132]. Allelochemical toxicity depends on the concentration and compounds involved. Several secondary metabolites of other natural substance groups from plants and microalgae, such as glycosides, polyphenols, polysaccharides, terpenes, flavones, and alkaloids, have shown to be ecologically safe and biodegradable [133–136]. Triketones are one of the most important herbicide classes developed from an allelochemical triketone. These compounds inhibit invasive plants. Another example is cryptophycin-1, a depsipeptide from the cyanobacterium *Nostoc* sp. Cryptophycin-1 has an antiproliferative and antimitotic activity, inhibiting the cell cycle during the metaphase of mitosis, mainly in yeasts of the genus *Cryptococcus* [74,137].

6. Microalgae Allelochemicals: Gaps and Opportunities

Even though the prospects for the application of microalgal allelochemicals as biopesticides seem promising, there are still many knowledge gaps about these compounds. For instance, the action mechanisms of most of them is poorly understood. Some allelochemicals are known to have modes of action similar to synthetic pesticides, such as the insecticidal compound anatoxin A (inhibition of acetylcholinesterase) and the herbicidal molecule cyanobacterin (inhibition of photosystem II). However, many others are likely to exhibit new mechanisms and multi-target actions, which should be elucidated [132].

Moreover, their persistence and biodegradation in the environment needs further study. Once released into the soil, allelochemicals are more readily microbially degraded than xenobiotic persistent synthetics compounds. Their biodegradation depends on the soil composition and on the structure and population of the soil microbial community. Most assays conducted to investigate pesticide properties of allelochemicals are conducted under artificial laboratory conditions. Thus, to further estimate their potential in agricultural applications and to assess their environmental impact, in-field bioassays are necessary [129,131]. Studies on their possible toxic effects to non-target organisms, including humans, are also mandatory. These steps are important for any biopesticide candidate and require a great amount of interdisciplinary research [129].

The research and application of plant allelochemicals in agriculture is undoubtedly far more advanced than those from microalgae. However, concerning the production process, microalgal allelochemicals may have multiple advantages. Microalgae outdoor cultivation systems can be implemented in non-arable lands employing wastewater from agricultural or industrial activities, thus avoiding competition with food production. Their cultivation is also possible in closed bioreactors. In this case, the most important advantage is the possibility of having good control of the chemical and physical conditions of the whole process, which enables a high reproducibility and high yield [30,138]. The mass cultivation of microalgae for the production of food and chemicals on a commercial scale has already been established. The carotenoid, astaxanthin, for instance, which can be obtained from different microalgae, annually moves approximately 2 billion dollars in the global market [138–140]. For the industrial production of astaxanthin, *Haematococcus pluvialis* is cultivated in a two-step process. In the first stage, the microalgae grow in photobioreactors with a culture-rich medium for biomass accumulation. The second stage, in which the production of the carotenoid is induced by nutrient starvation, can be carried out in larger bioreactors or in open ponds. This process is energy-intensive and costly; however, improvements have been intensively researched and their implementation can substantially reduce the prices of the product [140,141].

Thus, since it is already possible to achieve bioactive substances from microalgae on a large scale, the obtaining of other compounds, such as allelochemicals, employing similar biotechnological processes can be envisaged. Nevertheless, the costs of biopesticide production through microalgae cultivation are still higher than the production of synthetic pesticides. However, the production of microalgae-derived biopesticides could be integrated with the production of biofuels and other valuable compounds within a biorefinery concept. This integrated approach is expected to lower the overall costs and enable better energy utilization as well as a reduction in the generation of residues [30,142–145].

Synthetic biology also has the potential to modify microalgal allelochemicals to improve activity, increase productivity and decrease non-specific toxicity [146].

The literature shows progress is being made in developing natural products from microalgae as biopesticides. Studies are in progress to reach commercial biopesticides from microalgae; in the future, they can complement or be an alternative to synthetic pesticides. At the moment, several promising less toxic and biodegradable allelochemicals have been identified and are worthy of further investigation into their application.

7. Perspectives

Much has been reported on the ecological effects of allelochemicals from plants and microbes but less so from algae and microalgae. All studies have highlighted the importance of the secondary "special" metabolism that is crucial for plant productivity and the biocontrol of pests in natural, agricultural, and anthropogenic environments. Studies on microalgae, while they have not received as much attention as agricultural crop plants, are equally important and present a new and promising low-cost source of bioactive compounds of increasing economic importance. As previously discussed, microalgae can be employed as bio-factories not only as a source of green energy but also a low-cost source of bioactive molecules, which can be useful for agriculture with minimal negative environmental impact. In this light, the future for using algae within a circular economic context is strengthened and the value of the algal production is increased by separating the differing components of their metabolism. The mechanisms of algal allelopathy require further study and specifically at the genetic level. It seems highly likely that gene editing with CRISPR will only increase the production of molecules of interest in a directed way. Given that secondary metabolism often occurs in response to a stress, be that competitive or environmental, genetic studies involving the algal holobiont (algae + bacteria + virus) and the role of interference RNAs will help us manage the algal bio-factory. The increased use of bioactive compounds from algae is promising. Here, we have described the potential of the interesting allelopathic compounds from microalgae. We have made the case for more intensive research on microalgae and hinted that microalgal bio-factories warrant industrial investment.

Author Contributions: Conceptualization, L.M.C. and A.B.V.; writing—original draft preparation, L.M.C.; J.E.d.S.; A.N.J.; A.M. and A.B.V.; writing—review and editing, L.M.C.; J.E.d.S.; A.N.J.; A.M. and A.B.V. All authors have read and agreed to the published version of the manuscript.

Funding: This work was financially supported by the Brazilian National Council for Scientific and Technological Development (MCT-CNPq), Carlos Chagas Filho Foundation for Research Support in the State of Rio de Janeiro (FAPERJ), Coordination for the Improvement of Higher Education Personnel (CAPES), and the Brazilian Society of Microbiology (SBM).

Data Availability Statement: No new data were generated in this review article.

Conflicts of Interest: The authors declare no conflict of interest.

References

1. Molisch, H. Der Einfluss einer Pflanze auf die Andere, Allelopathie. *Nature* **1938**, *141*, 493. [CrossRef]
2. Tan, K.; Huang, Z.; Ji, R.; Qiu, Y.; Wang, Z.; Liu, J. A review of allelopathy on microalgae. *Microbiology* **2019**, *165*, 587–592. [CrossRef] [PubMed]
3. Gomes, M.P.; Garcia, Q.S.; Barreto, L.C.; Pimenta, L.P.S.; Matheus, M.T.; Figueredo, C.C. Allelopathy: An overview from micro- to macroscopic organisms, from cells to environments, and the perspectives in a climate-changing world. *Biologia* **2017**, *72*, 113–129. [CrossRef]
4. Inderjit; Dakshini, K.M.M. Algal allelopathy. *Bot. Rev.* **1994**, *60*, 182–196. [CrossRef]
5. Hemaiswarya, S.; Raja, R.; Ravikumar, R.; Carvalho, I.S. Microalgae Taxonomy and Breeding. In *Biofuel Crops: Production, Physiology and Genetics*; CABI: Wallingford, UK, 2013; pp. 44–53.
6. Sajjadi, B.; Chen, W.-Y.; Raman, A.A.A.; Ibrahim, S. Microalgae lipid and biomass for biofuel production: A comprehensive review on lipid enhancement strategies and their effects on fatty acid composition. *Renew. Sustain. Energy Rev.* **2018**, *97*, 200–232. [CrossRef]

7. Gallardo-Rodríguez, J.; Sánchez-Mirón, A.; García-Camacho, F.; López-Rosales, L.; Chisti, Y.; Molina-Grima, E. Bioactives from microalgal dinoflagellates. *Biotechnol. Adv.* **2012**, *30*, 1673–1684. [CrossRef]
8. Stevens, P.F. Angiosperm Phylogeny and Diversification. In *Encyclopedia of Evolutionary Biology*; Elsevier: Amsterdam, The Netherlands, 2016; pp. 78–83. ISBN 978-0-12-800426-5.
9. Watson, S.B.; Whitton, B.A.; Higgins, S.N.; Paerl, H.W.; Brooks, B.W.; Wehr, J.D. Harmful Algal Blooms. In *Freshwater Algae of North America*; Elsevier: Amsterdam, The Netherlands, 2015; pp. 873–920. ISBN 978-0-12-385876-4.
10. Mathieu, A. Advances in Bioengineering Study of Microalgae. *J. Pet. Environ. Biotechnol.* **2022**, *13*, 1000453.
11. Mendes, L.B.B.; Vermelho, A.B. Allelopathy as a potential strategy to improve microalgae cultivation. *Biotechnol. Biofuels* **2013**, *6*, 152. [CrossRef]
12. Leflaive, J.; Ten-Hage, L. Algal and cyanobacterial secondary metabolites in freshwaters: A comparison of allelopathic compounds and toxins. *Freshw. Biol.* **2007**, *52*, 199–214. [CrossRef]
13. Shahbaz, A.; Hussain, N.; Saba, S.; Bilal, M. Actinomycetes, Cyanobacteria, and Fungi: A Rich Source of Bioactive Molecules. In *Microbial Biomolecules*; Elsevier: Amsterdam, The Netherlands, 2023; pp. 113–133. ISBN 978-0-323-99476-7.
14. Luesch, H.; Yoshida, W.Y.; Moore, R.E.; Paul, V.J.; Mooberry, S.L. Isolation, Structure Determination, and Biological Activity of Lyngbyabellin A from the Marine Cyanobacterium *Lyngbya majuscula*. *J. Nat. Prod.* **2000**, *63*, 611–615. [CrossRef]
15. Legrand, C.; Rengefors, K.; Fistarol, G.O.; Granéli, E. Allelopathy in phytoplankton—Biochemical, ecological and evolutionary aspects. *Phycologia* **2003**, *42*, 406–419. [CrossRef]
16. Ali, S.; Ullah, M.I.; Sajjad, A.; Shakeel, Q.; Hussain, A. Environmental and Health Effects of Pesticide Residues. In *Sustainable Agriculture Reviews 48*; Inamuddin, Ahamed, M.I., Lichtfouse, E., Eds.; Sustainable Agriculture Reviews; Springer International Publishing: Cham, Switzerland, 2021; Volume 48, pp. 311–336. ISBN 978-3-030-54718-9.
17. Riyaz, M.; Ahmad Shah, R.; Sivasankaran, K. Pesticide Residues: Impacts on Fauna and the Environment. In *Biodegradation Technology of Organic and Inorganic Pollutants*; Ferreira Mendes, K., Nogueira de Sousa, R., Cabral Mielke, K., Eds.; IntechOpen: London, UK, 2022; ISBN 978-1-83968-895-9.
18. Gould, F.; Brown, Z.S.; Kuzma, J. Wicked evolution: Can we address the sociobiological dilemma of pesticide resistance? *Science* **2018**, *360*, 728–732. [CrossRef] [PubMed]
19. Chaïb, S.; Pistevos, J.C.A.; Bertrand, C.; Bonnard, I. Allelopathy and allelochemicals from microalgae: An innovative source for bio-herbicidal compounds and biocontrol research. *Algal Res.* **2021**, *54*, 102213. [CrossRef]
20. Molisch, H.; Narwal, S.S. *The Influence of One Plant on Another, Allelopathy*; Published by Scientific Publishers (India) for International Allelopathy Foundation, Hisar: Jodhpur, India, 2001; ISBN 978-81-7233-285-3.
21. Caruana, A.M.N.; Amzil, Z. Microalgae and Toxins. In *Microalgae in Health and Disease Prevention*; Elsevier: Amsterdam, The Netherlands, 2018; pp. 263–305. ISBN 978-0-12-811405-6.
22. Cembella, A.D. Chemical ecology of eukaryotic microalgae in marine ecosystems. *Phycologia* **2003**, *42*, 420–447. [CrossRef]
23. Śliwińska-Wilczewska, S.; Wiśniewska, K.; Konarzewska, Z.; Cieszyńska, A.; Felpeto, A.B.; Lewandowska, A.U.; Latała, A. The current state of knowledge on taxonomy, modulating factors, ecological roles, and mode of action of phytoplankton allelochemicals. *Sci. Total. Environ.* **2021**, *773*, 145681. [CrossRef]
24. Wang, Z.; Wang, C.; Li, W.; Wang, M.; Xiao, L. Interspecies competition between *Scrippsiella acuminata* and three marine diatoms: Growth inhibition and allelopathic effects. *Aquat. Toxicol.* **2021**, *237*, 105878. [CrossRef]
25. Kumari, P.; Cna'ani, A.; Didi-Cohen, S.; Tzin, V.; Khozin-Goldberg, I. Nitrogen Deprivation-Induced Production of Volatile Organic Compounds in the Arachidonic-Acid-Accumulating Microalga *Lobosphaera incisa* Underpins Their Role as ROS Scavengers and Chemical Messengers. *Front. Mar. Sci.* **2020**, *7*, 410. [CrossRef]
26. Sartori, R.B.; Siqueira, S.F.; Maroneze, M.M.; Fagundes, M.B.; Wagner, R.; Zepka, L.Q.; Jacob-Lopes, E. Microalgal secondary metabolites: Effect of climatic variables, seasons, and photocycles on the biogeneration of volatile organic compounds (VOCs). *J. Appl. Phycol.* **2021**, *33*, 1457–1472. [CrossRef]
27. Grabski, K.; Tukaj, Z. Autoinduction activity of a conditioned medium obtained from high density cultures of the green alga *Scenedesmus subspicatus*. *J. Appl. Phycol.* **2007**, *20*, 323–330. [CrossRef]
28. Chia, M.A.; Jankowiak, J.G.; Kramer, B.J.; Goleski, J.A.; Huang, I.-S.; Zimba, P.V.; do Carmo Bittencourt-Oliveira, M.; Gobler, C.J. Succession and toxicity of *Microcystis* and *Anabaena* (Dolichospermum) blooms are controlled by nutrient-dependent allelopathic interactions. *Harmful Algae* **2018**, *74*, 67–77. [CrossRef]
29. Ali, S.S.; Mastropetros, S.G.; Schagerl, M.; Sakarika, M.; Elsamahy, T.; El-Sheekh, M.; Sun, J.; Kornaros, M. Recent advances in wastewater microalgae-based biofuels production: A state-of-the-art review. *Energy Rep.* **2022**, *8*, 13253–13280. [CrossRef]
30. Casanova, L.M.; Mendes, L.B.B.; Corrêa, T.D.S.; da Silva, R.B.; Joao, R.R.; Macrae, A.; Vermelho, A.B. Development of Microalgae Biodiesel: Current Status and Perspectives. *Microorganisms* **2022**, *11*, 34. [CrossRef] [PubMed]
31. Oliveira, C.Y.B.; Jacob, A.; Nader, C.; Oliveira, C.D.L.; Matos, Â.P.; Araújo, E.S.; Shabnam, N.; Ashok, B.; Gálvez, A.O. An overview on microalgae as renewable resources for meeting sustainable development goals. *J. Environ. Manag.* **2022**, *320*, 115897. [CrossRef]
32. Loftus, S.E.; Johnson, Z.I. Cross-study analysis of factors affecting algae cultivation in recycled medium for biofuel production. *Algal Res.* **2017**, *24*, 154–166. [CrossRef]
33. Lu, Z.; Loftus, S.; Sha, J.; Wang, W.; Park, M.S.; Zhang, X.; Johnson, Z.I.; Hu, Q. Water reuse for sustainable microalgae cultivation: Current knowledge and future directions. *Resour. Conserv. Recycl.* **2020**, *161*, 104975. [CrossRef]

34. Thomas, P.K.; Hietala, D.C.; Cardinale, B.J. Tolerance to allelopathic inhibition by free fatty acids in five biofuel candidate microalgae strains. *Bioresour. Technol. Rep.* **2023**, *21*, 101321. [CrossRef]
35. Qixin, L.; Xuan, F.; Zhiya, S.; Wenxin, S.; Shuo, W.; Ji, L. Enhanced wastewater treatment performance by understanding the interaction between algae and bacteria based on quorum sensing. *Bioresour. Technol.* **2022**, *354*, 127161. [CrossRef]
36. Satake, M.; Honma, D.; Watanabe, R.; Oshima, Y. Alexandrolide, a diatom growth inhibitor isolated from the dinoflagellate *Alexandrium catenella*. *Tetrahedron Lett.* **2019**, *60*, 1341–1344. [CrossRef]
37. Pouvreau, J.-B.; Housson, E.; Le Tallec, L.; Morançais, M.; Rincé, Y.; Fleurence, J.; Pondaven, P. Growth inhibition of several marine diatom species induced by the shading effect and allelopathic activity of marennine, a blue-green polyphenolic pigment of the diatom *Haslea ostrearia* (Gaillon/Bory) Simonsen. *J. Exp. Mar. Biol. Ecol.* **2007**, *352*, 212–225. [CrossRef]
38. Kapoore, R.V.; Wood, E.E.; Llewellyn, C.A. Algae biostimulants: A critical look at microalgal biostimulants for sustainable agricultural practices. *Biotechnol. Adv.* **2021**, *49*, 107754. [CrossRef]
39. Sasso, S.; Pohnert, G.; Lohr, M.; Mittag, M.; Hertweck, C. Microalgae in the post-genomic era: A blooming reservoir for new natural products. *FEMS Microbiol. Rev.* **2012**, *36*, 761–785. [CrossRef] [PubMed]
40. Lichman, B.R. The scaffold-forming steps of plant alkaloid biosynthesis. *Nat. Prod. Rep.* **2020**, *38*, 103–129. [CrossRef] [PubMed]
41. Doan, N.T.; Rickards, R.W.; Rothschild, J.M.; Smith, G.D. Allelopathic actions of the alkaloid 12-epi-hapalindole E isonitrile and calothrixin A from cyanobacteria of the genera *Fischerella* and *Calothrix*. *J. Appl. Phycol.* **2000**, *12*, 409–416. [CrossRef]
42. Brunson, J.K.; McKinnie, S.M.K.; Chekan, J.R.; McCrow, J.P.; Miles, Z.D.; Bertrand, E.M.; Bielinski, V.A.; Luhavaya, H.; Oborník, M.; Smith, G.J.; et al. Biosynthesis of the neurotoxin domoic acid in a bloom-forming diatom. *Science* **2018**, *361*, 1356–1358. [CrossRef]
43. Quilliam, M.A.; Wright, J.L.C. The Amnesic Shellfish Poisoning Mystery. *Anal. Chem.* **1989**, *61*, 1053A–1060A. [CrossRef]
44. Takemoto, T.; Daigo, K. Constituents of *Chondria Armata*. *Chem. Pharm. Bull.* **1958**, *6*, 578–580. [CrossRef]
45. Orr, R.J.S.; Stüken, A.; Murray, S.A.; Jakobsen, K.S. Evolution and Distribution of Saxitoxin Biosynthesis in Dinoflagellates. *Mar. Drugs* **2013**, *11*, 2814–2828. [CrossRef]
46. Murray, S.A.; Mihali, T.K.; Neilan, B.A. Extraordinary Conservation, Gene Loss, and Positive Selection in the Evolution of an Ancient Neurotoxin. *Mol. Biol. Evol.* **2011**, *28*, 1173–1182. [CrossRef]
47. Schantz, E.J.; Lynch, J.M.; Vayvada, G.; Matsumoto, K.; Rapoport, H. The Purification and Characterization of the Poison Produced by *Gonyaulax catenella* in Axenic Culture. *Biochemistry* **1966**, *5*, 1191–1195. [CrossRef]
48. Jackim, E.; Gentile, J. Toxins of a Blue-Green Alga: Similarity to Saxitoxin. *Science* **1968**, *162*, 915–916. [CrossRef]
49. Chiang, I.-Z.; Huang, W.-Y.; Wu, J.-T. Allelochemicals of *Botryococcus braunii* (Chloroficeae). *J. Phycol.* **2004**, *40*, 474–480. [CrossRef]
50. DellaGreca, M.; Zarrelli, A.; Fergola, P.; Cerasuolo, M.; Pollio, A.; Pinto, G. Fatty Acids Released by *Chlorella vulgaris* and Their Role in Interference with *Pseudokirchneriella subcapitata*: Experiments and Modelling. *J. Chem. Ecol.* **2010**, *36*, 339–349. [CrossRef]
51. Benkendorff, K.; Davis, A.R.; Rogers, C.N.; Bremner, J.B. Free fatty acids and sterols in the benthic spawn of aquatic molluscs, and their associated antimicrobial properties. *J. Exp. Mar. Biol. Ecol.* **2005**, *316*, 29–44. [CrossRef]
52. Cózar, A.; Morillo-García, S.; Ortega, M.J.; Li, Q.P.; Bartual, A. Macroecological patterns of the phytoplankton production of polyunsaturated aldehydes. *Sci. Rep.* **2018**, *8*, 12282. [CrossRef] [PubMed]
53. Demay, J.; Bernard, C.; Reinhardt, A.; Marie, B. Natural Products from Cyanobacteria: Focus on Beneficial Activities. *Mar. Drugs* **2019**, *17*, 320. [CrossRef] [PubMed]
54. Yokoyama, A.; Murata, M.; Oshima, Y.; Iwashita, T.; Yasumoto, T. Some Chemical Properties of Maitotoxin, a Putative Calcium Channel Agonist Isolated from a Marine Dinoflagellate. *J. Biochem.* **1988**, *104*, 184–187. [CrossRef]
55. Reyes, J.G.; Sánchez-Cárdenas, C.; Acevedo-Castillo, W.; Leyton, P.; López-González, I.; Felix, R.; Gandini, M.A.; Treviño, M.B.; Treviño, C.L. Maitotoxin: An Enigmatic Toxic Molecule with Useful Applications in the Biomedical Sciences. In *Seafood and Freshwater Toxins*; Botana, L.M., Ed.; CRC Press: Boca Raton, FL, USA, 2014; pp. 677–694. ISBN 978-0-429-09621-1.
56. Baden, D.G.; Bourdelais, A.J.; Jacocks, H.; Michelliza, S.; Naar, J. Natural and Derivative Brevetoxins: Historical Background, Multiplicity, and Effects. *Environ. Health Perspect.* **2005**, *113*, 621–625. [CrossRef]
57. McManus, K.M.; Kirk, R.D.; Via, C.W.; Lotti, J.S.; Roduit, A.F.; Teta, R.; Scarpato, S.; Mangoni, A.; Bertin, M.J. Isolation of Isotrichophycin C and Trichophycins G–I from a Collection of *Trichodesmium thiebautii*. *J. Nat. Prod.* **2020**, *83*, 2664–2671. [CrossRef]
58. Bertin, M.J.; Wahome, P.G.; Zimba, P.V.; He, H.; Moeller, P.D.R. Trichophycin A, a Cytotoxic Linear Polyketide Isolated from a *Trichodesmium thiebautii* Bloom. *Mar. Drugs* **2017**, *15*, 10. [CrossRef]
59. Micallef, M.L.; D'Agostino, P.M.; Al-Sinawi, B.; Neilan, B.A.; Moffitt, M.C. Exploring cyanobacterial genomes for natural product biosynthesis pathways. *Mar. Genom.* **2014**, *21*, 1–12. [CrossRef]
60. Rastogi, R.P.; Sinha, R.P.; Incharoensakdi, A. The cyanotoxin-microcystins: Current overview. *Rev. Environ. Sci. Bio/Technol.* **2014**, *13*, 215–249. [CrossRef]
61. Sivonen, K.; Leikoski, N.; Fewer, D.P.; Jokela, J. Cyanobactins—Ribosomal cyclic peptides produced by cyanobacteria. *Appl. Microbiol. Biotechnol.* **2010**, *86*, 1213–1225. [CrossRef] [PubMed]
62. Del Mondo, A.; Smerilli, A.; Ambrosino, L.; Albini, A.; Noonan, D.M.; Sansone, C.; Brunet, C. Insights into phenolic compounds from microalgae: Structural variety and complex beneficial activities from health to nutraceutics. *Crit. Rev. Biotechnol.* **2021**, *41*, 155–171. [CrossRef]

63. Del Mondo, A.; Sansone, C.; Brunet, C. Insights into the biosynthesis pathway of phenolic compounds in microalgae. *Comput. Struct. Biotechnol. J.* **2022**, *20*, 1901–1913. [CrossRef]
64. Oldfield, E.; Lin, F.-Y. Terpene Biosynthesis: Modularity Rules. *Angew. Chem. Int. Ed.* **2011**, *51*, 1124–1137. [CrossRef]
65. Jaki, B.; Orjala, J.; Heilmann, J.; Linden, A.; Vogler, B.; Sticher, O. Novel Extracellular Diterpenoids with Biological Activity from the Cyanobacterium *Nostoc commune*. *J. Nat. Prod.* **2000**, *63*, 339–343. [CrossRef] [PubMed]
66. Höckelmann, C.; Becher, P.G.; von Reuß, S.H.; Jüttner, F. Sesquiterpenes of the Geosmin-Producing Cyanobacterium Calothrix PCC 7507 and their Toxicity to Invertebrates. *Z. Naturforsch C. J. Biosci.* **2009**, *64*, 49–55. [CrossRef] [PubMed]
67. Pozzer, A.C.; Gómez, P.A.; Weiss, J. Volatile organic compounds in aquatic ecosystems—Detection, origin, significance and applications. *Sci. Total. Environ.* **2022**, *838*, 156155. [CrossRef] [PubMed]
68. Zuo, Z. Why Algae Release Volatile Organic Compounds—The Emission and Roles. *Front. Microbiol.* **2019**, *10*, 491. [CrossRef]
69. Achyuthan, K.E.; Harper, J.C.; Manginell, R.P.; Moorman, M.W. Volatile Metabolites Emission by In Vivo Microalgae—An Overlooked Opportunity? *Metabolites* **2017**, *7*, 39. [CrossRef]
70. Xie, Y.; Tian, L.; Han, X.; Yang, Y. Research Advances in Allelopathy of Volatile Organic Compounds (VOCs) of Plants. *Horticulturae* **2021**, *7*, 278. [CrossRef]
71. Liu, X.; Cao, A.; Yan, D.; Ouyang, C.; Wang, Q.; Li, Y. Overview of mechanisms and uses of biopesticides. *Int. J. Pest Manag.* **2021**, *67*, 65–72. [CrossRef]
72. Kumar, J.; Ramlal, A.; Mallick, D.; Mishra, V. An Overview of Some Biopesticides and Their Importance in Plant Protection for Commercial Acceptance. *Plants* **2021**, *10*, 1185. [CrossRef] [PubMed]
73. Farooq, M.; Jabran, K.; Cheema, Z.A.; Wahid, A.; Siddique, K.H. The role of allelopathy in agricultural pest management. *Pest Manag. Sci.* **2011**, *67*, 493–506. [CrossRef] [PubMed]
74. Costa, J.A.V.; Freitas, B.C.B.; Cruz, C.G.; Silveira, J.; Morais, M.G. Potential of microalgae as biopesticides to contribute to sustainable agriculture and environmental development. *J. Environ. Sci. Health Part B* **2019**, *54*, 366–375. [CrossRef]
75. Berthon, J.-Y.; Michel, T.; Wauquier, A.; Joly, P.; Gerbore, J.; Filaire, E. Seaweed and microalgae as major actors of blue biotechnology to achieve plant stimulation and pest and pathogen biocontrol—A review of the latest advances and future prospects. *J. Agric. Sci.* **2021**, *159*, 523–534. [CrossRef]
76. Pattanayak, S.; Das, S.; Biswal, G. Crop Disease Control through Application of Algal Biopesticides: Alternate Route Map in Organic Farming. In *Biopesticides in Organic Farming*; CRC Press: Boca Raton, FL, USA, 2021; pp. 123–126. ISBN 978-0-367-46017-4.
77. Gonçalves, A.L. The Use of Microalgae and Cyanobacteria in the Improvement of Agricultural Practices: A Review on Their Biofertilising, Biostimulating and Biopesticide Roles. *Appl. Sci.* **2021**, *11*, 871. [CrossRef]
78. Bratchkova, A.; Alexander, D. Kroumov Microalgae as Producers of Biologically Active Compounds with Anti-bacterial, Antiviral, Antifungal, Antialgal, Antiprotozoal, Antiparasitic and Anticancer Activity. *Acta Microbiol. Bulg.* **2020**, *36*, 79–89.
79. Stirk, W.A.; van Staden, J. Bioprospecting for bioactive compounds in microalgae: Antimicrobial compounds. *Biotechnol. Adv.* **2022**, *59*, 107977. [CrossRef]
80. Swain, S.S.; Paidesetty, S.K.; Padhy, R.N. Antibacterial, antifungal and antimycobacterial compounds from cyanobacteria. *Biomed. Pharmacother.* **2017**, *90*, 760–776. [CrossRef]
81. Hernández-Carlos, B.; Gamboa-Angulo, M.M. Metabolites from freshwater aquatic microalgae and fungi as potential natural pesticides. *Phytochem. Rev.* **2011**, *10*, 261–286. [CrossRef]
82. Asthana, R.K.; Srivastava, A.; Singh, A.P.; Deepali; Singh, S.P.; Nath, G.; Srivastava, R.; Srivastava, B.S. Identification of an antimicrobial entity from the cyanobacterium *Fischerella* sp. isolated from bark of *Azadirachta indica* (Neem) tree. *J. Appl. Phycol.* **2006**, *18*, 33–39. [CrossRef]
83. Kajiyama, S.-I.; Kanzaki, H.; Kawazu, K.; Kobayashi, A. Nostofungicidine, an antifungal lipopeptide from the field-grown terrestrial blue-green alga *Nostoc commune*. *Tetrahedron Lett.* **1998**, *39*, 3737–3740. [CrossRef]
84. Desbois, A.P.; Mearns-Spragg, A.; Smith, V.J. A Fatty Acid from the Diatom *Phaeodactylum tricornutum* is Antibacterial Against Diverse Bacteria Including Multi-resistant *Staphylococcus aureus* (MRSA). *Mar. Biotechnol.* **2009**, *11*, 45–52. [CrossRef]
85. Senousy, H.H.; El-Sheekh, M.M.; Saber, A.A.; Khairy, H.M.; Said, H.A.; Alhoqail, W.A.; Abu-Elsaoud, A.M. Biochemical Analyses of Ten Cyanobacterial and Microalgal Strains Isolated from Egyptian Habitats, and Screening for Their Potential against Some Selected Phytopathogenic Fungal Strains. *Agronomy* **2022**, *12*, 1340. [CrossRef]
86. Schmid, B.; Coelho, L.; Schulze, P.S.C.; Pereira, H.; Santos, T.; Maia, I.B.; Reis, M.; Varela, J. Antifungal properties of aqueous microalgal extracts. *Bioresour. Technol. Rep.* **2022**, *18*, 101096. [CrossRef]
87. Almalki, M.A.; Khalifa, A.Y.Z.; Alkhamis, Y.A. In vitro Antibiosis of Chlorella vulgaris Extract against the Phytopathogen, *Stenotrophomonas maltophilia*. *J. Pure Appl. Microbiol.* **2022**, *16*, 630–637. [CrossRef]
88. Ragaa, A.; Hamouda, M.S.M. El-Ansary Potential of Plant-Parasitic Nematode Control in Banana Plants by Micro-algae as a New Approach Towards Resistance. *Egypt. J. Biol. Pest Control.* **2017**, *27*, 165–172.
89. Righini, H.; Francioso, O.; Quintana, A.M.; Roberti, R. Cyanobacteria: A Natural Source for Controlling Agricultural Plant Diseases Caused by Fungi and Oomycetes and Improving Plant Growth. *Horticulturae* **2022**, *8*, 58. [CrossRef]
90. Scaglioni, P.T.; Pagnussatt, F.A.; Lemos, A.C.; Nicolli, C.P.; Del Ponte, E.M.; Badiale-Furlong, E. *Nannochloropsis* sp. and *Spirulina* sp. as a Source of Antifungal Compounds to Mitigate Contamination by *Fusarium graminearum* Species Complex. *Curr. Microbiol.* **2019**, *76*, 930–938. [CrossRef]

91. Natarajan, C.; Prasanna, R.; Gupta, V.; Dureja, P.; Nain, L. Characterization of the fungicidal activity of *Calothrix elenkinii* using chemical methods and microscopy. *Appl. Biochem. Microbiol.* **2012**, *48*, 51–57. [CrossRef]
92. Bang, K.-H.; Lee, D.-W.; Park, H.-M.; Rhee, Y.-H. Inhibition of Fungal Cell Wall Synthesizing Enzymes by *trans*-Cinnamaldehyde. *Biosci. Biotechnol. Biochem.* **2000**, *64*, 1061–1063. [CrossRef] [PubMed]
93. Wang, S.-Y.; Chen, P.-F.; Chang, S.-T. Antifungal activities of essential oils and their constituents from indigenous cinnamon (*Cinnamomum osmophloeum*) leaves against wood decay fungi. *Bioresour. Technol.* **2005**, *96*, 813–818. [CrossRef] [PubMed]
94. Huang, Q.-S.; Zhu, Y.-J.; Li, H.-L.; Zhuang, J.-X.; Zhang, C.-L.; Zhou, J.-J.; Li, W.-G.; Chen, Q.-X. Inhibitory Effects of Methyl *trans*-Cinnamate on Mushroom Tyrosinase and Its Antimicrobial Activities. *J. Agric. Food Chem.* **2009**, *57*, 2565–2569. [CrossRef] [PubMed]
95. Amborabé, B.-E.; Fleurat-Lessard, P.; Chollet, J.-F.; Roblin, G. Antifungal effects of salicylic acid and other benzoic acid derivatives towards *Eutypa lata*: Structure–activity relationship. *Plant Physiol. Biochem.* **2002**, *40*, 1051–1060. [CrossRef]
96. Berne, S.; Kovačič, L.; Sova, M.; Kraševec, N.; Gobec, S.; Križaj, I.; Komel, R. Benzoic acid derivatives with improved antifungal activity: Design, synthesis, structure–activity relationship (SAR) and CYP53 docking studies. *Bioorganic Med. Chem.* **2015**, *23*, 4264–4276. [CrossRef] [PubMed]
97. Nehela, Y.; Taha, N.A.; Elzaawely, A.A.; Xuan, T.D.; Amin, M.A.; Ahmed, M.E.; El-Nagar, A. Benzoic Acid and Its Hydroxylated Derivatives Suppress Early Blight of Tomato (*Alternaria solani*) via the Induction of Salicylic Acid Biosynthesis and Enzymatic and Nonenzymatic Antioxidant Defense Machinery. *J. Fungi* **2021**, *7*, 663. [CrossRef] [PubMed]
98. D'agostino, P.M.; Seel, C.J.; Ji, X.; Gulder, T.; Gulder, T.A.M. Biosynthesis of cyanobacterin, a paradigm for furanolide core structure assembly. *Nat. Chem. Biol.* **2022**, *18*, 652–658. [CrossRef]
99. Berry, J. Marine and Freshwater Microalgae as a Potential Source of Novel Herbicides. In *Herbicides and Environment*; Kortekamp, A., Ed.; InTech: London, UK, 2011; ISBN 978-953-307-476-4.
100. Zhang, Y.; Whalen, J.K.; Sauvé, S. Phytotoxicity and bioconcentration of microcystins in agricultural plants: Meta-analysis and risk assessment. *Environ. Pollut.* **2020**, *272*, 115966. [CrossRef]
101. Koodkaew, I.; Sunohara, Y.; Matsuyama, S.; Matsumoto, H. Isolation of ambiguine D isonitrile from *Hapalosiphon* sp. and characterization of its phytotoxic activity. *Plant Growth Regul.* **2012**, *68*, 141–150. [CrossRef]
102. Biondi, N.; Piccardi, R.; Margheri, M.C.; Rodolfi, L.; Smith, G.D.; Tredici, M.R. Evaluation of *Nostoc* Strain ATCC 53789 as a Potential Source of Natural Pesticides. *Appl. Environ. Microbiol.* **2004**, *70*, 3313–3320. [CrossRef]
103. Abdel-Baky, H.H.; Shallan, M.A.; Baroty, G.E.; El-Baz, F.K. Volatile Compounds of the Microalga *Chlorella vulgaris* and Their Phytotoxic Effect. *Pak. J. Biol. Sci.* **2001**, *5*, 61–65. [CrossRef]
104. Berry, J.P.; Gantar, M.; Perez, M.H.; Berry, G.; Noriega, F.G. Cyanobacterial Toxins as Allelochemicals with Potential Applications as Algaecides, Herbicides and Insecticides. *Mar. Drugs* **2008**, *6*, 117–146. [CrossRef] [PubMed]
105. Marten, G.G. Larvicidal algae. *J. Am. Mosq. Control. Assoc.* **2007**, *23*, 177–183. [CrossRef]
106. Rankic, I.; Zelinka, R.; Ridoskova, A.; Gagic, M.; Pelcova, P.; Huska, D. Nano/microparticles in conjunction with microalgae extract as novel insecticides against Mealworm beetles, *Tenebrio molitor*. *Sci. Rep.* **2021**, *11*, 17125. [CrossRef] [PubMed]
107. Becher, P.G.; Jüttner, F. Insecticidal compounds of the biofilm-forming cyanobacterium *Fischerella* sp. (ATCC 43239). *Environ. Toxicol.* **2005**, *20*, 363–372. [CrossRef] [PubMed]
108. Becher, P.G.; Keller, S.; Jung, G.; Süssmuth, R.; Jüttner, F. Insecticidal activity of 12-epi-hapalindole J isonitrile. *Phytochemistry* **2007**, *68*, 2493–2497. [CrossRef] [PubMed]
109. Kiviranta, J.; Abdel-Hameed, A.; Sivonen, K.; Niemelä, S.I.; Carlberg, G. Toxicity of cyanobacteria to mosquito larvae—Screening of active compounds. *Environ. Toxicol. Water Qual.* **1993**, *8*, 63–71. [CrossRef]
110. Delaney, J.; Wilkins, R. Toxicity of microcystin-LR, isolated from *Microcystis aeruginosa*, against various insect species. *Toxicon* **1995**, *33*, 771–778. [CrossRef]
111. Saario, E.; Abdel-Hameed, A.; Kiviranta, J. Larvicidal microcystin toxins of cyanobacteria affect midgut epithelial cells of *Aedes aegypti* mosquitoes. *Med. Veter. Èntomol.* **1994**, *8*, 398–400. [CrossRef]
112. Cook, W.; Beasley, V.; Dahlem, A.; Dellinger, J.; Harlin, K.; Carmichael, W. Comparison of effects of anatoxin-a(s) and paraoxon, physostigmine and pyridostigmine on mouse brain cholinesterase activity. *Toxicon* **1988**, *26*, 750–753. [CrossRef]
113. dos Santos, D.S.; Rosa, M.E.; Zanatta, A.P.; Oliveira, R.S.; de Almeida, C.G.M.; Leal, A.P.; Sanz, M.; Fernandes, K.A.; de Souza, V.Q.; de Assis, D.R.; et al. Neurotoxic effects of sublethal concentrations of cyanobacterial extract containing anatoxin-a(s) on *Nauphoeta cinerea* cockroaches. *Ecotoxicol. Environ. Saf.* **2018**, *171*, 138–145. [CrossRef] [PubMed]
114. Hassan, M.E.; Mohafrash, S.M.M.; Fallatah, S.A.; El-Sayed, A.E.-K.B.; Mossa, A.-T.H. Eco-friendly larvicide of *Amphora coffeaeformis* and *Scenedesmus obliquus* microalgae extracts against *Culex pipiens*. *J. Appl. Phycol.* **2021**, *33*, 2683–2693. [CrossRef]
115. Dolma, S.K.; Singh, P.P.; Reddy, S.G.E. Insecticidal and Enzyme Inhibition Activities of Leaf/Bark Extracts, Fractions, Seed Oil and Isolated Compounds from *Triadica sebifera* (L.) Small against *Aphis craccivora* Koch. *Molecules* **2022**, *27*, 1967. [CrossRef] [PubMed]
116. Punia, A.; Chauhan, N.S.; Singh, D.; Kesavan, A.K.; Kaur, S.; Sohal, S.K. Effect of gallic acid on the larvae of *Spodoptera litura* and its parasitoid *Bracon hebetor*. *Sci. Rep.* **2021**, *11*, 531. [CrossRef]
117. Joshi, R.S.; Wagh, T.P.; Sharma, N.; Mulani, F.A.; Sonavane, U.; Thulasiram, H.V.; Joshi, R.; Gupta, V.S.; Giri, A.P. Way toward "Dietary Pesticides": Molecular Investigation of Insecticidal Action of Caffeic Acid against *Helicoverpa armigera*. *J. Agric. Food Chem.* **2014**, *62*, 10847–10854. [CrossRef]

118. Zhang, X.-Y.; Shen, J.; Zhou, Y.; Wei, Z.-P.; Gao, J.-M. Insecticidal Constituents from *Buddlej aalbiflora* Hemsl. *Nat. Prod. Res.* **2017**, *31*, 1446–1449. [CrossRef]
119. Saber, A.A.; Moussa, S.; Abdel-Rahim, E.F.M.; Cantonati, M. Insecticidal Prospects of Algal and Cyanobacterial Extracts against the Cotton Leaf Worm *Spodoptera littoralis*. *Vie Milieu Life Environ.* **2018**, *68*, 199–212.
120. Sigamani, S.; Chinnasamy, R.; Dharmaraj, R.K.; Ramamurthy, D.; Devarajan, N.; Narayanasamy, M.; Natarajan, H. Larvicidal potency of the extracts from *Chlorella* sp. against *Aedes aegypti*. *Biocatal. Agric. Biotechnol.* **2020**, *27*, 101663. [CrossRef]
121. Rahuman, A.A.; Gopalakrishnan, G.; Ghouse, B.S.; Arumugam, S.; Himalayan, B. Effect of *Feronia limonia* on mosquito larvae. *Fitoterapia* **2000**, *71*, 553–555. [CrossRef]
122. Aguilar-Marcelino, L.; Pineda-Alegría, J.A.; Salinas-Sánchez, D.O.; Hernández-Velázquez, V.M.; Silva-Aguayo, G.I.; Navarro-Tito, N.; Sotelo-Leyva, C. In Vitro Insecticidal Effect of Commercial Fatty Acids, β-Sitosterol, and Rutin against the Sugarcane Aphid, *Melanaphis sacchari* Zehntner (Hemiptera: Aphididae). *J. Food Prot.* **2022**, *85*, 671–675. [CrossRef] [PubMed]
123. Perumalsamy, H.; Jang, M.J.; Kim, J.-R.; Kadarkarai, M.; Ahn, Y.-J. Larvicidal activity and possible mode of action of four flavonoids and two fatty acids identified in *Millettia pinnata* seed toward three mosquito species. *Parasites Vectors* **2015**, *8*, 237. [CrossRef] [PubMed]
124. Ghatrehsamani, S.; Jha, G.; Dutta, W.; Molaei, F.; Nazrul, F.; Fortin, M.; Bansal, S.; Debangshi, U.; Neupane, J. Artificial Intelligence Tools and Techniques to Combat Herbicide Resistant Weeds—A Review. *Sustainability* **2023**, *15*, 1843. [CrossRef]
125. Rani, L.; Thapa, K.; Kanojia, N.; Sharma, N.; Singh, S.; Grewal, A.S.; Srivastav, A.L.; Kaushal, J. An extensive review on the consequences of chemical pesticides on human health and environment. *J. Clean. Prod.* **2020**, *283*, 124657. [CrossRef]
126. Zhang, A.; Luo, W.; Sun, J.; Xiao, H.; Liu, W. Distribution and uptake pathways of organochlorine pesticides in greenhouse and conventional vegetables. *Sci. Total. Environ.* **2015**, *505*, 1142–1147. [CrossRef]
127. Balmer, J.E.; Morris, A.D.; Hung, H.; Jantunen, L.; Vorkamp, K.; Rigét, F.; Evans, M.; Houde, M.; Muir, D.C. Levels and trends of current-use pesticides (CUPs) in the arctic: An updated review, 2010–2018. *Emerg. Contam.* **2019**, *5*, 70–88. [CrossRef]
128. Liu, L.; Pohnert, G.; Wei, D. Extracellular Metabolites from Industrial Microalgae and Their Biotechnological Potential. *Mar. Drugs* **2016**, *14*, 191. [CrossRef]
129. Hickman, D.T.; Rasmussen, A.; Ritz, K.; Birkett, M.A.; Neve, P. Review: Allelochemicals as multi-kingdom plant defence compounds: Towards an integrated approach. *Pest Manag. Sci.* **2020**, *77*, 1121–1131. [CrossRef]
130. Qureshi, H.; Anwar, T.; Ali, Q.; Haider, M.Z.; Habib, N.; Fatima, S.; Waseem, M.; Bibi, Y.; Arshad, M.; Adkins, S.W. Isolation of natural herbicidal compound from *Lantana camara*. *Int. J. Environ. Anal. Chem.* **2019**, *101*, 631–638. [CrossRef]
131. Cheng, F.; Cheng, Z. Research Progress on the use of Plant Allelopathy in Agriculture and the Physiological and Ecological Mechanisms of Allelopathy. *Front. Plant Sci.* **2015**, *6*, 1020. [CrossRef]
132. Soltys, D.; Krasuska, U.; Bogatek, R.; Gniazdowsk, A. Allelochemicals as Bioherbicides—Present and Perspectives. In *Herbicides—Current Research and Case Studies in Use*; Price, A., Ed.; InTech: London, UK, 2013; ISBN 978-953-51-1112-2.
133. Balaji-Prasath, B.; Wang, Y.; Su, Y.P.; Hamilton, D.P.; Lin, H.; Zheng, L.; Zhang, Y. Methods to control harmful algal blooms: A review. *Environ. Chem. Lett.* **2022**, *20*, 3133–3152. [CrossRef]
134. Yu, S.; Zhang, Y.; Ran, Y.; Lai, W.; Ran, Z.; Xu, J.; Zhou, C.; Yan, X. Characterization of Steryl Glycosides in Marine Microalgae by Gas Chromatography-Triple Quadrupole Mass Spectrometry (GC-QQQ-MS): Steryl Glycosides in Microalgae Analyzed by GC-QQQ-MS. *J. Sci. Food Agric.* **2018**, *98*, 1574–1583. [CrossRef] [PubMed]
135. Cichoński, J.; Chrzanowski, G. Microalgae as a Source of Valuable Phenolic Compounds and Carotenoids. *Molecules* **2022**, *27*, 8852. [CrossRef] [PubMed]
136. Latif, S.; Chiapusio, G.; Weston, L.A. Allelopathy and the Role of Allelochemicals in Plant Defence. In *Advances in Botanical Research*; Elsevier: Amsterdam, The Netherlands, 2017; Volume 82, pp. 19–54. ISBN 978-0-12-801431-8.
137. Fidor, A.; Konkel, R.; Mazur-Marzec, H. Bioactive Peptides Produced by Cyanobacteria of the Genus *Nostoc*: A Review. *Mar. Drugs* **2019**, *17*, 561. [CrossRef]
138. Mendes, L.B.B.; Viegas, C.V.; Joao, R.R.; da Silva, R.B. Microalgae Production: A Sustainable Alternative for a Low-carbon Economy Transition. *Open Microalgae Biotechnol.* **2021**, *1*, 1–7. [CrossRef]
139. Grand View Research GVR Report CoverAstaxanthin Market Size, Share & Trends Report Astaxanthin Market Size, Share & Trends Analysis Report by Product (Dried Algae Meal or Biomass, Oil, Softgel, Liquid), by Source (Natural, Synthetic), by Application, by Region, and Segment Forecasts, 2023–2030. Available online: https://www.grandviewresearch.com/industry-analysis/global-astaxanthin-market (accessed on 1 March 2023).
140. Villaró, S.; Ciardi, M.; Morillas-España, A.; Sánchez-Zurano, A.; Acién-Fernández, G.; Lafarga, T. Microalgae Derived Astaxanthin: Research and Consumer Trends and Industrial Use as Food. *Foods* **2021**, *10*, 2303. [CrossRef]
141. Li, X.; Wang, X.; Duan, C.; Yi, S.; Gao, Z.; Xiao, C.; Agathos, S.N.; Wang, G.; Li, J. Biotechnological production of astaxanthin from the microalga *Haematococcus pluvialis*. *Biotechnol. Adv.* **2020**, *43*, 107602. [CrossRef]
142. Kumar, D.; Singh, B. Algal biorefinery: An integrated approach for sustainable biodiesel production. *Biomass Bioenergy* **2019**, *131*, 105398. [CrossRef]
143. Zhu, L. Biorefinery as a promising approach to promote microalgae industry: An innovative framework. *Renew. Sustain. Energy Rev.* **2015**, *41*, 1376–1384. [CrossRef]
144. Branco-Vieira, M.; Martin, S.S.; Agurto, C.; Freitas, M.A.V.; Martins, A.A.; Mata, T.M.; Caetano, N.S. Biotechnological potential of *Phaeodactylum tricornutum* for biorefinery processes. *Fuel* **2020**, *268*, 117357. [CrossRef]

145. Sivaramakrishnan, R.; Suresh, S.; Kanwal, S.; Ramadoss, G.; Ramprakash, B.; Incharoensakdi, A. Microalgal Biorefinery Concepts' Developments for Biofuel and Bioproducts: Current Perspective and Bottlenecks. *Int. J. Mol. Sci.* **2022**, *23*, 2623. [CrossRef] [PubMed]
146. Reijnders, M.J.M.F.; van Heck, R.G.A.; Lam, C.M.C.; Scaife, M.A.; dos Santos, V.A.P.M.; Smith, A.G.; Schaap, P.J. Green genes: Bioinformatics and systems-biology innovations drive algal biotechnology. *Trends Biotechnol.* **2014**, *32*, 617–626. [CrossRef] [PubMed]

MDPI AG
Grosspeteranlage 5
4052 Basel
Switzerland
Tel.: +41 61 683 77 34

Plants Editorial Office
E-mail: plants@mdpi.com
www.mdpi.com/journal/plants